Soil Biology

Volume 21

Series Editor
Ajit Varma, Amity Institute of Microbial Technology,
Amity University Uttar Pradesh, Noida, UP, India

For further volumes:
http://www.springer.com/series/5138

Patrice Dion
Editor

Soil Biology and Agriculture in the Tropics

Springer

Editor
Professor Dr. Patrice Dion
Département de phytologie
Pavillon Charles-Eugène-Marchand
1030, avenue de la Médecine
Université Laval
Québec (Québec) G1V 0A6
Canada
Patrice.Dion@fsaa.ulaval.ca

ISSN: 1613-3382
ISBN: 978-3-642-05075-6 e-ISBN: 978-3-642-05076-3
DOI 10.1007/978-3-642-05076-3
Springer Heidelberg Dordrecht London New York

Soil Biology

Library of Congress Control Number: 2010920235

Cover design: SPI Publisher Services

Printed on acid-free paper

Springer is part of Springer Science+Business Media (www.springer.com)

Preface

All human actions are a response. Some are elicited by need, others by fantasy or curiosity, or else by anxiety or particular feelings. Some actions are performed on the spot and others are indefinitely postponed or brought to existence only in our dreams. Much to our relief, this ultimate fate will not be that of the present book, which has been prepared under a sense of urgency that both the scientific and day-to-day experience inspire. The criticality of tropical agriculture is made manifest everywhere, through a crucial economic, social, and ecological North–South interplay. Although the soil is often explicitly ignored in this dialog, it always remains an issue, since the development and survival of civilizations depend on the state of the soils they exploit. Humanity's future may lie at skin-of-earth depth, and nowhere is this skin so fragile and bruised than in the tropics.

Whereas the Ancients maintained a rich tradition of agricultural representation in poems, paintings, and other works of fine art, Christianity conceived of God as being uninterested in agriculture. What is more, the labors of the soil were attributed a negative value, as they have been viewed as a punishment to those who were reckless enough to commit the original sin. As a result and with some notable exceptions, a world was created by artists and then embraced by an increasingly dominant culture, where food often did not exist or, when it did, came as a magical, unproduced entity. Now food production is being reinserted in the collective imaginary of the developed world, as the fragility of our civilization's biological foundations is being insistently brought to our collective conscience.

Some readers of this Preface will remember with nostalgia the fictions of Edgar Rice Burroughs, where the tropical "jungle" was endowed with the power to overwhelm human nature and reversibly mould an English Lord into a gracious and largely innocuous tree dweller. More prosaically, the reverse is happening today, as the various tropical milieus have become an ever-regressing frontier of human expansion. Soils of the tropics, in particular, yield with amazing ease to the destructive power of our agricultural processes. It is the purpose of this book to present the soil component of tropical agriculture, and to document some tools and strategies that will contribute to its preservation. Deforestation, erosion, salinization, and other forms of soil degradation arise as a consequence of our increasing needs and purchasing power. Nowhere are these effects made more evident than under tropical latitudes. The authors of this book have highlighted why this is so, and what solutions can be envisioned.

By showing how they are shaped through the interplay of parent materials and climates, Chap.1 illustrates the peculiarities of the soils of the tropics. This analysis naturally leads to a description of the main types of soils found in the tropics, and is pursued in Chap. 2 by an extensive survey and compilation of the literature dealing with organic matter and soil biomass. In a precise and concise manner, these two introductory chapters together provide a foundation for discussions on soil biology to be found in the following contributions.

The next three chapters deal with various aspects of soil management. The conversion of forested land to agricultural use brings about profound changes in the structure and function of microbial communities. This effect is unevenly distributed among microbial taxonomical or physiological types. The coexistence of crop plants and soil microorganisms can be oriented and exploited so as to promote the sustainability of agricultural practices and the preservation of biodiversity: this is the objective of conservation agriculture. By contrast, slash and burn agriculture appears as a rather intrusive and even destructive process, especially when accompanied by heavy population pressures. However, in this case also, microorganisms can be used so as to mitigate impacts and promote sustainability.

Desert soils and inundated rice fields are sequentially presented in Chaps. 7 and 8, offering illuminating contrast. In arid soils, some plants define a perimeter, or resource island, within which microbial and plant life thrives with enhanced intensity. While being virtually aquatic environments, inundated rice fields also provide examples of mutual dependencies, with interrelated life forms relying on co-metabolism, associations, and symbioses.

Being privileged objects of microbiological research performed on temperate soils, microbial associations with plants are of major impact in tropical settings: their essentiality to tropical agriculture is eloquently demonstrated in Chaps. 8–11. Rhizobia interact both with legume and nonlegume tropical hosts, often showing little host specificity and a broad range of effects. Likewise, mycorrhizas exert profound influences on plant communities assembled under agroforestry management, establishing large networks involved in nutrient circulation, maintenance of plant biodiversity, and determination of soil structure. Although they do not elicit the formation of visible root structures, the phosphate-solubilizing bacteria are an integral part of the plant microbiome and act through a variety of well-characterized mechanisms. These various plant-associated microorganisms may be formulated and commercialized as biofertilizers, which are particularly appealing to smallholder farmers of the tropics.

Problems related to pesticide use and wastewater irrigation are of paramount importance considering the fragility and instability of tropical ecosystems. Pesticides, which are applied worldwide and hence provide a virtually universal and specific selection pressure, offer a unique opportunity to those wishing to explore the largely uncharted terrain of microbial biogeography. By contrast, wastewater irrigation is a practice specific to areas with lesser economical resource; it is enmeshed in social and cultural values and presents health risks which call for adapted solutions.

The book closes with a discussion of traditional soil knowledge, highlighting its significance as a component of the human experience and proposing that it should become integrated with all forms of knowledge so as to produce a renewed framework for our interaction with the world.

In assembling and ordering the chapters of this book, we have deemed it best to delineate a continuous trajectory through the various subdisciplines and concepts of soil biology: hence, the chapters come in a continuous flow, rather than being grouped in sections. The intention here is to provide a sense of integration, as the soil itself, with its various horizons and its organic and mineral components, constitutes a complex and multiply integrated whole. While we hope this notion will be conveyed to the reader, certainly it has been shared by all the contributors of the present volume, who have demonstrated over the last 18 months a community of thought and objectives. I thank them here for their professionalism and dedication. My thanks also go to Professor Ajit Varma, Soil Biology Series Editor, and Dr. Jutta Lindenborn, Springer Managing Editor, for their careful guidance and patience, to Marie-Claude Julien for preparing the index, and to Cécile Gauthier for checking the references.

Québec City, Canada Patrice Dion

Contents

Contributors

Abaidoo, Robert C. International Institute of Tropical Agriculture, Ibadan, Nigeria; Provost, College of Agriculture and Natural Resources, Kwame Nkrumah University of Science and Technology (KNUST), University Post Office, Kumasi, Ghana, rcabaidoo@yahoo.com

Abossolo Angue, Monique Department of Earth Sciences, Faculty of Science, University of Yaoundé I, Cameroon

Arbeli, Ziv Facultad Agronomía, Universidad Nacional de Colombia, Sede Bogotá, A. A. 14490, Bogotá, D.C., Colombia, aziv@bt.unal.edu.co

Bashan, Yoav Environmental Microbiology Group, Northwestern Center for Biological Research (CIBNOR), La Paz, B.C.S., Mexico; Department of Soil, Water and Environmental Science, The University of Arizona, Tucson, Arizona, USA; Bashan Foundation, Corvallis, Oregon, USA, bashan@cals.arizona.edu

Cardoso, Irene Maria Department of Soil Science, Federal University of Viçosa, Viçosa, Minas Gerais, Brazil, 36570-000, irene@ufv.br

de Aquino, Adriana Maria Embrapa Agrobiologia, BR 465, km 7, P.O. Box 74505, Seropédica, RJ, Brazil, Cep 23.890-000, adriana@cnpab.embrapa.br

de-Bashan, Luz Estela Environmental Microbiology Group, Northwestern Center for Biological Research (CIBNOR), La Paz, B.C.S., Mexico; Bashan Foundation, Corvallis, Oregon, USA, luzb@cals.arizona.edu

de Carvalho, André Mundstock Xavier Department of Soil Science, Federal University of Viçosa, Viçosa, Minas Gerais, Brazil, 36570-000, andremundstock@yahoo.com.br

Dion, Patrice Département de phytologie, Pavillon Charles-Eugène Marchand, 1030, avenue de la Médecine, Université Laval, Québec (Québec) Canada G1V 0A6, Patrice.Dion@fsaa.ulaval.ca

Dissanayake, Priyanka International Water Management Institute, P.O. Box 2075, Colombo, Sri Lanka, p.dissanayake@cgiar.org

Drechsel, Pay International Water Management Institute, P.O. Box 2075, Colombo, Sri Lanka, p.drechsel@cgiar.org

Fernandes Correia, Maria Elizabeth Embrapa Agrobiologia, BR 465, km 7, P.O. Box 74505, Seropédica, RJ, Brazil, Cep 23.890-000, ecorreia@cnpab.embrapa.br

Frossard, Emmanuel ETH Zurich, Institute of Plant Sciences, FMG C 23 CH-8315 Lindau (ZH), Switzerland

Fuentes, Cilia L. Facultad Agronomía, Universidad Nacional de Colombia, Sede Bogotá, A. A. 14490, Bogotá, D.C., Colombia, clfuentesd@unal.edu.co

Jansa, Jan ETH Zurich, Institute of Plant Sciences, FMG C 23 CH-8315 Lindau (ZH), Switzerland

Joergensen, Rainer Georg Department of Soil Biology and Plant Nutrition, University of Kassel, Nordbahnhofstr. 1a, 37213 Witzenhausen, Germany, joerge@uni-kassel.de

Keraita, Bernard International Water Management Institute (IWMI), West Africa Office, PMB CT 112, Accra, Ghana; Department of International Health, University of Copenhagen, Denmark, b.keraita@cgiar.org

Kuyper, Thomas W. Department of Soil Quality, Wageningen University, P.O. BOX 47, 6700 AA Wageningen, The Netherlands, Thom.Kuyper@wur.nl

López-López, Aline Ecological Genomics Program, Centro de Ciencias Genómicas, UNAM, Avenida Universidad SN, Chamilpa, Cuernavaca, Mor. 62210, Mexico

Martínez, Julio Ecological Genomics Program, Centro de Ciencias Genómicas, UNAM, Avenida Universidad SN, Chamilpa, Cuernavaca, Mor. 62210, Mexico

Martínez-Romero, Esperanza Ecological Genomics Program, Centro de Ciencias Genómicas, UNAM, Avenida Universidad SN, Chamilpa, Cuernavaca, Mor. 62210, Mexico, emartine@ccg.unam.mx

Maxwell, Akple S. International Water Management Institute (IWMI), KNUST liaison office, Kumasi, Ghana, iwmi-kumasi@cgiar.org

Nain, Lata Division of Microbiology, Indian Agricultural Research Institute, New Delhi 110012, lata_rajat@yahoo.co.in

Nayak, Saswati National Institute of Science, Technology and Developmental Studies, New Delhi 110012, nayaksaswati@gmail.com

Nortcliff, Stephen Department of Soil Science, University of Reading, Reading, RG6 6DW, United Kingdom, s.nortcliff@reading.ac.uk

Nwaga, Dieudonné Laboratory of Soil Microbiology, Faculty of Science/Biotechnology Centre, P.O. Box 812, University of Yaoundé I, Yaoundé, Cameroon, dnwaga@yahoo.fr

Pandey, Alok Kumar Division of Microbiology, Indian Agricultural Research Institute, New Delhi 110012, alok2k2@gmail.com

Prasanna, Radha Division of Microbiology, Indian Agricultural Research Institute, New Delhi 110012, radhapr@gmail.com

Rosenblueth, Mónica Ecological Genomics Program, Centro de Ciencias Genómicas, UNAM, Avenida Universidad SN, Chamilpa, Cuernavaca, Mor. 62210, Mexico

Rumjanek, Norma Gouvêa Embrapa Agrobiologia, BR 465, km 7, P.O. Box 74505, Seropédica, RJ, Brazil, Cep 23.890-000, norma@cnpab.embrapa.br

Sánchez-Nieves, Jimena Facultad de Ciencias, Departamento de Biología, Universidad Nacional de Colombia Bogotá, Colombia A. A 14-490, Bogotá, D.C., Colombia, jsanchezn@unal.edu.co

Tavares, Rodrigo de Castro Department of Soil Science, Federal University of Viçosa, Viçosa, Minas Gerais, Brazil, 36570-000, rocatavares@yahoo.com.br

Uribe, Daniel Instituto de Biotecnología, Universidad Nacional de Colombia Bogotá, Colombia A. A 14-490, Bogotá, D.C., Colombia, duribev@unal.edu.co

Vanegas, Javier Instituto de Biotecnología, Universidad Nacional de Colombia Bogotá, Colombia A. A 14-490, Bogotá, D.C., Colombia, jvanegasg@unal.edu.co

Verchot, Louis V. Center for International Forestry Research (CIFOR), P.O. Box 0113 BOCBD, Bogor 16000, Indonesia, L.verchot@cgiar.org

Xavier, Gustavo Ribeiro Embrapa Agrobiologia, BR 465, km 7, P.O. Box 74505, Seropédica, RJ, Brazil, Cep 23.890-000, gustavo@cnpab.embrapa.br

Yarzábal, Luis Andrés Laboratorio de Microbiología Molecular y Biotecnología, Departamento de Biología, Facultad de Ciencias, Núcleo de La Hechicera, Universidad de Los Andes, Mérida 5101, Estado Mérida, Venezuela, yluis@ula.ve

Zilli, Jerri Édson Embrapa Roraima, BR 174, km 8, Boa Vista, RR, Brazil, Cep 69301-970, zilli@cpafrr.embrapa.br

Chapter 1
Soils of the Tropics

Stephen Nortcliff

1.1 Introduction

The Tropics of Cancer and of Capricorn, located 23½°N and S respectively, broadly define the area we recognise as the tropics and within which we find tropical environments. Whilst these limits are to a degree arbitrary, they do provide recognisable boundaries. Natural boundaries also occur in some parts of the world, for example in South Asia, where the Himalayas form a natural boundary, although this actually reaches 34°N. Within this "tropical" area, some have argued that there are no such things as tropical soils, merely soils found within this inter-tropical zone. Whilst there are soils within this zone which have strong similarities with soils found beyond these tropical limits, there are a number of soils and soil development processes which are specific to this zone. This is particularly the case within the humid and sub-humid zones of this region, where the soil environment is characterised by periods of intense weathering and leaching. In many cases there is a further distinguishing feature about these soils; their development in this region has been uninterrupted for many hundreds of thousands of years and possibly millions of years.

1.2 Factors of Soil Formation

Soils form in response to their environmental context. Historically, this relationship has often been expressed in the form of a soil-forming equation, where the nature of the soil is considered as a function of the interactions between various environmental factors. This relationship between soils and the environment was first introduced

S. Nortcliff
Department of Soil Science, University of Reading, Reading, Berkshire RG6 6DW, UK
e-mail: s.nortcliff@reading.ac.uk

P. Dion (ed.), *Soil Biology and Agriculture in the Tropics*, Soil Biology 21,
DOI 10.1007/978-3-642-05076-3_1,

in the second half of the nineteenth Century by a Russian soil scientist, Dokuchaev, but it is probably most widely known by the presentation of Jenny (1941).

Jenny suggested that the nature of soils and their development was influenced by what he described as the *Factors of Soil Formation*:

1. Climate
2. Parent material
3. Relief or topography
4. Organisms (vegetation and soil fauna)
5. Time

Latterly, many have suggested that mankind should be added as a further and often very significant factor, because our impact on the soils has in many situations dramatically altered both the nature of the soils and their development, and perhaps more importantly their use.

1.2.1 Climate

By definition, in the tropics there are relatively small differences in temperature during the year. In the lowland tropics (approximately 87% of the land area of the tropical zone), mean annual temperatures are normally above 25°C. Where the mean annual soil temperatures are above 22°C and the temperature range throughout the year is greater than 6°C, the soil temperature regime is described as *hyperthermic*, and where the range in soil temperatures is less than 6°C it is described as *isohyperthermic*.

There are marked differences in terms of the rainfall distribution within this zone and consequently in the soil moisture regimes. Approximately 24% of this zone has a high annual rainfall, frequently in excess of 3,500 mm, well distributed through the year, and is often characterised, under natural conditions, by closed canopy forest. The soil moisture regime under these conditions is described as *udic*, in which the soil is not dry for as long as 90 cumulative days. This zone is often described as the *Humid Tropics,* with a natural vegetation frequently characterised by a forest cover. For much (approximately 49%) of the tropical zone, the climate is distinguished by pronounced wet and dry season(s). The soil moisture regime characteristic of this zone is *ustic*. The broad concept of this moisture regime is that the soil remains dry for part of the year (90 or more consecutive days), but when moisture is present, conditions are suitable for plant growth. This zone, often described as the *Seasonally Moist* or *Seasonally Arid Tropics*, is often characterised by a natural vegetation cover of grass and non-continuous, often low-stature tree cover. The remainder of the zone has a semi-arid or desert climate where evaporation exceeds precipitation in most months; here the soil moisture regime is described as *torric*, where soils are moist for 90 or less days per year.

Whilst the broad context of soil development is determined by the climatic conditions, local variations will occur because of the modifying influence of the other environmental factors.

1.2.2 Parent Material

Parent material is the material from which the soil is derived. Whilst in many cases the original material from which the soil has developed will be found unweathered below this soil itself, there are frequently circumstances where the soil is developed in a relatively thin layer of superficial material which has been deposited over the underlying rock by transporting processes involving water, wind or mass movement.

The properties of the parent material which significantly affect the nature of the soil are the degree of consolidation, the grain size and the composition. Coherent and impermeable rock materials which are not fissured will offer only a limited surface area for weathering, and consequently these processes will occur only at the surface. In contrast, fissured and loosely consolidated material will expose a large surface area, and weathering will take place more extensively throughout the material. Where the soils are developed on transported, unconsolidated materials, soil development can begin immediately. Where soil development takes place on consolidated fresh rock, there must be weathering of this rock material before soil development can begin. The grain size of the parent material determines the texture (the relative proportions of sand, silt and clay) of the soil. For example, a parent material such as sandstone, with a composition including a substantial proportion of sand-sized quartz, will weather to give a sandy soil; in contrast, a fine grained material such as basalt will weather to produce a predominantly fine-textured soil. Parent material composition strongly influences the nature of tropical soils and the soil processes which operate, together with the nature of the overlying vegetation and the agricultural potential of the soil. Whilst there are detailed classifications of rock materials, a relatively straightforward approach is to use the proportion of silica present in the material as the basis for classification:

- *Felsic* parent materials have not less than 66% total silica, which includes quartz and combined silica. These soils often contain free silica as quartz and orthoclase and plagioclase feldspars, and muscovite and biotite. When freshly exposed, these materials are often pale in colour.
- The *Intermediate* group of parent materials have between 55% and 66% total silica, and contain quartz, plagioclase, less orthoclase and some ferromanganese minerals such as hornblende.
- The *Basic* (or Mafic) parent materials have less than 55% total silica, and contain ferromanganese minerals such as hornblende, olivine and augite and limited amounts of plagioclase. These materials are often dark coloured when newly exposed.

In broad terms, Basic rocks will weather more rapidly than Felsic rocks. It is important to note that soil development in the tropics has in most cases been taking place for long periods of time, and the parent materials have been subject to sustained weathering and leaching during this development.

Because of the intensity and in many cases the long duration of the weathering processes which the parent materials have been subjected to, the mantle of weathered material at the earth's surface is often of considerable thickness. This weathered mantle is generally referred to as the *Regolith* and, particularly in the tropics, what we recognise as soil development may often only be clearly evident in the upper parts of the regolith.

1.2.3 Relief or Topography

There are many strong influences on the nature of soil through relief or topography. There is the indirect influence that for every 1,000 m increase in altitude there is an adiabatic fall in temperature of 6°C. This fall in temperature will greatly affect the rate of most chemical and biological processes in the soil, and hence influence the rate of weathering and the decomposition of organic matter. The hydrological variations within a landscape will also influence the nature and pattern of the soils, with the soils in the upper part of the slope/topographic system generally being freely drained and soils in the lower parts often receiving solutes and particulate materials from upslope, and often being characterised by poorer drainage conditions.

1.2.4 Organisms

Organisms as a soil-forming factor include both vegetation and soil fauna. Of the soil organisms, termites, earthworms, bacteria and fungi are particularly important in the key task of the incorporation of organic residues from the surface of the soil and dead root material from within the soil, breaking these materials down physically and chemically to release plant nutrients, and to produce relatively stable organic by-products which may have key roles in affecting other soil properties such as aggregate stability.

There is a strong two-way relationship between soil and vegetation in both natural and cultivated environments. Soils exert an influence on the type of vegetation present, and the vegetation will influence the soils and soil processes through the addition of organic residues to the surface, which play a key role in nutrient cycling. Vegetation also induces changes in soil moisture because of its demands arising from plant respiration, its action as a "protective" cover at the surface and, where the plants are leguminous and able to establish nitrogen-fixing rhizobial symbioses, its input of nitrogen-rich residues to the soil. Most important amongst these effects is the input of organic residues to the soil surface.

1.2.5 Time

Time is very different from the other soil-forming factors, because its influence is in determining the duration of the interactions between the soil and the other environmental factors. Through time, the nature of these relationships will change. In much of the tropics, there is strong evidence to suggest that many soils have been developing at a site for many hundreds of thousands of years and possibly millions of years, although not necessarily under the same combinations of environmental factors. Time is a factor which distinguishes tropical soils and soil development in the tropics from the cool temperate climates where during the Pleistocene the landscape was glaciated or subject to periglacial conditions and the soil cover removed or disrupted, with the consequence that soil development in these areas has often occurred over much shorter periods of time, possibly as little as a few tens of thousands of years.

1.2.6 Man

Increasingly, man is recognised as a major influence on many ecosystems. Crutzen (2002) suggested that the influence of man has been so great that we should distinguish the period of this influence as a separate geological epoch, the Anthropocene. If we consider soils in general, it is evident that man has had a major influence on the nature of many soils through soil management practices, and in broader context through environmental changes. This is certainly true in the tropics, although the period of major human intervention in many parts of the tropics is much shorter than in other parts of the globe. There have been major soil changes and extensive degradation of soil as a result of inappropriate management, and many tropical soils have been degraded as a result of broader environmental changes. Often because of the inherent properties of the soils in this region and the nature of the tropical environment, tropical soils have less resilience to these changes and are often more rapidly degraded.

1.3 Soil-Forming Processes

Soil-forming processes are often associated with the type of soil that results when the process is operating, for example a Podzol is the product of the Podzolisation soil-forming process. These processes which may be described as *Composite Processes* represent different combinations of a range of basic or *Specific Processes*. The composite processes will have some of the specific processes in common, but the key differences are in their relative magnitudes and duration.

The specific processes include a range of processes which might be considered as transformation processes in which minerals (weathering) and organic materials (decomposition) are altered. In transfer processes, materials are moved within the soil (the transfer may in some cases follow from the transformation of materials). In removal processes, materials are lost from the soil system completely. While the specific processes will occur in all regions of the globe, under tropical conditions their absolute and relative intensities may be greater because of the higher temperatures and duration of soil wetness.

Water plays a key role in many of these processes. Within the soil, water occurs as the soil solution, and the volume of water present in the soil at a particular time will depend on the climatic regime of the soil and its capacity for retaining water and allowing water to flow through the soil, either vertically or laterally. In an udic moisture regime the soil will be moist for much of the time, and differences in the nature and intensity of specific processes will depend upon the water retention characteristics and the flow of water through the soil. In contrast, under a torric moisture regime the soil will be moist only infrequently. Soils with an ustic moisture regime will have periods when the soil is moist and periods when the soil is relatively dry. Of importance are not only the presence of water, but also its nature, principally its temperature, pH and solute content.

Chemical transformation of minerals (weathering) includes the breakdown of primary minerals and the synthesis of secondary minerals (important amongst these are the clay minerals such as smectite, illite and kaolinite). In cases where these processes have occurred at high rate and intensities or for long periods of time, the secondary minerals will also be subject to strong weathering, with transformation of smectite clays to kaolinite, illite to kaolinite and kaolinite to gibbsite. Minerals vary markedly in their susceptibility to this chemical transformation. In broad terms, quartz and muscovite are the most resistant to chemical weathering; feldspars and biotite form part of an intermediate group and orthoclase is the least resistant. This sequence of weathering susceptibility does not differ significantly between tropical and other climatic zones, but the rates of weathering tend to be higher and the duration of the weathering longer in the tropical zones because of the higher temperatures of the water and possibly because the soil solutions are often more acidic.

Transfer of materials within the soil will be as solutes in the water moving through the soil. The soluble products of the weathering processes are removed from the soil through the process of leaching. The effectiveness of the leaching depends on the volume of water passing through the soil, the temperature and pH of the water, and its rate of flow through the soil. The materials released during the weathering process also vary in the ease with which they can be leached. The following considers broad groups of materials in terms of the relative ease with which they can be leached under normal conditions in the soil.

- *Soluble salts* (e.g. chloride and sulphates) are readily removed when leaching occurs.
- *Exchangeable cations* (e.g. Ca^{++}, Mg^{++}, K^{+} and Na^{+}) are very soluble and readily lost under moderate leaching. Silica has a low solubility in the pH range 3.5–8.0, and quartz has a very low solubility.

- The solubility of *iron* is dependent on its speciation; in broad terms, ferrous iron (FeO) is relatively soluble, but ferric iron (Fe_2O_3 or FeOOH) is relatively insoluble. Under free drainage, iron is predominantly found in the ferric forms, so is considered to be only of very limited solubility. Where there is high groundwater, the iron may be reduced to the ferrous form, and hence its solubility (and mobility) increases.
- *Aluminium* occurring in the form Al_2O_3 is soluble only at very low pH values, and in most soils is considered immobile.

The silica:sesquioxide ratio has been used as an indicator of the degree of weathering and leaching. In many soils, particularly those in temperate regions, this ratio is well above 2.0, but in the tropics, as the weathering releases silica and this is removed, there is a relative accumulation of sesquioxides (iron and aluminium oxides). In many tropical soils this ratio is below 2.0. When weathering is intense and iron is removed, the silica:alumina ratio is used. This will fall below 2.0 only where weathering and leaching are intense.

Under freely draining conditions where the dominant processes are strong weathering and leaching, and where these processes have been operating uninterrupted for long periods of time, the soil that develops will be deeply weathered, with diffuse boundaries between the horizons. The subsurface horizon under these conditions is often known as an oxic (Soil Taxonomy — see Sect. 1.4) or ferralic (WRB — see Sect. 1.4) horizon. These horizons are often relatively thick (greater than 30 cm), have a low percentage of weatherable minerals, a low ability to retain and exchange cations (cation exchange capacity), and a very small proportion of the fine earth of the soil in the silt fraction, and are often clay-textured. These horizons are often characterised by a fine, sand-sized stable aggregation of the clay, sometimes referred to as pseudosands. Because of this fine stable structure, the horizons often have a high hydraulic conductivity. Water flow through these soil horizons is often described as biphasic, with an initial rapid percolation through the macropores and a second flow, much slower in the microporosity. Nortcliff and Thornes (1989) suggested that this may explain why the chemical composition of the groundwater under these horizons does not reflect the composition of the water in the micropores. A distinctive feature of these horizons is that, particularly when the iron content is high, phosphate and other anions may be strongly adsorbed.

In addition to the transfers and removals in solution, there may also be transfers as suspensions. A key process involving suspension is the vertical translocation of clays. The clays are removed from upper layers as a result of physical or chemical dispersion (eluviation), transported in colloidal suspension in the water flowing vertically through the soil pores and fissures, and deposited on the surfaces of the pores and fissures at depth within the soil (illuviation). This processes is particularly marked where the clays are smectitic or illitic, and less prominent where the clays are predominantly kaolinitic. This process of clay translocation is favoured by processes which facilitate the dispersion of the clays: these may be physical, such as the rapid wetting of soil material at the end of a pronounced dry period, or chemical, where the soil solution contains mineral and organic material which facilitates dispersion. Where these processes are actively operating, there may be a marked vertical contrast in the distribution of clay, with distinctive decreases in the upper part of the soil from the zone from which the clay is being eluviated, and an increase in clay (sometimes referred to as a *clay bulge*) in the lower part of the soil. Where deposition has occurred it is possible, using soil thin sections and microscopic analysis, to identify clay skins (also referred to as argillans or clay cutans) which have distinct characteristics, such as a narrow range of particle size compared to the bulk of the soil, and orientation in the clays which provides evidence of a transporting process acting on materials in suspension. Where the deposition is clearly evidenced, the horizon is often described as an argillic (Soil Taxonomy) or argic (WRB) horizon.

Lateral translocations include both movements in solution through soils at different positions in the landscape and movements in solution and suspension over the surface. These processes will strongly influence the nature and patterns of soil found at different positions in the landscape.

1.4 Soil Classification

Soil classification is to many soil scientists the starting point for a disagreement! In the past, the classifications were often based on soils within national boundaries and divisions were made to accommodate the soils that were found nationally and the relative balance between different soil types. Within the tropics, particularly the African tropics, the soil classification in use in a particular country often reflected that of the Colonial "ruler", hence there were systems that had broad principles derived from British, French and Belgian soil scientists. The countries of former French Africa used in the past a broadly common system; there was a markedly different system within what was the Belgian Congo; within the former British Africa there were often locally relevant classifications. Initially, there were few attempts to relate national classifications to any form of international standard, or to see any form of cross-referencing to aid interpretation across national boundaries.

In 1960, the USDA introduced a system of classification which sought to classify all soils of the world (Soil Survey Staff 1960). Initially called "Soil Classification: A Comprehensive System, 7th Approximation", this system revolutionised soil classification in that it sought to introduce a whole new set of terminology and established limits for class units within the classification. This has been widely used and much revised, and has changed its name to Soil Taxonomy, the most recent version of which was published in 1999 (Soil Survey Staff 1999). Within Africa and as a result of the collaboration of Belgian and French soil scientists, D'Hoore (1964) produced a monograph which described the soils included in the 1: 5,000,000 Soil Map of Africa, and was a product of the Commission for Technical Co-operation in Africa (CCTA). In 1974, FAO produced the legend to the Soil Map of the World (FAO-UNESCO 1974), which whilst not a true soil classification provided broad indications of classes and soil types.

Today there are two widely used international soil classifications; the above mentioned Soil Taxonomy and the World Reference Base for Soils (WRB), which has its roots in the Soil Map of the World produced by FAO-UNESCO. Both classification systems use the identification of horizons which are diagnostic of the operation of particular soil-forming processes or combinations of processes influenced by the nature of the soil-forming factors. They recognise the importance of soil climates in determining the nature of the soils, and seek to set limits for the membership classes. Both are the subject of ongoing appraisal and revision, and the most recent version of WRB was published in 2007 (IUSS Working Group WRB 2007). Whilst there are some similarities between these two widely used soil classification systems, and there have been attempts in recent years to seek to reduce the differences in the diagnostic characteristics chosen and the limits set for membership, there is still no universal agreement on which should be used. Because of this, I have chosen to approach the description of the soils using the broad climatic zones present in the lowland tropical regions, and seek to refer to similar soils classified within these two schemes.

1.5 The Soils

Within the tropics there are probably occurrences of soils also found in extra-tropical regions, but there are a limited number of soils which are predominantly found in the tropical zone and are to a degree characteristic of this zone.

1.5.1 Soils of the Humid Tropics

The zonal soils of the humid tropics, often associated with a natural vegetation of rainforest, are characterised as deeply weathered, with low inherent fertility and strong acidity. The soils have an udic moisture regime, and as a consequence are moist throughout most of the year; they are characterised by intense weathering and rapid leaching. The soils are very varied in terms of colour, but red, dark red, reddish yellow and yellow colours predominate. Within the CCTA Soil Map of Africa (D'Hoore 1964) these soils were described as *ferrallitic soils*, and have broad equivalents in *Oxisols* (Soil Survey Staff 1999) and *Ferralsols* (IUSS Working Group WRB 2007).

These soils exhibit weathering to depths of 10 m or more, particularly where development is on crystalline rocks. The soil profiles are deep, but often lack clearly recognisable horizons. There are differences with depth within these soils, but the differences are transitional rather than abrupt. The soils are often described as structureless with textures often described as clayey, but where the clays are finely aggregated into pseudosands they may feel loamy-textured. Hydrologically, the soils often behave as loamy-textured soils with high hydraulic conductivities, but may also show the biphasic nature of water flow referred to above (see Sect. 1.3).

The intense weathering and rapid leaching results in the rapid breakdown of most minerals and a residual concentration of resistant primary minerals, iron and aluminium oxides and hydrous oxides and quartz. Quartz will often remain as the sand fraction, but most other minerals will be weathered to clay-size materials, with very little silt-size material remaining. These minerals have low cation exchange capacities, which is a characteristic feature of these soils. The soils have organically enriched layers at the surface. Because of the high productivity of natural vegetation systems developed in these soils, there is a large input of organic debris at the surface. Nonetheless, as a consequence of the environmental conditions which are conducive to rapid degradation of these materials, the organically enriched surface layers are not deep. Because of the importance of the nutrients cycled through the vegetation, dense networks of fine roots are often found in the upper layers of these highly weathered soils and within the organic litter at the surface; these roots may access the nutrients released during organic matter decomposition. Stark (1971) and Jordan (1985) have suggested this as evidence for an almost closed nutrient cycle in natural tropical forested environments, where the bulk of the nutrient pool is held in the biomass rather than in the mineral soil.

In summary, these soils, particularly in their subsurface horizons, often have the following broad characteristics:

- Acidic to strongly acidic pH values: pH below 5.5 and often below 5.0
- Low cation exchange capacity: less than 20 $cmol_c/kg$, but often much lower
- Low to very low base saturation: usually below 20%
- Silt: Clay ratio of less than 0.15 and often lower
- Low content of weatherable minerals in the silt and fine sand fraction
- Silica:sesquioxide ratio below 2.0
- Silica:alumina ratio may be below 2.0
- Kaolinite (1:1 type) is the dominant clay mineral with rare 2:1 minerals
- Gibbsite [$Al(OH)_3$] is present
- High exchangeable Al may occur in some soils

Soil Taxonomy identifies the *Oxic* horizon and the WRB the *Ferralic* horizon, with some or all of the above characteristics, as diagnostic of these soil-forming environments.

These soils generally have good physical properties; they are deep, with rapid permeability and strong microstructure. Because of the good structure and rapid permeability, these soils are not normally susceptible to erosion unless poorly managed. Because of the high hydraulic conductivity characteristic of these soils, plants may suffer moisture stress during dry periods. This may be particularly marked if, because of aluminium toxicity in some soil layers, the plant roots are not able to fully exploit the moisture held in the full depth of the soil.

As mentioned above, most nutrients are held in the biomass; for this reason the nutrient pool is rapidly depleted, and the soils will sustain only low levels of renewed plant growth if the process of nutrient cycling is interrupted by removal of the vegetation. On clearance of the natural vegetation, it is therefore essential that strategies are introduced to maintain soil organic matter levels and to recycle nutrients through the biomass. These may involve mulching, manuring, agroforestry or periods of fallow.

In addition to the low levels of plant nutrients provision from the mineral fraction, these soils have a high potential to fix phosphorus. In addition, given the low pH conditions found in these soils, aluminium toxicity may be a constraint on plant growth (see Chap. 10). The conditions of low nutrient status are further exacerbated by the rapid leaching common in these soils. If nutrients are released through normal weathering or added as organic or inorganic fertilisers, they will be rapidly leached from the system.

1.5.2 Soils of the Seasonally Moist Tropics

The zonal soils of the seasonally moist tropics are often associated with vegetation comprising grassland and low-stature, often sparse trees. Such vegetation is known locally as Savanna (Africa) and Cerrado (Brazil). The corresponding soils are characterised by moderate to intense weathering and leaching, with a well-developed clay-enriched B horizon (known as an argillic horizon in Soil Taxonomy and an argic horizon in WRB, which also identifies a *nitic* horizon as a clay-enriched subsurface horizon where the enrichment may be due to illuviation and in situ weathering). There is more input of moisture from precipitation than loss through transpiration and evaporation in some seasons, and during this time water and materials in solution are leached from the soil. Within the CCTA Soil Map of Africa (D'Hoore 1964) these soils were described as *ferruginous soils*, and have broad equivalents in *Ultisols* where the soils have a relatively low base status, and *Alfisols* where the soils have higher base status (Soil Survey Staff 1999), and in *Alisols*, *Nitisols* and some *Lixisols* (IUSS Working Group WRB 2007). The soils exhibit a wide range of colours, but are commonly red, reddish brown or yellowish red. These reddish colours are caused by the separation and dehydration of iron compounds in the sharply contrasting wet and dry seasons.

In contrast to the Ferrallitic soils of the humid tropics, these soils are relatively shallow, with depth to weathered rock not normally exceeding 2.5 m. The less intense weathering and leaching will normally result in higher contents of weatherable minerals. The distinctive feature of these soils is the development of the textural B horizon. In addition to showing the clear "clay bulge" referred to above, this horizon will normally have clear evidence of translocated clays in the form of clay cutans or argillans.

Many of the soils of this climatic zone have higher base status than the zonal soils of the humid tropics, but the range of base saturation is large, from below 30% to as high as 90%. Because the soil is moist for only part of the year, weathering is moderately intense. The contrast between wet and dry parts of the year results in the release of iron oxide during the wetter parts of the year and its dehydration during drier parts. This process, known as rubefication, gives the often characteristic red colours commonly found in these soils. During the wet season(s) the soil will be close to field capacity. During and immediately after rain, excess water will drain rapidly, leaching soluble constituents. In the dry part of the year, the upper horizons dry out and leaching ceases.

Whilst the soils of this climatic zone are diverse, there are some broad characteristics which summarise many of these soils:

- The soils are weakly to moderately acid, with a pH generally above 5.5.
- The silt:clay ratio is low but often above 0.15.
- Clay minerals are predominantly kaolinite and the oxides and hydrous oxides of iron, but some 2:1 minerals are present.
- The cation exchange capacity is normally above 15 $cmol_c$/kg.
- Silica:sesquioxide ratio is normally below 2.0.
- Silica:alumina ratio is above 2.0.
- Gibbsite is very uncommon.

The range of soils developed in this zone will vary in response both to the number and duration of the periods of soil moisture and to the nature of the parent material from which the soils have developed. Soils with lower base saturation, developed on low base status parent materials or where leaching has been more intense, include the *Ultisols* of Soil Taxonomy with a base saturation in the B horizon less than 35%, and the *Alisols* in WRB. Both these soils may also exhibit low pH surface layers, and Al toxicity may occur at relatively shallow depth, inhibiting the growth of some plants. Aluminium toxicity is often more common in the low base status soils in the seasonally moist tropics than under humid tropical conditions. *Alfisols* (Soil Taxonomy) are characteristic of soils developed on higher base status parent materials and where weathering may be less intense. Whilst the WRB allows no direct comparison, the WRB *Lixisols* and *Nitisols* tend to be of higher base status, and correspond to some of the *Alfisols* found in tropical regions.

The soils found in this region vary considerably in their agronomic potential, but most have a higher fertility than soils of the humid tropics. Where the soils are developed on base-rich parent materials, the soils may be considered locally very fertile. The supply of water is markedly seasonal, and plants must be capable of surviving periods of drought or be able to grow during the periods when the soil is moist. Soil organic matter levels are lower in these soils than under rainforest, and a key task when these soils are managed is to maintain soil organic matter levels through additions of manures and plant residues. Because of the relatively high levels of free iron, phosphorus may be fixed and as such be a limiting nutrient for plant growth unless carefully managed. During the dry season, the surface layers from which clay has been lost may harden and, although this is temporary and generally disappears once the rain starts, the initiation of normal land preparation may be delayed as a consequence. If the loss of clay from surface layers is combined with marked declines in soil organic matter, the surface soil may become unstable and the soil may become vulnerable to erosion, particularly at the onset of the rains following the dry season, when the soil is afforded limited protection by plant cover. Particularly vulnerable are poorly managed soils on sloping lands. Management must ensure that the erodibility of the surface soil material is reduced and that infiltration capacity is maintained or improved both to prevent the loss of soil and to ensure that rain infiltrates into the soil.

1.5.3 Other Soils

The two broad groups of soils described above are those which are found predominantly in the tropics and subtropics. Where the climate is *torric*, the soil is moist for less than 90 days. Under these conditions, weathering is severely restricted, and leaching is often a very rare event. Materials which are brought into solution during periods when the soil is moist may be transferred within the soil profile rather than leached out at the base. With this limited leaching, salts may accumulate in the soil profile, giving rise to halomorphic soils; where carbonates accumulate, calcimorphic soils develop.

A widely occurring but not extensive soil in the tropics is the *Vertisol*. The Vertisols are dark, churning, heavy clay soils which develop on relatively flat environments on base-rich parent materials. They are subject to limited leaching losses because of arid conditions and impeded drainage. The soils have a high proportion of clays which swell on wetting and shrink on drying, forming wide cracks from the surface as they dry out. When dry, they are extremely hard and massive, and may break into large prismatic structures. When wet, the soils are sticky and plastic and have a very low permeability. These soils occur within the Tropics under *udic, ustic* and *torric* moisture regimes, with consequent very marked differences in the periods when the soils are wet. Their base status will be strongly influenced by the nature of their parent materials.

Laterite, frequently referred to in the context of the soils of the tropics, is a material for which there are many local vernacular terms. Unfortunately, the materials covered by these terms are exceptionally varied. The first documented description of laterite is by Buchanan (1807). Buchanan was working in India, and described a material that when found in situ could be cut easily, but once left exposed to the atmosphere hardened irreversibly. In its hardened form it was being used locally as bricks, hence the name, which was coined from the Latin for brick, *later*. This contrast between the soft nature of the material when freshly exposed and the hardened nature after exposure has resulted in the material being widely discussed, although it may not be particularly extensive in some parts of the tropics. Since the initial description by Buchanan, the term laterite has been used for a very wide range of both soft and hardened materials.

More recently, the term has been replaced in Soil Taxonomy by the term *plinthite* and in WRB by the *plinthic horizon*. Whilst there are differences in details between the two definitions, they do have broadly similar meanings. They derive not from the Latin for brick but from the Greek! Plinthite is described as an iron-rich, humus-poor mixture of clay with quartz and other minerals. It changes irreversibly to an ironstone hardpan or to irregular aggregates on exposure to repeated wetting and drying, especially if exposed to the heat of the sun.

Plinthite forms by the segregation of iron. The iron may be derived in situ or may have been transferred vertically from other horizons or laterally from other soils. Plinthite normally forms in a horizon that is saturated with water for part of the year. Under these conditions, iron is segregated in the form of soft, more or less clayey, red or dark red redox concretions, formed as a result of iron migration under anaerobic conditions and subsequent deposition in localised concentrations when aerobic conditions prevail; these concretions will irreversibly harden on exposure to repeated wetting and drying. Plinthite is firm when the soil is moist and hard when the soil is dry. Plinthite includes much of what has previously been described as laterite.

1.6 Conclusions

Whilst the tropics cover a diverse range of environments and environmental conditions, in terms of soils and soil development the key characteristics are the relatively high temperatures of the lowland tropics and the rainfall patterns and distributions. Many soil processes operate only when the soil is moist: consequently, the characteristics of the rain regime, with some regions experiencing year-round rainfall and other regions marked wet and dry seasons, will strongly influence the nature of soil development. The relatively high temperatures experienced in the lowland tropics will also markedly affect the rate of the processes which operate.

Many of the temperate landscapes of the globe were markedly affected by Ice Ages during the Pleistocene period, when soil development was severely curtailed. In many cases, soils were stripped from the surface of the land, and soil development had to commence on freshly deposited materials after the ice had retreated. By contrast, the tropics have probably had uninterrupted soil developments in some parts of the region for many hundreds of thousands of years. It is unlikely that all of the conditions of soil formation which prevail today have persisted throughout this period, but nevertheless soils have been subject to suites of development processes over very long time periods, and the nature of the soils is likely to reflect this. As a consequence, many of the soils of the lowland tropics exhibit features which are not seen in soils in other regions.

References

Buchanan FH (1807) A journey from Madras through the countries of Mysore, Canara and Malabar, vol 2. East India Company, London

Crutzen PJ (2002) Geology of mankind–the Anthropocene. Nature 415:23

D'Hoore J (1964) Soil map of Africa scale 1:5000000. Explanatory Monograph. CCTA Publication 93, Lagos

FAO-UNESCO (1974) FAO-UNESCO Soil map of the world, 1: 5000000, vol 1, Legend. UNESCO, Paris

IUSS Working Group WRB (2007) World reference base for soil resources 2006, First update 2007. World Soil Resources Reports No. 103. FAO, Rome

Jenny H (1941) The factors of soil formation: a system of quantitative pedology. McGraw-Hill, San Francisco, CA

Jordan CF (1985) Nutrient cycling in tropical forest ecosystems: principles and their application in conservation and management. Wiley, Chichester

Nortcliff S, Thornes JB (1989) Variations in soil nutrients in relation to soil moisture status. In: Proctor J (ed) Mineral nutrients in tropical forest and savanna ecosystems, Special Publication No. 9 of the British Ecological Society. Blackwell Scientific, Oxford, pp 43–54

Soil Survey Staff (1999) Soil taxonomy — A basic system of soil classification for making and interpreting soil surveys. USDA Agriculture Handbook 436. US Department of Agriculture, Washington, DC

Soil Survey Staff (1960) Soil classification: a comprehensive system, 7th Approximation. US Department of Agriculture, Washington, DC

Stark N (1971) Nutrient cycling: 1. Nutrient distribution in some Amazonian soils. Trop Ecol 12:24–50

Chapter 2
Organic Matter and Micro-Organisms in Tropical Soils

Rainer Georg Joergensen

2.1 Introduction

Organic matter and soil micro-organisms are key components for soil fertility, i.e., the ability of a soil to provide nutrients for plant growth, and for soil quality, i.e., the capacity of a soil to maintain key ecological functions such as decomposition of organic residues (Craswell and Lefroy 2001; Joergensen and Castillo 2001; Cerri et al. 2006). Soil micro-organisms encompass archaea, bacteria, fungi, and protozoa. They maintain the majority of enzymatic processes in soil and preserve energy and nutrients in their biomass (Jenkinson and Ladd 1981). The diversity of soil micro-organisms is enormous and the majority of species are still unknown, although microbial diversity has received much research interest over the past decade (Torsvik and Øvreås 2007).

The maintenance of soil fertility and soil quality is especially important in tropical regions (Bationo et al. 2007). Since temperature controls many processes in soil, especially those mediated by soil micro-organisms, the higher temperature in tropical regions leads to faster turnover rates of microbial biomass and soil organic matter in comparison to temperate climatic conditions (Jenkinson and Ayanaba 1977; Diels et al. 2004; see Chap. 1), shortening the time taken for ecosystems to respond to changes in management practices and increasing the risk of permanent damage (Cerri et al. 2003). In contrast to the nearly constant air temperature, humidity and soil water content are characterized by extreme changes between dry and rainy seasons under tropical savannah (Gijsman et al. 1997; Andersson et al. 2004; Ndiaye et al. 2004) and tropical monsoon climatic conditions (Srivastava and Singh 1991; Manna et al. 2007; Vineela et al. 2008).

R. Georg Joergensen
Department of Soil Biology and Plant Nutrition, University of Kassel, Nordbahnhofstr. 1a, 37213 Witzenhausen, Germany
e-mail: joerge@uni-kassel.de

P. Dion (ed.), *Soil Biology and Agriculture in the Tropics*, Soil Biology 21,
DOI 10.1007/978-3-642-05076-3_2,

Only under equatorial rainforest climatic conditions is microbial activity not limited throughout the year by humidity and soil water content (Henrot and Robertson 1994; Cleveland et al. 2003; Li et al. 2006a,b). Tropical equatorial rainforests have been subjected to heavy logging and agricultural clearance throughout the twentieth century and up to the present day, and the area covered by rainforests around the world is rapidly shrinking (Luizao et al. 1992; Henrot and Robertson 1994; Adachi et al. 2006). In these equatorial areas, agricultural land-use systems are new and may not be sustainable (Motavalli et al. 2000; Dinesh et al. 2003; Bossio et al. 2005). In other tropical regions, especially those with tropical monsoon climatic (Witt et al. 2000) conditions, sustainable agricultural land-use systems have existed for several hundreds of years.

Human impact on soil micro-organisms by changes in land use may lead to adverse effects on soil processes, which is the driving force behind many research activities. Important and typical research objectives have been land-use conversion (Ayanaba et al. 1976; Luizao et al. 1992; Basu and Behera 1993; Prasad et al. 1995; Gijsman et al. 1997; Motavalli et al. 2000; Dinesh et al. 2004a; Bossio et al. 2005; Nogueira et al. 2006), crop rotation (Wick et al. 1998; Oberson et al. 2001; Bünemann et al. 2004; Izquierdo et al. 2003; Adeboye et al. 2006; Vineela et al. 2008), tillage (Salinas-García et al. 2002; Balota et al. 2004; Agele et al. 2005), agroforestry (Mazzarino et al. 1993; Tornquist et al. 1999; Marschner et al. 2002), and conversion of rainforest to plantation (Behera and Sahani 2003; Dinesh et al. 2003; Adachi et al. 2006).

The variability of soils is much higher in tropical regions than in the Luvisol- and Cambisol-dominated landscape of temperate regions. The range of tropical soils goes from young and nutrient rich volcanic Andisols (Joergensen and Castillo 2001; Rabary et al. 2008) to Oxisols (Zhang and Zhang 2003; Bünemann et al. 2004; Xu et al. 2007), Ultisols (Haron et al. 1998; Cantarutti et al. 2002; Cerri et al. 2003) and old landscapes devoid of nutrients after elution by tropical rainfall for millions of years. This variability may mask damage to important soil functions. In addition, the habitat soil needs long periods not only for formation, but also for recovery and restoration. In contrast, soil organisms are generally recognized as highly dynamic, reacting rapidly to environmental conditions (Winding et al. 2005; Bastida et al. 2008). It has been reported that microbial biomass sensitively indicates changes due to soil management, long before other parameters such as soil organic C or total nitrogen (Powlson et al. 1987). Knowledge about soil micro-organisms and soil biological processes may improve the scientific basis of management decisions, e.g., determination of the type of species to be used for planting and cropping in tropical areas.

The aim of the present review is to examine to what extent differences in tropical soil properties and land-use management affect soil organic matter, soil microbial biomass and microbial activity, particularly basal respiration, and modify the relationships between these elements. In preparing the current chapter, due consideration was given to the fact that our present knowledge on this subject is mainly derived from soils developed under temperate humid climatic conditions. After some methodological remarks, the progress of our knowledge on soil micro-organisms in tropical soils is examined, as are the severe gaps in our understanding that need to be addressed by future research.

2.2 Methods Used in Preparation of the Present Review

2.2.1 Data Collection

Data were obtained from the literature for sites located in the tropics between latitude of 24°00′N and 23°40′S, with an altitude not exceeding 2,200 m elevation above sea level (ASL) and a mean annual temperature not falling below 15.3°C. This definition is somewhat arbitrary, as tropics and tropical climate have sometimes been defined differently. One site was considered as an independent sample, i.e., the results of different plots in short-term and small-scale field experiments were combined to obtain one average value (Wick et al. 1998; Balota et al. 2004; Adeboye et al. 2006). Also, the results of different sampling dates throughout the year were averaged to one value (Basu and Behera 1993; Behera and Sahani 2003; Cleveland et al. 2003), as the present review is not focused on seasonal variations of the microbial biomass. Strong seasonal variations of the soil organic C and total N content indicate the occurrence of nondetected spatial variability (Basu and Behera 1993; Sahani and Behera 2001; Saynes et al. 2005; Galicia and García-Oliva 2008).

Basic requirements for including the data in the present review were information on the location of sampling and on the contents of microbial biomass C and soil organic C in $\mu g\ g^{-1}$ soil. Data presented on a per area basis ($g\ m^{-2}$ or $kg\ ha^{-1}$) were recalculated into the respective contents if bulk densities were provided (Cerri et al. 2003; Saynes et al. 2005; Li et al. 2006a). Other soil data collected from the references were soil pH, contents of clay, sand, total N, total P, microbial biomass N, microbial biomass P, basal respiration, ergosterol, ATP, and soil type (USDA classification system). Soils classified according to the WRB-FAO classification system were converted to the USDA classification system (see Chap. 1 for further information on the USDA and WRB classification systems).

Information on land-use systems, mean annual temperature, mean annual precipitation and were collected as important environmental variables. If latitude and ASL were not presented in a particular reference, these data were obtained from Google Earth. Also, data on mean annual temperature and mean annual precipitation were obtained from the Internet if not provided by the respective reference. All sites were separated into four classes according to their land-use management: (1) arable (Table 2.1), (2) plantation (Table 2.2), (3) pasture (Table 2.3), and (4) forest sites (Table 2.4). Plantation sites include agroforestry sites. Pasture sites include meadows, grass savannah, tree savannah, and fallow sites. Forest sites include natural primary and secondary forest sites.

Table 2.1 Arable soils of the tropics: basic parameters and carbon-related microbial properties

Reference, number of sites	Latitude	ASL (m)	Temperature (°C)	Rainfall (mm)	pH (H_2O)	Total N (mg g^{-1} soil)	Microbial biomass C (μg g^{-1} soil)	Microbial biomass C/soil organic C (%)
Islam and Weil (2000), 1	24°00′N	20	26.0	1,800	5.3	0.8	160	2.1
Srivastava and Singh (1991), 1	23°45′N	360	26.5	1,040	6.7	1.1	250	2.3
Shen et al. (2001), 5	23°08′N	10	21.5	1,680	4.5	0.7	290	3.7
Izquierdo et al. (2003), 1	23°02′N	30	24.6	1,300	5.7	3.0	450	2.2
Manna et al. (2007), 6	22°41′N	310	25.5	830	8.1	0.5	310	5.9
Tripathi et al. (2006), 9	22°15′N	3	22.1	1,900	6.2	ND	290	3.0
Bhattacharyya et al. (2005), 1	22°02′N	10	26.5	1,600	5.5	1.7	180	1.3
Das et al. (2007), 2	21°51′N	10	26.5	1,715	7.2	0.9	320	3.2
Prasad et al. (1995), 1	21°22′N	150	25.0	1,400	6.7	1.3	310	2.4
Basu and Behera (1993), 1	19°59′N	260	24.0	1,500	6.7	3.2	350	1.1
Zhang and Zhang (2003), 1	19°31′N	150	23.3	1,830	4.9	ND	220	2.3
Ndiaye et al. (2004), 1	14°47′N	70	26.2	440	ND	0.5	50	0.8
Villenave et al. (2004), 1	14°41′N	20	26.2	460	6.9	0.6	70	0.8
Ndour et al. (2008), 2	14°41′N	20	26.2	460	6.5	1.0	100	1.2
Witt et al. (2000), 16	14°10′N	20	30.0	2,100	6.4	1.0	220[a,b]	2.1
Villenave et al. (2003), 2	13°47′N	20	26.2	440	5.3[c]	0.3	70	1.8
Sall et al. (2006), 1	13°45′N	10	26.2	650	5.5	0.3	50	0.9
Rajashekhara Rao and Siddaramappa (2008), 1	12°58′N	930	23.9	870	6.4	ND	80	1.8
Vineela et al. (2008), 4	12°58′N	930	23.8	920	5.5	ND	140	3.3
Joergensen and Castillo (2001), 25	12°04′N	50	28.5	1,800	6.5	1.5	120	0.7
Bilgo et al. (2007), 4	11°51′N	330	29.0	850	6.6[c]	0.3	42	1.1
Adeboye et al. (2006), 1	11°11′N	680	27.0	1,060	5.7	0.7	260[b]	4.5
Mazzarino et al. (1993), 1	9°53′N	600	22.0	2,640	5.0	2.6	130[a,b]	0.3
Lulu and Insam (2000a, b), 13	9°01′N	2200	16.5	1,110	6.6	1.9	320	1.7
Wick et al. (1998), 3	7°30′N	210	26.2	1,250	6.2	0.5	120	2.2
Ayanaba et al. (1976), 12	7°23′N	210	26.5	1,120	6.4	0.9	190	1.8

Mulongoy (1986), 1	7°23′N	210	26.5	1,300	5.6	1.9	170	1.0
Tian (1998), 1	7°23′N	210	26.5	1,300	5.8	ND	240	2.4
Agele et al. (2005), 1	7°05′N	270	26.2	1,250	6.7	2.7	370[a]	1.1
Oberson et al. (2001), 1	4°30′N	150	27.0	2,240	4.7	1.7	160[a,b,d]	0.6
Gijsman et al. (1997), 5	4°00′N	200	27.0	2,200	4.4	1.3	450[a,b]	1.5
Bünemann et al. (2004), 4	0°09′N	1480	18.0	960	5.0	1.8	280[d]	1.1
Bossio et al. (2005), 7	0°02′N	1540	18.5	1,700	5.1	ND	160[a]	0.8
Rabary et al. (2008), 4	19°46′S	1600	16.9	1,450	5.7	3.8	340	0.7
Fernandes et al. (2005), 2	22°47′S	570	22.0	1,430	5.8	1.1	140[a,b]	0.9
Nogueira et al. (2006), 4	23°27′S	580	20.9	1,620	6.0[c]	ND	660[a]	2.7
Balota et al. (2004), 6	23°40′S	580	22.4	1,990	4.6	ND	234	1.4

Sequentially shown are: reference, number of sites analyzed, site parameters (latitude, elevation above sea level (ASL), mean annual temperature, mean annual rainfall), soil pH, soil total N, soil microbial biomass C, soil microbial biomass C/soil organic C ratio; *ND* not determined

[a]corrected k_{EC} value

[b]corrected k_{EN} value

[c]adjusted soil pH

[d]corrected k_{EP} value

Table 2.2 Plantation soils of the tropics: basic parameters and carbon-related microbial properties

Reference, number of sites	Latitude	ASL (m)	Temperature (°C)	Rainfall (mm)	pH (H_2O)	Total N (mg g^{-1} soil)	Microbial biomass C (µg g^{-1} soil)	Microbial biomass C/soil organic C (%)
Xu et al. (2007), 1	21°56′N	720	22.0	1,400	4.6	1.0	230	2.3
Zhang and Zhang (2003), 3	19°31′N	150	23.3	1,830	4.5	ND	180	1.9
Zhang et al. (2007), 2	19°28′N	150	23.3	1,830	4.5	0.7	200	2.0
Templer et al. (2005), 1	19°00′N	50	25.2	2,000	5.5	ND	1,510[a,b]	1.9
Dinesh et al. (2003), 4	11°37′N	40	30.1	3,100	5.1	0.8	160	2.2
Dinesh (2004), 5	11°37′N	40	30.1	3,100	5.2	0.7	290	3.9
Dinesh et al. (2004a), 5	11°37′N	40	30.1	3,100	5.9[e]	1.3	390	2.9[d]
Dinesh et al. (2004b), 3	11°37′N	40	30.1	3,100	5.8[e]	0.9	190	2.2
Tornquist et al. (1999), 1	10°26′N	40	26.5	3,910	4.8	3.7	1,260[a,b]	3.4
Adachi et al. (2006), 1	2°50′N	20	27.5	1,900	4.7[e]	1.2	80	0.5
Haron et al. (1998), 3	1°02′N	50	27.0	2,000	4.4	1.4	260	1.5
Bossio et al. (2005), 1	0°02′N	1540	18.5	1,700	4.1	ND	290[c]	1.0
Förster et al. (2006), 1	3°07′S	90	26.9	2,000	4.0	2.1	220	0.8
Marschner et al. (2002), 2	3°08′S	50	25.8	2,530	4.3	ND	480[c]	1.6
Turgay et al. (2002), 4	5°08′S	650	24.0	1,750	5.3	2.6	660	2.1
Waldrop et al. (2000), 4	17°30′S	270	26.0	1,730	4.1	1.9	220	0.6

Sequentially shown are: reference, number of sites analyzed, site parameters (latitude, elevation ASL, mean annual temperature, mean annual rainfall), soil pH, soil total N, soil microbial biomass C, soil microbial biomass C/soil organic C ratio; *ND* not determined

[a]corrected k_C value

[b]corrected k_N value

[c]corrected k_{EC} value

[d]corrected soil organic C content

[e]adjusted soil pH

Table 2.3 Savannah soils of the tropics: basic parameters and carbon-related microbial properties

Reference, number of sites	Latitude	ASL (m)	Temperature (°C)	Rainfall (mm)	pH (H_2O)	Total N (mg g^{-1} soil)	Microbial biomass C (μg g^{-1} soil)	Microbial biomass C/soil organic C (%)
Islam and Weil (2000), 1	24°00′N	20	26.0	1,800	5.6	1.1	390	3.1
Srivastava and Singh (1991), 1	23°45′N	360	26.5	1,040	7.0	1.1	400	3.3
Izquierdo et al. (2003), 2	23°02′N	30	24.6	1,300	5.9	3.1	550	2.3
Sahani and Behera (2001), 1	22°00′N	170	25.0	1,670	ND	1.4	150	2.7
Behera and Sahani (2003), 1	22°00′N	170	25.0	1,960	6.3	2.4	250	1.7
Prasad et al. (1995), 1	21°22′N	150	25.0	1,400	6.6	1.5	420	2.8
Basu and Behera (1993), 1	19°59′N	260	24.0	1,500	6.6	3.6	400	1.1
Zhang and Zhang (2003), 1	19°31′N	150	23.3	1,830	5.1	ND	120	1.4
Galicia and García-Oliva (2004), 6	19°30′N	100	24.6	750	7.0	2.0	340[c,d]	1.1
Galicia and García-Oliva (2008), 3	19°30′N	100	24.6	750	7.2	2.2	490	1.6
Toledo-Aceves and Garcia-Oliva (2008), 4	19°29′N	240	24.6	750	7.0	8.0	1,960[a,b]	3.0
Zhang et al. (2007), 1	19°28′N	150	23.3	1,830	4.5	0.7	220	2.0
Geissen and Morales Guzman (2006), 1	17°29′N	600	25.4	3,900	5.6[e]	3.0	170	0.5
Sierra and Marbán (2000), 2	16°15′N	40	25.5	4,000	5.5	1.9	190[c]	1.1
Sall et al. (2006), 1	13°45′N	10	26.2	650	6.1	1.0	120	0.8
Bilgo et al. (2007), 8	11°51′N	330	29.0	850	6.6[e]	0.4	70	1.2
Groffman et al. (2001), 2	10°30′N	90	27.5	1,600	5.4	3.9	1,140[d]	2.5
Tornquist et al. (1999), 1	10°26′N	40	26.5	3,910	5.1	4.2	490[a,b]	1.3
Henrot and Robertson (1994), 2	9°50′N	670	26.0	4,300	5.5	2.1	950	2.9
Cleveland et al. (2003), 2	8°43′N	160	26.0	5,100	5.5	4.5	780	1.5
Michelsen et al. (2004), 3	8°06′N	580	28.0	1,100	6.8	ND	520	1.3
Andersson et al. (2004), 1	8°01′N	550	28.5	900	7.0[e]	1.0	1,360	3.4
Ayanaba et al. (1976), 3	7°23′N	210	26.5	1,120	6.7	1.1	330	2.8

(continued)

Table 2.3 (continued)

Reference, number of sites	Latitude	ASL (m)	Temperature (°C)	Rainfall (mm)	pH (H_2O)	Total N (mg g^{-1} soil)	Microbial biomass C (µg g^{-1} soil)	Microbial biomass C/soil organic C (%)
Agele et al. (2005), 1	7°05′N	270	26.2	1,250	7.3	3.4	560[a]	1.1
Oberson et al. (2001), 2	4°30′N	150	27.0	2,240	4.9	1.7	350[a,b,f]	1.3
Gijsman et al. (1997), 1	4°00′N	200	27.0	2,200	4.4	1.2	440[a,b]	1.7
Luizao et al. (1992), 1	3°08′S	30	24.6	2,100	3.7	ND	990	4.0
Cerri et al. (2003), 1	10°10′S	140	25.6	2,200	6.2	1.2	260[a]	1.5
Cerri et al. (2006), 5	10°10′S	140	25.6	2,200	5.9	1.2	560[a]	3.9
Cantarutti et al. (2002), 1	16°40′S	110	24.0	1,300	5.5	0.8	130[a,b]	1.4
de Oliveira et al. (2004), 4	17°13′S	660	25.7	1,750	5.1	1.2	150[a,b]	0.8
Waldrop et al. (2000), 1	17°30′S	270	26.0	1,730	4.6	4.0	860	1.1
Rabary et al. (2008), 4	19°46′S	1600	16.9	1,450	5.7	5.0	390	0.6
Northup et al. (1999), 5	20°11′S	220	24.7	530	6.7	0.5	220	2.5

Sequentially shown are: reference, number of sites analyzed, site parameters (latitude, elevation ASL, mean annual temperature, mean annual rainfall), soil pH, soil total N, soil microbial biomass C, soil microbial biomass C/soil organic C ratio; *ND* not determined

[a]corrected k_{EC} value

[b]corrected k_{EN} value

[c]corrected k_C value

[d]corrected k_N value

[e]adjusted soil pH

[f]corrected k_{EP} value

Table 2.4 Forest soils of the tropics: basic parameters and carbon-related microbial properties

Reference, number of sites	Latitude	ASL (m)	Temperature (°C)	Rainfall (mm)	pH (H_2O)	Total N (mg g^{-1} soil)	Microbial biomass C (μg g^{-1} soil)	Microbial biomass C/soil organic C (%)
Islam and Weil (2000), 2	24°00′N	20	26.0	1,800	5.0	1.1	300	2.9
Srivastava and Singh (1991), 1	23°45′N	360	26.5	1,040	6.4	2.2	610	2.8
Sahani and Behera (2001), 1	22°00′N	170	25.0	1,670	ND	2.6	900	3.8
Behera and Sahani (2003), 2	22°00′N	170	25.0	1,960	6.3	2.2	740	3.5
Xu et al. (2007), 1	21°56′N	720	22.0	1,400	4.2	1.4	380	2.4
Mao et al. (1992), 7	21°28′N	40	23.0	1,600	4.6	0.3	120	1.4
Prasad et al. (1995), 1	21°22′N	150	25.0	1,400	6.9	2.0	640	3.2
Basu and Behera (1993), 1	19°59′N	260	24.0	1,500	6.7	6.7	820	1.3
Montano et al. (2007), 2	19°29′N	240	24.6	750	6.5	2.2	530	1.8
Toledo-Aceves and Garcia-Oliva (2008), 1	19°29′N	240	24.6	750	7.4	4.0	1,250[a,b]	3.0
Templer et al. (2005), 1	19°0′N	50	25.2	2,000	6.0	ND	2,830[a,b]	1.9
Saynes et al. (2005), 4	18°28′N	1070	24.5	850	7.1	2.1	570[c]	1.4
Li et al. (2006a), 2	18°18′N	400	22.1	3,500	4.9	ND	400	0.8
Salamanca et al. (2002, 2006), 8	16°06′N	30	28.0	2,300	6.5	1.3	380	2.9
Duponnois et al. (2005), 1	14°10′N	10	25.0	700	5.2	3.9	80	0.2
Motavalli et al. (2000), 1	13°31′N	110	27.3	2,160	7.3	ND	1,400[d]	2.7
Dinesh et al. (2006), 16	12°10′N	1	30.1	3,100	5.5	1.4	410	2.6
Dinesh et al. (2003), 4	11°37′N	40	30.1	3,100	5.2	1.9	600	2.5
Dinesh et al. (2004b), 3	11°37′N	40	30.1	3,100	5.9[e]	1.7	390	2.9
Groffman et al. (2001), 4	10°30′N	90	27.5	1,600	5.8	3.3	1,060	3.9
Vance and Nadkarni (1990), 1	10°18′N	1550	20.0	2,000	4.7	9.8	1,950	1.3
Mazzarino et al. (1993), 1	9°53′N	600	22.0	2,640	5.0	2.8	160[c,d]	0.4
Henrot and Robertson (1994), 2	9°50′N	670	26.0	4,300	5.7	2.4	1,590	4.3
Cleveland et al. (2003), 2	8°43′N	160	26.0	5,100	5.7	6.0	1,300	2.0
Jensen et al. (2001), 1	8°09′N	550	28.5	950	6.2	0.8	230[c]	1.4
Michelsen et al. (2004), 2	8°09′N	580	28.5	900	6.8	ND	270	1.1
Wick et al. (1998), 1	7°30′N	210	26.2	1,250	6.7	0.9	360	3.2
Ayanaba et al. (1976), 5	7°23′N	210	26.5	1,120	6.4	0.9	220	2.0

(continued)

Table 2.4 (continued)

Reference, number of sites	Latitude	ASL (m)	Temperature (°C)	Rainfall (mm)	pH (H_2O)	Total N (mg g^{-1} soil)	Microbial biomass C (μg g^{-1} soil)	Microbial biomass C/soil organic C (%)
Tian (1998), 1	7°23′N	210	26.5	1,300	6.4	ND	550	3.5
Priess and Fölster (2001), 3	5°00′N	1180	20.6	2,200	4.9[e]	6.9	1,260	1.1
Adachi et al. (2006), 4	2°50′N	20	27.5	1,900	4.0[e]	1.4	60	0.3
Bossio et al. (2005), 2	0°02′N	1540	18.5	1,800	4.6	ND	380	0.8
Luizao et al. (1992), 1	3°08′S	30	24.6	2,100	3.7	ND	940	3.1
Turgay et al. (2002), 4	5°08′S	650	24.0	1,750	5.3	3.2	950	2.6
Cerri et al. (2003), 1	10°10′S	140	25.6	2,200	4.9	1.1	200[d]	1.7
Cantarutti et al. (2002), 1	16°40′S	110	24.0	1,300	5.5	0.8	180[c,d]	2.0
de Oliveira et al. (2004), 1	17°13′S	660	25.7	1,750	5.0	1.0	150[c,d]	1.0
Nogueira et al. (2006), 5	23°27′S	580	20.9	1,620	6.0[e]	ND	810[d]	2.9

Sequentially shown are: reference, number of sites analyzed, site parameters (latitude, elevation ASL, mean annual temperature, mean annual rainfall), soil pH, soil total N, soil microbial biomass C, soil microbial biomass C/soil organic C ratio; *ND* not determined

[a]corrected k_C value

[b]corrected k_N value

[c]corrected k_{EN} value

[d]corrected k_{EC} value

[e]adjusted soil pH

2.2.2 Soil Chemical Properties

The soil pH was usually measured in water suspension. If the pH was measured in 0.01 M $CaCl_2$ or 1 M KCl solution, the data were converted to pH in water by multiplication by 1.1, this ratio having been obtained from those references where two pH values were presented (Mao et al. 1992; Joergensen and Castillo 2001; Cerri et al. 2006). The majority of soil organic C values were obtained by the Walkley-Black method (Srivastava and Singh 1991; Mazzarino et al. 1993; Motavalli et al. 2000), using potassium dichromate sulphuric acid oxidation as described, for example, by Nelson and Sommers (1982). Loss of ignition was sometimes used to determine soil organic C (Northup et al. 1999; Michelsen et al. 2004). Total N was measured in the majority of cases by the Kjeldahl procedure (Srivastava and Singh 1991; Northup et al. 1999; Motavalli et al. 2000). On the other hand, gas chromatographic determination of total C and total N were often used by different authors (Joergensen and Castillo 2001; Ndiaye et al. 2004; Li et al. 2006a, b). A variety of other methods were also applied for determining organic C and total N in soil.

2.2.3 Microbial Biomass Indices

Microbial biomass C was obtained by fumigation extraction (FE) (Vance et al. 1987) at 280 sites or 75.3% of the data, by fumigation incubation (FI) (Jenkinson and Powlson 1976) at 67 sites or 18.0% of the data, by substrate-induced respiration (Anderson and Domsch 1978; Kaiser et al. 1992; Lulu and Insam 2000a, b) at 18 sites or 4.9% of the data, by ATP (Jenkinson and Oades 1979; Adachi et al. 2006) at five sites or 1.3% of the data, and by direct microscopy (Babiuk and Paul 1970; Lodge and Ingham 1991; Li et al. 2006a) at two sites or 0.5% of the data. The microbial biomass C values obtained by FE were corrected if the conversion value of 0.38 for dichromate oxidation (Vance et al. 1987) or 0.45 for ultraviolet or oven oxidation (Wu et al. 1990; Joergensen 1996a) had not been used. At 45 sites, the microbial biomass C data were obtained by measuring ninhydrin-reactive N according to Amato and Ladd (1988) or Joergensen and Brookes (1990). The microbial biomass C values obtained by FI were corrected if the conversion value of 0.45 (Jenkinson and Powlson 1976; Joergensen 1996a) had not been used. Data were not used if the FI method was obviously incorrectly applied (Salinas-García et al. 2002).

The soil microbial biomass is not completely extractable (FE method) or decomposable after fumigation (FI method). For this reason, conversion values must be used to account for the nonextractable (FE method) or the nondecomposable (FI method) fraction (Jenkinson 1988). The conversion of measured data into microbial biomass values is an important step for assessing the flux of energy and elements (Jenkinson and Ladd 1981). Biomass values serve as an internal control by relating microbial growth and CO_2 evolution data, and they help to prove the soundness of the values in comparison with other data. As the calibration of conversion values is difficult and erroneous, the use of a single standard conversion value for each element is recommended also for tropical soils (Joergensen 1996a; Joergensen and Mueller 1996). In particular, the calibration of the FE method was not affected by soil pH and clay content (Kaiser et al. 1992). Only the conversion of ninhydrin-reactive N to microbial biomass C leads to underestimation below pH 4.8 according to Joergensen (1996b), using the constant factors of Amato and Ladd (1988) and Joergensen and Brookes (1990).

ATP was converted to microbial biomass C on the basis of the ATP to microbial biomass C ratio of 10.5 $\mu mol\ g^{-1}$ obtained by Contin et al. (2001). Direct microscopic counts were converted to microbial biomass C assuming an average C content of 47% (Jenkinson and Ladd 1981).

Microbial biomass N was obtained by FE (Vance et al. 1987) at 141 sites or 89.8% of the data and by FI (Jenkinson and Powlson 1976) at 16 sites or 10.2% of the data. The microbial biomass N values obtained by FE were corrected if the conversion value of 0.54 had not been used (Brookes et al. 1985; Joergensen and Mueller 1996). The microbial biomass N values obtained by FI were corrected if the conversion value of 0.68 (Jenkinson and Powlson 1976) had not been used. Microbial biomass P can only be measured by FE (Brookes et al. 1982). The microbial biomass P values were corrected if the conversion value of 0.40 had not been used (Brookes et al. 1982).

2.2.4 Basal Respiration

Basal respiration was usually measured in a closed glass vessel for between 10 h and 10 days at between 20 and 30°C. In some cases, the data were unit-corrected if they were suspiciously high or low. All data were adjusted, if necessary, to a 25°C rate per day using the rate-modifying factor (f) for temperature (= T in °C) effects of the Rothamsted Carbon model (Jenkinson et al. 1987): $f = 47.9/(1 + e^{(106/(T + 18.3)})$.

2.2.5 Statistical Evaluation

The results presented in Tables 2.1–2.4 are arithmetic means and expressed on an oven-dry basis. Those in Tables 2.5 and 2.6 are arithmetic or geometric means. The significance of land-use or soil-type-specific effects was tested by one-way analysis of variance using the PLSD-test (Fisher's protected least significant difference). Relationships between selected soil biological and chemical properties were analyzed by regression analysis. Data were ln-transformed if they did not fit a normal distribution. All statistical analyses were performed using StatView 5.0 (SAS Institute Inc.).

Table 2.5 General and land use-specific means of site characteristics and chemical properties in soils of the tropics covered in the references used in this review

Characteristic	Type of soil				
	General	Arable	Plantation	Pasture	Forest
Latitude	13°20′ (372)	13°28′ *ab* (152)	11°17′ *b* (42)	13°50′ *a* (73)	13°39′ *ab* (105)
Mean annual temperature (°C)	25.4 (372)	24.4 *b* (152)	27.0 *a* (42)	25.9 *a* (73)	26.0 *a* (105)
Mean annual precipitation (mm)	1,800 (372)	1,510 *c* (152)	2,420 *a* (42)	1,660 *c* (73)	2,070 *b* (105)
Sampling depth (mm)	160 (371)	180 *b* (152)	230 *a* (42)	110 *c* (73)	130 *c* (104)
Clay (%)	30 (263)	35 *a* (127)	30 *ab* (31)	21 *c* (45)	26 *bc* (61)
pH (H_2O)	5.9 (369)	6.1 *a* (151)	5.0 *c* (42)	6.1 *a* (72)	5.7 *b* (104)
Soil organic C (mg g^{-1} soil)[a]	16.3 (372)	12.5 *c* (152)	14.9 *bc* (42)	17.7 *b* (73)	23.2 *a* (105)
Total N (mg g^{-1} soil)[a]	1.3 (305)	1.0 *c* (112)	1.3 *bc* (35)	1.3 *b* (68)	1.7 *a* (90)
Total P (mg g^{-1} soil)[a]	0.31 (89)	0.48 *a* (41)	0.32 *ab* (2)	0.20 *b* (32)	0.22 *b* (14)
Soil organic C/total N[a]	12.3 (305)	11.9 *bc* (112)	11.1 *c* (35)	13.4 *a* (68)	12.5 *ab* (90)

For each combination of characteristic and soil type, the number of observations is shown in parentheses; figures within a row followed by different letters in italics are significantly different according to the PLSD-test
[a]geometric mean

Table 2.6 General and land use-specific means of biological properties in soils of the tropics covered in the references used in this review

Characteristic	Type of soil				
	General	Arable	Plantation	Pasture	Forest
Microbial biomass C (μg g^{-1} soil)[a]	270 (372)	190 *c* (152)	280 *b* (42)	280 *b* (73)	450 *a* (105)
Microbial biomass N (μg g^{-1} soil)[a]	39 (157)	24 *b* (46)	32 *b* (24)	49 *a* (26)	56 *a* (61)
Microbial biomass P (μg g^{-1} soil)[a]	3.9 (46)	2.9 *b* (37)	ND	14.3 *a* (5)	14.8 *a* (4)
Basal respiration (CO_2-C g^{-1} soil d^{-1})[a]	13 (192)	11 *b* (89)	10 *b* (22)	12 *b* (19)	20 *a* (62)
Ergosterol (μg g^{-1} soil)[a]	0.60 (61)	0.29 *c* (26)	0.42 *b* (16)	ND	1.39 *a* (19)
ATP (nmol g^{-1} soil)[a]	1.6 (53)	ND	1.0 *b* (18)	ND	2.1 *a* (35)
Microbial biomass C/N	10.5 (157)	12.0 *a* (46)	11.0 *ab* (24)	7.5 *c* (26)	10.4 *b* (61)
Microbial biomass C/P[a]	49 (46)	52 *a* (37)	ND	27 *a* (5)	61 *a* (4)
Ergosterol/microbial biomass C (%)[a]	0.27 (61)	0.28 *a* (26)	0.27 *a* (16)	ND	0.27 *a* (19)
ATP/microbial biomass C (μmol g^{-1})	5.3 (48)	ND	4.3 *b* (17)	ND	5.8 *a* (31)
Microbial biomass C/soil organic C (%)[a]	1.7 (374)	1.5 *b* (152)	1.9 *a* (42)	1.6 *ab* (73)	1.9 *a* (105)
qCO_2 (mg CO_2-C g^{-1} biomass C d^{-1})[a]	52 (192)	59 *a* (89)	35 *b* (22)	61 *a* (19)	48 *ab* (62)

For each combination of characteristic and soil type, number of observations is shown in parentheses; figures within a row followed by different letters in italics are significantly different according to the PLSD-test; *ND* not determined
[a]geometric mean

2.3 Environmental Variables

Data were obtained from 26 countries in tropical regions. About 38% of the data were obtained from Asia (11% of the total number from the Andaman Islands and 10% from India), 34% from Latin America (11% from Brazil and 7% from Nicaragua), 24% from Africa (8% from Nigeria and 5% from Ethiopia), and 3% from other tropical regions such as Australia, Guam and Tahiti. However, no information is available on soil micro-organisms for many tropical countries in refereed international journals.

The mean latitude of all sampling sites was 13°20′, which is close to the midpoint between the equator and the 24°00′ latitude, which were set as northern and southern boundaries (Table 2.5). The soils of most land-use forms are evenly distributed over the tropics, except the plantation soils, which were on average closest to the equator and had the highest mean annual temperature of 27.0°C and the highest mean annual precipitation of 2,420 mm. The plantation soils also had the lowest mean soil pH of 5.0. In general, the tropical soils were moderately acidic, with a mean soil pH of 5.9. They had a considerable clay content of 30%, which was highest in the group of arable soils. The lowest soil pH of around 5.0 was observed in Oxisols and Inceptisols, the highest soil pH of around 6.6 in Vertisols and Alfisols. Soil types were assigned for 290 sites or 78% of the total number, which decreased in percent of the total number from Oxisols (22%) > Entisols (18%) > Ultisols (11%) > Vertisols (9%) > Andisols (8%) > Alfisols (5%) > Inceptisols (4%) > Mollisols (1%).

The mean contents of soil organic C and total N were 16.3 and 1.3 mg g^{-1} soil respectively (Table 2.5). These contents declined in the order forest > pasture > plantation > arable soils. The soil C/N ratio varied in most soils between 12 and 14 and was lowest at 11.1 in the plantation soils. It is obvious that tropical forest soils contain considerable amounts of organic matter in the A-horizon of the mineral soil. Decomposition processes not only occur in the litter layer, which was only analyzed by a minority of references (Vance and Nadkarni 1990; Priess and Fölster 2001; Li et al. 2006a, b), but also in the A horizon of the mineral soil. Soil organic matter decreased apparently with increasing intensity of human land use and decreasing C input by the vegetation. However, only a minority of references presented data on the annual production of aboveground plant biomass (Duponnois et al. 2005; Li et al. 2006a, b), which was usually related to the C input by roots (Marschner et al. 2002; Galicia and García-Oliva 2008). The belowground plant biomass is an important source of organic matter in most tropical soils (Sierra and Nygren 2005; Kaonga and Coleman 2008), as harvest residues are often used as energy sources or removed by burning (Castillo and Joergensen 2001; Bationo et al. 2007). The use of crop residue mulch and compost as C and nutrient source for arable soils is still rare (Izquierdo et al. 2003; Bhattacharyya et al. 2005).

Information on total P is rare and only available for 24% of the soils. This is surprising, as total P is very low in many tropical soils, especially in Oxisols (Gijsman et al. 1997; Northup et al. 1999; Andersson et al. 2004; Bilgo et al. 2007; Zhang et al. 2007). P deficiency not only limits plant growth in tropical soils, it also has strong negative effects on microbial growth (Duah-Yentumi et al. 1998; Salamanca et al. 2006; Gnankambary et al. 2008).

2.4 Soil Microbial Biomass

The mean content of all microbial biomass C data was 270 μg g^{-1} soil (Table 2.7), ranging from 20 to 2,830 μg g^{-1} soil (Fig. 2.1a). Microbial biomass C was significantly highest in the forest soils, on a medium level in pasture and plantation soils

Table 2.7 Correlation coefficients between soil organic matter content, soil total N content and soil microbial properties in soils of the tropics covered in the references used in this review

	Total N	Microbial biomass C	Microbial biomass N	Basal respiration	Ergosterol	ATP
Soil organic C	0.92 (305)	0.66 (372)	0.78 (157)	0.58 (192)		0.52 (53)
Total N		0.68 (305)	0.85 (148)	0.63 (159)		0.62 (53)
Microbial biomass C			0.85 (157)	0.67 (192)	0.73 (61)	0.88 (48)
Microbial biomass N				0.60 (77)		0.85 (48)
Basal respiration					0.72 (52)	0.89 (48)
Ergosterol						0.82 (48)

For each combination of characteristics analyzed, the number of observations is indicated in parentheses. All correlation coefficients are significant at $P < 0.0001$

and significantly lowest in the arable soils, reaching only 40% of the forest soils (Table 2.7). The mean content of all microbial biomass N data was 39 μg g^{-1} soil, ranging from 7 to 353 μg g^{-1} soil (Fig. 2.1b). However, the differences between the four groups of land-use systems were less significant, as the amount of data was 60% lower (Table 2.7). The mean of all microbial biomass C/N ratios was 10.5; it was about 7.5 in the pasture soils and was, consequently, significantly lower for pasture that for the other three groups of land-use management, where the mean C/N ratio was 11.1. Microbial biomass C and microbial biomass N were closely interrelated and also showed significant correlations with soil organic C and total N (Table 2.7). However, microbial biomass N showed closer correlations to soil organic C and total N than microbial biomass C. Microbial biomass P followed microbial biomass C with an average C/P ratio of 49 (Table 2.7). The differences between the land-use systems were not significant.

The total geometric mean for microbial biomass C in tropical soils is only 20% below the worldwide average of 330 μg g^{-1} soil reported by Wardle (1998). This is amazing, considering that the decomposition of plant residues and the turnover of soil organic matter is much higher under tropical conditions (Jenkinson and Ayanaba 1977; Cerri et al. 2003; Diels et al. 2004). This means that the C input by plants must be much higher in tropical ecosystems or that the turnover is only increased for a short time after fresh organic residues have entered the soil. However, low pH values may counteract positive temperature effects on the microbial turnover of organic matter (Wardle 1998). The relative difference between arable and forest soils was markedly stronger for microbial biomass C than for soil organic C. This indicates that the microbial biomass content reacts much more strongly on the C input by plants than does the soil organic content, as suggested by Powlson et al. (1987). Another reason might be that tillage leads to a dilution of microbial biomass on a larger soil volume with a higher bulk density, which is accentuated by a larger sampling depth of the arable soils in comparison to that of the forest soils. Water and wind erosion, especially when the soils are bare or the crops are small and developing, may additionally lead to losses of organic matter in the surface horizon

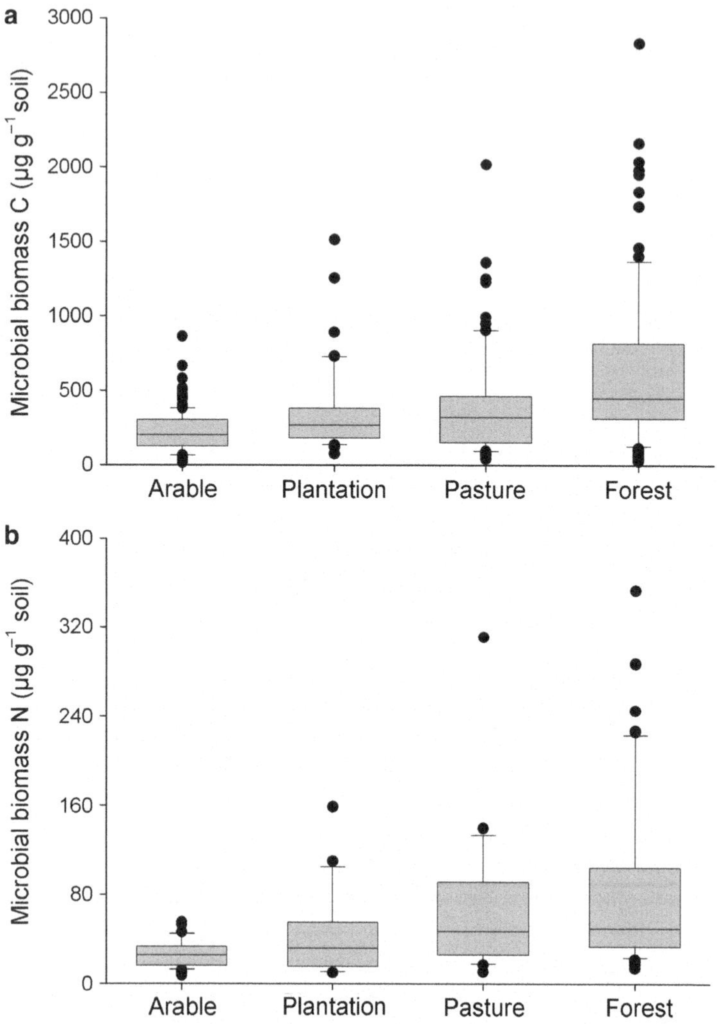

Fig. 2.1 Land use-specific box plots for the content of (**a**) microbial biomass C and (**b**) microbial biomass N in soils of the tropics. For each type of soil, the box plots show the smallest observation, the lower quartile, the median, the upper quartile, and the largest observation, with observations considered as outliers being indicated as *dots*

(Bationo et al. 2007). Irrigation of arable sites may increase the turnover of microbial biomass and soil organic matter during dry periods, but the reverse is also possible, depending on the interaction of water and oxygen supply (Yavitt et al. 1993; Witt et al. 2000; Ndour et al. 2008). Wildfires and prescribed burning reduce the aboveground biomass, but may have positive effects on the belowground C inputs, as suggested by the observed increases of the soil microbial biomass in surface horizons (Jensen et al. 2001; Andersson et al. 2004; Michelsen et al. 2004).

Fires may also increase the soil organic C content, due the incorporation of charcoaled organic residues (Glaser et al. 2002).

The reasons for the relatively large microbial biomass C/N ratios exceeding 10 in tropical soils compared to arable soils from a humid climate are not fully understood (Joergensen and Emmerling 2006). A sometimes stated explanation is the possibility of fungal dominance in soils with a large microbial biomass C/N ratio (Dilly et al. 2003; Moore et al. 2005). However, as explained below, a shift towards fungi is not likely to be the only explanation for the observed wide microbial biomass C/N ratios. In some cases, P limitation has been demonstrated as a possible explanation (Salamanca et al. 2006). However, limitation of other nutrients such as S and Zn might also be responsible for the large C/N ratio of the microbial biomass in tropical soils. Changes in the microbial biomass C/N ratio may also indicate segregation of the microbial community according to their enzymatic and catabolic potential, along with organic matter quality and energy flux, as discussed below (Lulu and Insam 2000a)

The mean microbial biomass C/P ratio of the tropical soils was more than four times the mean of 11 measured in 44 German agricultural soils (Jörgensen 1995) and roughly twice the mean value of around 25 found in sub-tropical India (Singh et al. 1991; Srivastava and Lal 1994). The microbial biomass C/P ratio shows considerably more variation than the microbial biomass C/N ratio (Anderson and Domsch 1980), contrasting the view of Cleveland and Liptzin (2007) that rather uniform microbial biomass C/P ratios exist in soil. Its variability is caused by the possibility for bacteria, and also for fungi, to store excess P in the form of cell-internal polyphosphates and cell wall components, e.g., teichoic acids (Grant 1979; Nielsen et al. 2002; Oberson and Joner 2005). These storage components can be used as a phosphate source in times of P limitation (Oberson and Joner 2005). The application of inorganic P and organic P components to soil significantly decreased the microbial biomass C/P ratio and strongly reflects P availability to soil micro-organisms (Oehl et al. 2004; Muhammad et al. 2007). The microbial biomass C/P ratio is increased by low P availability, but also by low N availability in combination with high C availability (Anderson and Domsch 1980). However, it is possible that microbial biomass P has been underestimated due to methodological problems, e.g., with alkaline extractants in acidic soils, especially in the presence of aluminum and iron oxides (Wu et al. 2000).

2.5 Basal Respiration Rate

The mean basal respiration was 13 $\mu g\ CO_2\ g^{-1}$ soil d^{-1} and increased in the order plantation < arable < pasture < forest soils. However, only the differences between forest and arable soils and forest and plantation soils were significant, due to the large variation within the four soil groups. Basal respiration is highly affected by the incubation conditions, e.g., temperature and the sample pretreatment, especially sieving (Joergensen 1996a). The mean basal respiration rate for tropical soils was only slightly above the mean basal respiration rate of 11 μg CO_2-C g^{-1} soil measured in 62 German soils (Jörgensen 1995). This indicates that most data obtained from the references listed in Tables 2.1–2.4 were accurate and suggests that similar levels of mineralizable soil organic matter exist in tropical soils and soils from a temperate humid climate.

2.6 Microbial Biomass C-to-Soil Organic C Ratio and qCO_2

The mean microbial biomass C-to-soil organic C ratio was 1.7% in all soils (Table 2.7), ranging from 0.1 to 8.2% (Fig. 2.2a). It was lowest in the arable soils and highest in the plantation and forest soils (Table 2.7). The microbial biomass C-to-soil organic C ratio is an important index for organic matter availability to the soil microbial community (Anderson and Domsch 1989). This suggests a better availability of organic matter in soils with permanent tree vegetation than in arable and grassland soils. However, the microbial biomass C-to-soil organic C ratios are generally lower in tropical soils than in soils from a humid temperate climate, which was 2.6% in 26 arable soils (Anderson and Domsch 1989) or 2.2% in 27 arable and 2.3% in 17 grassland soils (Jörgensen 1995). The microbial biomass C-to-soil organic C was exceptionally low in Andisols, with a mean of 0.6%, due to aluminium oxides fixing organic matter by covalent bondings, which make this organic matter unavailable to soil micro-organisms (Zunino et al. 1982). An even more important reason is most likely the generally low pH level of most soils (Anderson and Domsch 1993; Anderson and Joergensen 1997). Highest microbial biomass C-to-soil organic C ratios were measured in Vertisols, with a mean of 2.2% and in Inceptisols with a mean of 2.5%. The microbial biomass C-to-soil organic C ratio in the present tropical soils was negatively correlated with the soil organic C-to-total N ratio, which is in accordance with Jörgensen (1995), and positively correlated with the microbial biomass C/N ratio (Table 2.8). This indicates that the soil organic matter quality had a strong effect on the substrate availability to soil micro-organisms, on the microbial biomass C/N ratio, and probably also on the microbial biomass C/P ratio (see above).

The mean metabolic quotient was 52 mg CO_2-C g^{-1} microbial biomass C d^{-1} (Table 2.7), ranging from 8 to 520 (Fig. 2.2b). The land-use specific qCO_2 values increased in the order plantation < forest < arable soils < pasture soils (Table 2.7, Fig. 2.2b). Only the differences between arable soils and plantation soils and between pasture soils and plantation soils were significant (Table 2.7). The ratio of basal respiration to microbial biomass, the metabolic quotient qCO_2, is an important index for the age structure of the microbial biomass, i.e., the higher the qCO_2, the younger the mean age of the microbial population (Anderson and Domsch 1990; Dilly 2005). The significant inverse relationship of qCO_2 with the microbial biomass C-to-soil organic C ratio (Table 2.8) reveals the interdependence of catabolism and anabolism (Anderson and Domsch 1990). Mean qCO_2 value of the tropical soils was more than twice as high as the mean qCO_2 value of 21 in 99

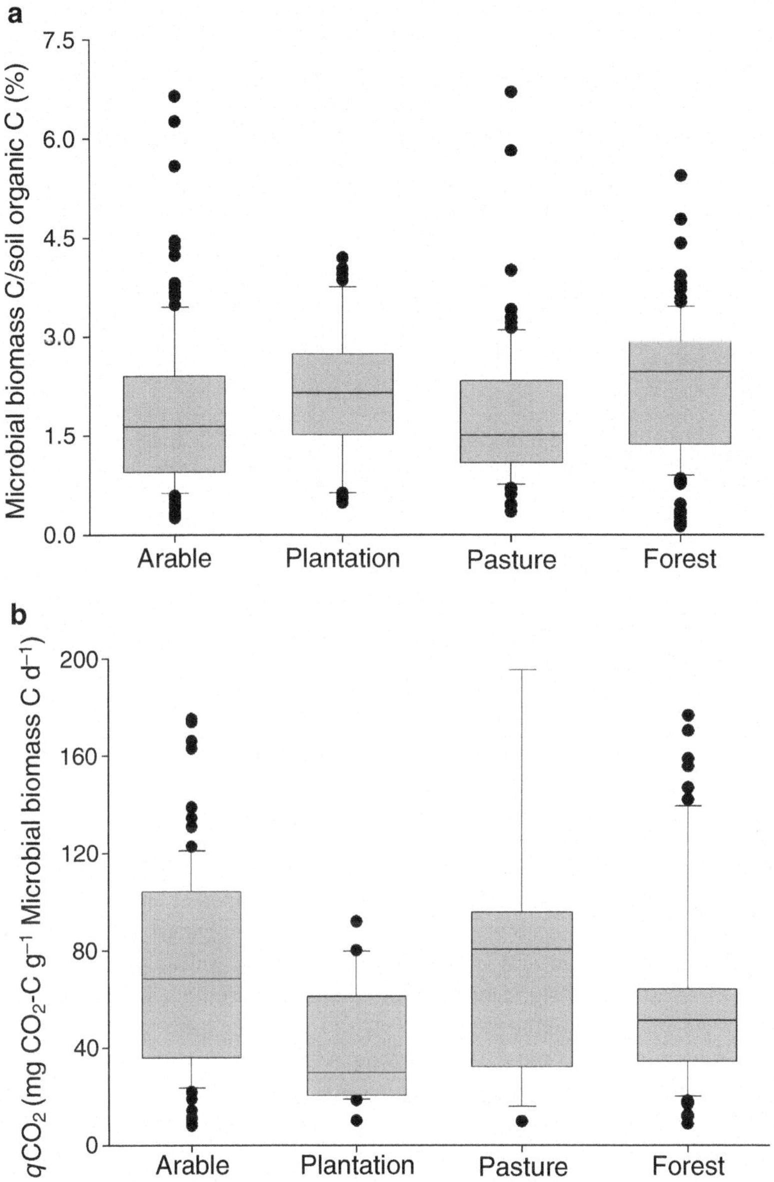

Fig. 2.2 Land use-specific box plots for (**a**) the microbial biomass C/soil organic C ratio and (**b**) the metabolic quotient qCO_2 in soils of the tropics. For each type of soil, the box plots show the smallest observation, the lower quartile, the median, the upper quartile, and the largest observation, with observations considered as outliers being indicated as *dots*

Table 2.8 Correlation coefficients between ratios of values for properties related to soil organic matter content and soil microbial biomass, in soils of the tropics covered in the references used in this review

	Microbial biomass C/soil organic C	qCO_2	ATP/microbial biomass C
Soil organic C/total N	−0.31 (305)****		
Microbial biomass C/N	0.49 (157)****		
Microbial biomass C/soil organic C		−0.33 (192)****	
qCO_2			0.52 (48)***
Ergosterol/microbial biomass C			0.57 (27)**

For each combination of ratios analyzed, the number of observations is indicated in parentheses
$^{**}P < 0.01$, $^{***}P < 0.001$, $^{****}P < 0.0001$

German soils under different land-use forms (Anderson and Domsch 1990; Jörgensen 1995). The generally lower pH (Anderson and Domsch 1993), and most probably also the presence of iron and aluminium oxides in Andisols (Joergensen and Castillo 2001), seem to increase the metabolic quotient considerably. Sub-lethal stress lowers the efficiency of substrate use, i.e., more substrate must be catabolized to CO_2 and less substrate can be incorporated into the microbial biomass (Anderson and Domsch 1993; Anderson and Joergensen 1997). This relationship seems to hold for soils all over the world.

2.7 ATP and Ergosterol

ATP and ergosterol contents were only measured in a minority of soils (Table 2.7). ATP was mainly determined in the Philippines (Salamanca et al. 2002, 2006) and the Andaman Islands (Dinesh et al. 2003, 2004a, b, 2006; Dinesh 2004), ergosterol in Nicaragua (Joergensen and Castillo 2001) and again in the Andaman Islands (Dinesh et al. 2003, 2004a; Dinesh 2004). The ATP content of the forest soils markedly exceeded that of the plantation soils (Table 2.7), leading to a significantly larger ATP/microbial biomass C ratio. The ATP concentration in the microbial biomass of all tropical soils was in the lower part of the range presented by Dyckmans et al. (2003). However, a negative relationship between the ATP/microbial biomass C ratio and the microbial biomass C/N ratio repeatedly observed by Salamanca et al. (2002, 2006) could not be confirmed in the present set of soils. In contrast, positive linear relationships were observed between qCO_2 and the ergosterol/microbial biomass C ratio as observed by Chander et al. (2001) and Salamanca et al. (2006), but not with the microbial biomass C/N ratio. This suggests a negative relationship between substrate use efficiency and presence of saprotrophic fungi. The present results are in line with Scheller and Joergensen (2008), but contradict the view that fungi are more efficient in the use of organic substrates (Sakamoto and Oba 1994; Jastrow et al. 2007).

ATP was significantly correlated to all soil organic matter and microbial biomass indices analyzed (Table 2.7). In contrast, ergosterol was only correlated with ATP, microbial biomass C, and basal respiration. The ergosterol content always increased significantly in the order arable < plantation < forest soils. The mean ergosterol/microbial biomass C ratio was 0.28%, without significant effects of the land-use system. However, the mean ergosterol/microbial biomass C ratio, which is an indicator for living saprotrophic fungi (Joergensen and Wichern 2008), was only half that of soils from a temperate humid climate (Djajakirana et al. 1996). If ergosterol is recalculated into fungal biomass C by multiplication by 90 (Djajakirana et al. 1996), fungi represent only 25% of total microbial biomass C in tropical soils. In contrast, the few data on the ratio of fungal to bacterial biomass indicate that fungi dominate the microbial biomass in tropical soils by roughly 60% (Prasad et al. 1995; Sahani and Behera 2001) or by 87% (Li et al. 2006a, b), and do not fall below 50% (Henrot and Robertson 1994).

This is in line with data on the ratio of fungal residue C to bacterial residue C, calculated on the basis of fungal glucosamine and bacterial muramic acid, which suggest that microbial residue C in soils from semi-arid West Africa consists of 77% fungal residues. This is close to the worldwide average of 75% for all forms of land-use management (Joergensen and Wichern 2008). The present data suggest that the fungi of tropical soils generally have very low ergosterol contents, perhaps indicating a large contribution of arbuscular mycorrhizal fungi, which do not contain ergosterol (Olsson et al. 2003), to the total fungal biomass of soils (Rillig et al. 2001).

2.8 Conclusions

For soils of the tropics, the relationships between microbial biomass C, basal respiration, and soil organic matter are generally similar to those of soils from a temperate humid climate. This is amazing, considering the large variety of soil types and the fact that all these soil biological and chemical properties were measured using a wide variety of methods. This gives confidence that microbial biomass C, basal respiration, and soil organic C are useful indices for soil fertility in tropical soils. Those soils are characterized by low pH, low microbial biomass C-to-soil organic C ratios and high metabolic quotients qCO_2. This is especially true for arable soils, indicating the negative impact of intensive land use on soil fertility. Compost addition, mulching, alley cropping, and reduced tillage are recommended to improve soil biological quality indices at arable sites. Increased efforts should be made to monitor the spatial heterogeneity of soil biological indices, and sampling should be carried out down to a 20- or 30-cm depth in 10-cm steps. The calculation of C budgets requires information on bulk density, and on the amount of organic C stored in the litter layer as well as in the aboveground plant biomass. Despite the importance of phosphorus in tropical soils, the relationship of microbial biomass P to the contents of total P and soil organic matter is a severely neglected area of soil biological research. Likewise, little information is available on the respective contribution of fungi and bacteria to the microbial biomass or to the microbial residue fraction of tropical soils.

References

Adachi M, Bekku YS, Rashidah W, Okuda T, Koizumi H (2006) Differences in soil respiration between different tropical ecosystems. Appl Soil Ecol 34:258–265

Adeboye MKA, Iwuafor ENO, Agbenin JO (2006) The effects of crop rotation and nitrogen fertilization on soil chemical and microbial properties in a Guinea Savanna Alfisol of Nigeria. Plant Soil 281:97–107

Agele SO, Ewulo BS, Oyewusi IK (2005) Effects of some soil management systems on soil physical properties, microbial biomass and nutrient distribution under rainfed maize production in a humid rainforest Alfisol. Nutr Cycl Agroecosyst 72:121–134

Amato M, Ladd JN (1988) Assay for microbial biomass based on ninhydrin-reactive nitrogen in extracts of fumigated soil. Soil Biol Biochem 20:107–114

Anderson JPE, Domsch KH (1978) A physiological method for quantitative measurement of microbial biomass in soils. Soil Biol Biochem 10:519–525

Anderson JPE, Domsch KH (1980) Quantities of plant nutrients in the microbial biomass of selected soils. Soil Sci 130:211–216

Anderson TH, Domsch KH (1989) Ratios of microbial biomass C to total organic carbon in arable soils. Soil Biol Biochem 21:471–479

Anderson TH, Domsch KH (1990) Application of eco-physiological quotients (qCO_2 and qD) on microbial biomasses from soils of different cropping histories. Soil Biol Biochem 22:251–255

Anderson TH, Domsch KH (1993) The metabolic quotient for CO_2 (qCO_2) as a specific activity parameter to assess the effects of environmental conditions, such as pH, on the microbial biomass of forest soils. Soil Biol Biochem 25:393–395

Anderson TH, Joergensen RG (1997) Relationship between SIR and FE estimates of microbial biomass C in deciduous forest soils at different pH. Soil Biol Biochem 29:1033–1042

Andersson M, Michelsen A, Jensen M, Kjøller A (2004) Tropical savannah woodland: effects of experimental fire on soil microorganisms and soil emissions of carbon dioxide. Soil Biol Biochem 36:849–858

Ayanaba A, Tuckwell SB, Jenkinson DS (1976) The effects of clearing and cropping on the organic reserves and biomass of tropical forest soils. Soil Biol Biochem 8:519–525

Babiuk LA, Paul EA (1970) The use of fluorescein isothiocyanate in the determination of the bacterial biomass of grassland soil. Can J Microbiol 16:57–62

Balota EL, Filho AC, Andrade DS, Dick RP (2004) Long-term tillage and crop rotation effects on microbial biomass and C and N mineralization in a Brazilian Oxisol. Soil Till Res 77:137–145

Bastida F, Zsolnay A, Hernández T, García C (2008) Past, present and future of soil quality indices: a biological perspective. Geoderma 147:159–171

Basu S, Behera N (1993) The effect of tropical forest conversion on soil microbial biomass. Biol Fertil Soils 16:302–304

Bationo A, Kihara J, Vanlauwe B, Waswa B, Kimetu J (2007) Soil organic carbon dynamics, functions and management in West African agro-ecosystems. Agric Syst 94:13–25

Behera N, Sahani U (2003) Soil microbial biomass and activity in response to *Eucalyptus* plantation and natural regeneration on tropical soil. For Ecol Manage 174:1–11

Bhattacharyya P, Chakrabarti K, Chakraborty A (2005) Microbial biomass and enzyme activities in submerged rice soils amended with municipal solid waste compost and decomposed cow manure. Chemosphere 60:1513–1522

Bilgo A, Masse D, Sall S, Serpantie G, Chotte JL, Hien V (2007) Chemical and microbial properties of semiarid tropical soils of short-term fallows in Burkina Faso, West Africa. Biol Fertil Soils 43:313–320

Bossio DA, Girvan MS, Verchot L, Bullimore J, Borelli T, Albrecht A, Scow KM, Ball AS, Pretty JN, Osborn AM (2005) Soil microbial community response to land use change in an agricultural landscape of Western Kenya. Microb Ecol 49:50–62

Brookes PC, Powlson DS, Jenkinson DS (1982) Measurement of microbial biomass phosphorus in soil. Soil Biol Biochem 14:319–329

Brookes PC, Landman A, Pruden G, Jenkinson DS (1985) Chloroform fumigation and the release of soil nitrogen: a rapid direct extraction method for measuring microbial biomass nitrogen in soil. Soil Biol Biochem 17:837–842

Bünemann EK, Bossio DA, Smithson PC, Frossard E, Oberson A (2004) Microbial community composition and substrate use in a highly weathered soil as affected by crop rotation and P fertilization. Soil Biol Biochem 36:889–901

Cantarutti RB et al (2002) The effect of grazing intensity and the presence of a forage legume on nitrogen dynamics in Brachiaria pastures in the Atlantic forest region of the south of Bahia, Brazil. Nutr Cycl Agroecosyst 64:257–271

Castillo X, Joergensen RG (2001) Impact of ecological and conventional arable management systems on chemical and biological soil quality indices in Nicaragua. Soil Biol Biochem 33:1591–1597

Cerri CEP, Coleman K, Jenkinson DS, Bernoux M, Victoria R, Cerri CC (2003) Modeling soil carbon from forest and pasture ecosystems of Amazon, Brazil. Soil Sci Soc Am J 67:1879–1887

Cerri CEP, Piccolo MC, Feigl BJ, Paustian K, Cerri CC, Victoria RL, Melillo JM (2006) Interrelationships among soil total C and N, microbial biomass, trace gas fluxes and internal N-cycling in soils under pasture of the Amazon region. J Sustain Agric 27:45–69

Chander K, Dyckmans J, Joergensen RG, Meyer B, Raubuch M (2001) Different sources of heavy metals and their long-term effects on soil microbial properties. Biol Fertil Soils 34:241–247

Cleveland CC, Liptzin D (2007) C:N:P stoichiometry in soil: is there a "Redfield ratio" for the microbial biomass. Biogeochemistry 85:235–252

Cleveland CC, Townsend AR, Schmidt SK, Constance BC (2003) Soil microbial dynamics and biogeochemistry in tropical forests and pastures, southwestern Costa Rica. Ecol Appl 13:314–326

Contin M, Todd A, Brookes PC (2001) The ATP concentrations in the soil microbial biomass. Soil Biol Biochem 33:701–704

Craswell ET, Lefroy RDB (2001) The role and function of organic matter in tropical soils. Nutr Cycl Agroecosyst 61:7–18

Das P, Pal R, Chowdhury A (2007) Effect of novaluron on microbial biomass, respiration, and fluorescein diacetate-hydrolyzing activity in tropical soils. Biol Fertil Soils 44:387–391

de Oliveira OC, de Oliveira IP, Alves BJR, Urquiaga S, Boddey RM (2004) Chemical and biological indicators of decline/degradation of *Brachiaria* pastures in the Brazilian Cerrado. Agric Ecosyst Environ 103:289–300

Diels J, Vanlauwe B, van der Meersch MK, Sanginga N, Merckx R (2004) Long-term soil organic carbon dynamics in a subhumid tropical climate: ^{13}C data in mixed C_3/C_4 cropping and modeling with ROTHC. Soil Biol Biochem 36:1739–1750

Dilly O (2005) Microbial energetics in soils. In: Buscot F, Varma A (eds) Microorganisms in soils: Roles in genesis and functions. Springer, Berlin, pp 123–138

Dilly O, Blume HP, Sehy U, Jimenez M, Munch JC (2003) Variation of stabilised, microbial and biologically active carbon and nitrogen in soil under contrasting land use and agricultural management practices. Chemosphere 52:557–569

Dinesh R (2004) Long-term effects of leguminous cover crops on microbial indices and their relationships in soils of a coconut plantation of a humid tropical region. J Plant Nutr Soil Sci 167:189–195

Dinesh R, Chaudhuri SG, Ganeshamurthy AN, Dey C (2003) Changes in soil microbial indices and their relationships following deforestation and cultivation in wet tropical forests. Appl Soil Ecol 24:17–26

Dinesh R, Chaudhuri SG, Sheeja TE (2004a) Soil biochemical and microbial indices in wet tropical forests: effects of deforestation and cultivation. J Plant Nutr Soil Sci 167:24–32

Dinesh R, Suryanarayana MA, Chaudhuri SG, Sheeja TE (2004b) Long-term influence of leguminous cover crops on the biochemical properties of a sandy clay loam Fluventic Sulfaquent in a humid tropical region of India. Soil Till Res 77:69–77

Dinesh R, Chaudhuri SG, Sheeja TE (2006) ATP levels and adenylate energy charge in soils of mangroves in the Andamans. Curr Sci 90:1258–1263

Djajakirana G, Joergensen RG, Meyer B (1996) Ergosterol and microbial biomass relationship in soil. Biol Fertil Soils 22:299–304

Duah-Yentumi S, Rønn R, Christensen S (1998) Nutrients limiting microbial growth in a tropical forest soil of Ghana under different management. Appl Soil Ecol 8:19–24

Duponnois R, Founoune H, Masse D, Pontanier R (2005) Inoculation of *Acacia holosericea* with ectomycorrhizal fungi in a semiarid site in Senegal: growth response and influences on the mycorrhizal soil infectivity after 2 years plantation. For Ecol Manag 207:351–362

Dyckmans J, Chander K, Joergensen RG, Priess J, Raubuch M, Sehy U (2003) Adenylates as an estimate of microbial biomass C in different soil groups. Soil Biol Biochem 35:1485–1491

Fernandes SAP, Bettiol W, Cerri CC (2005) Effect of sewage sludge on microbial biomass, basal respiration, metabolic quotient and soil enzymatic activity. Appl Soil Ecol 30:65–77

Förster B, Muroya K, Garcia M (2006) Plant growth and microbial activity in a tropical soil amended with faecal pellets from millipedes and woodlice. Pedobiology 50:281–290

Galicia L, García-Oliva F (2004) The effects of C, N and P additions on soil microbial activity under two remnant tree species in a tropical seasonal pasture. Appl Soil Ecol 26:31–39

Galicia L, García-Oliva F (2008) Remnant tree effects on soil microbial carbon and nitrogen in tropical seasonal pasture in western Mexico. Eur J Soil Biol 44:290–297

Geissen V, Morales Guzman G (2006) Fertility of tropical soils under different land use system — a case study of soils in Tabasco, Mexico. Appl Soil Ecol 31:169–178

Gijsman AJ, Oberson A, Friesen DK, Sanz JI, Thomas RJ (1997) Nutrient cycling through microbial biomass under rice-pasture rotations replacing native savanna. Soil Biol Biochem 29:1433–1441

Glaser B, Lehmann J, Zech W (2002) Ameliorating physical and chemical properties of highly weathered soils in the tropics with charcoal — a review. Biol Fertil Soils 35:219–230

Gnankambary Z, Ilstedt U, Nyberg G, Hien V, Malmer A (2008) Nitrogen and phosphorus limitation of soil microbial respiration in two tropical agroforestry parklands in the south-Sudanese zone of Burkina Faso: the effects of tree canopy and fertilization. Soil Biol Biochem 40:350–359

Grant WD (1979) Cell wall teichoic acid as a reserve phosphate source in *Bacillus subtilis*. J Bacteriol 137:35–68

Groffman PM, McDowell WH, Myers JC, Merriam JL (2001) Soil microbial biomass and activity in tropical riparian forests. Soil Biol Biochem 33:1339–1348

Haron K, Brookes PC, Anderson JM, Zakaria ZZ (1998) Microbial biomass and soil organic matter dynamics in oil palm (*Elaeis guineensis* Jacq.) plantations, West Malaysia. Soil Biol Biochem 30:547–552

Henrot J, Robertson GP (1994) Vegetation removal in two soils of the humid tropics: effect on microbial biomass. Soil Biol Biochem 26:111–116

Islam KR, Weil RR (2000) Land use effects on soil quality in a tropical forest ecosystem of Bangladesh. Agric Ecosyst Environ 79:9–16

Izquierdo I, Caravaca F, Alguacil MM, Roldan A (2003) Changes in physical and biological soil quality indicators in a tropical crop system (Havana, Cuba) in response to different agroecological management practices. Environ Manag 32:639–645

Jastrow JD, Amonette JE, Bailey VL (2007) Mechanisms controlling soil carbon turnover and their potential application for enhancing carbon sequestration. Clim Change 80:5–23

Jenkinson DS (1988) The determination of microbial biomass carbon and nitrogen in soil. In: Wilson JR (ed) Advances in nitrogen cycling in agricultural ecosystems. CABI, Wallingford, pp 368–386
Jenkinson DS, Ayanaba A (1977) Decomposition of carbon-14 labeled plant material under tropical conditions. Soil Sci Soc Am J 41:912–915
Jenkinson DS, Ladd JN (1981) Microbial biomass in soil: measurement and turnover. In: Paul EA, Ladd JN (eds) Soil biochemistry, vol 5. Marcel Dekker, New York, pp 415–471
Jenkinson DS, Oades JM (1979) A method for measuring adenosine triphosphate in soil. Soil Biol Biochem 11:193–199
Jenkinson DS, Powlson DS (1976) The effects of biocidal treatments on metabolism in soil V. A method for measuring soil biomass. Soil Biol Biochem 8:209–213
Jenkinson DS, Hart PBS, Rayner JH, Parry LC (1987) Modelling the turnover of organic matter in long-term experiments at Rothamsted. INTECOL Bull 15:1–8
Jensen M, Michelsen A, Gashaw M (2001) Responses in plant, soil inorganic and microbial nutrient pools to experimental fire, ash and biomass addition in a woodland savanna. Oecologia 128:85–93
Jörgensen RG (1995) Die quantitative Bestimmung der mikrobiellen Biomasse in Böden mit der Chloroform-Fumigations-Extraktions-Methode. Gött Bodenkundl Ber 104:1–229
Joergensen RG (1996a) The fumigation-extraction method to estimate soil microbial biomass: calibration of the k_{EC} value. Soil Biol Biochem 28:25–31
Joergensen RG (1996b) Quantification of the microbial biomass by determining ninhydrin-reactive N. Soil Biol Biochem 28:301–306
Joergensen RG, Brookes PC (1990) Ninhydrin-reactive nitrogen measurements of microbial biomass in 0.5 M K_2SO_4 soil extracts. Soil Biol Biochem 22:1023–1027
Joergensen RG, Castillo X (2001) Interrelationships between microbial and soil properties in young volcanic ash soils of Nicaragua. Soil Biol Biochem 33:1581–1589
Joergensen RG, Emmerling C (2006) Human impact on soil organisms and on current measurement methods. J Plant Nutr Soil Sci 169:295–309
Joergensen RG, Mueller T (1996) The fumigation-extraction method to estimate soil microbial biomass: calibration of the k_{EN} value. Soil Biol Biochem 28:33–37
Joergensen RG, Wichern F (2008) Quantitative assessment of the fungal contribution to microbial tissue in soil. Soil Biol Biochem 40:2977–2991
Kaiser EA, Mueller T, Joergensen RG, Insam H, Heinemeyer O (1992) Evaluation of methods to estimate the soil microbial biomass and the relationship with soil texture and organic matter. Soil Biol Biochem 24:675–683
Kaonga ML, Coleman K (2008) Modelling soil organic carbon turnover in improved fallows in eastern Zambia using the RothC-26.3 model. For Ecol Manage 256:1160–1166
Li Y, Xu M, Zou X (2006a) Effects of nutrient additions on ecosystem carbon cycle in a Puerto Rican tropical wet forest. Global Change Biol 12:284–293
Li Y, Xu M, Zou X (2006b) Heterotrophic soil respiration in relation to environmental factors and microbial biomass in two wet tropical forests. Plant Soil 281:193–201
Lodge DJ, Ingham ER (1991) A comparison of agar film techniques for estimating fungal biovolumes in litter and soil. Agric Ecosys Environ 5:31–37
Luizao RCC, Bonde TA, Rosswall T (1992) Seasonal variation of soil microbial biomass – the effects of clearfelling a tropical rainforest and establishment of pasture in the central Amazon. Soil Biol Biochem 24:805–813
Lulu B, Insam H (2000a) Medium-term effects of a single application of mustard residues on soil microbiota and C content of vertisols. Biol Fertil Soils 31:108–113
Lulu B, Insam H (2000b) Effects of cropping systems and manuring on soil microbial and C status of a Vertisol in Ethiopia. Bodenkultur 51:107–114
Manna MC, Swarup A, Wanjari RH, Ravankar HN (2007) Long-term effects of NPK fertiliser and manure on soil fertility and a sorghum-wheat farming system. Aust J Exp Agric 47:700–711
Mao DM, Min YW, Yu LL, Martens R, Insam H (1992) Effect of afforestation on microbial biomass and activity in soils of tropical China. Soil Biol Biochem 24:865–872

Marschner P, Marino W, Lieberei R (2002) Seasonal effects on microorganisms in the rhizosphere of two tropical plants in a polyculture agroforestry system in Central Amazonia, Brazil. Biol Fertil Soils 35:68–71

Mazzarino MJ, Szott L, Jimenez M (1993) Dynamics of soil total C and N, microbial biomass, and water-soluble C in tropical agroecosystems. Soil Biol Biochem 25:205–214

Michelsen A, Andersson M, Jensen M, Kjøller A, Gashew M (2004) Carbon stocks, soil respiration and microbial biomass in fire-prone tropical grassland, woodland and forest ecosystems. Soil Biol Biochem 36:1707–1717

Montano NM, Garcia-Oliva F, Jaramillo VJ (2007) Dissolved organic carbon affects soil microbial activity and nitrogen dynamics in a Mexican tropical deciduous forest. Plant Soil 295: 265–277

Moore JC, McCann K, de Ruiter PC (2005) Modeling trophic pathways, nutrient cycling, and dynamic stability in soils. Pedobiologia 49:499–510

Motavalli PP, Discekici H, Kuhn J (2000) The impact of land clearing and agricultural practices on soil organic C fractions and CO_2 efflux in the Northern Guam aquifer. Agric Ecosyst Environ 79:17–27

Muhammad S, Müller T, Joergensen RG (2007) Compost and phosphorus amendments for stimulating microorganisms and maize growth in a saline Pakistani soil in comparison with a non-saline German soil. J Plant Nutr Soil Sci 170:745–751

Mulongoy K (1986) Microbial biomass and maize nitrogen uptake under a *Psophocarpus palustris* live-mulch grown on a tropical Alfisol. Soil Biol Biochem 18:395–398

Ndiaye D, Lepage M, Sall CE, Brauman A (2004) Nitrogen transformations associated with termite biogenic structures in a dry savanna ecosystem. Plant Soil 265:189–196

Ndour NYB, Baudoin E, Guisse A, Seck M, Khouma M, Brauman A (2008) Impact of irrigation water quality on soil nitrifying and total bacterial communities. Biol Fertil Soils 44:797–803

Nelson DW, Sommers LE (1982) Total carbon, organic carbon and organic matter. In: Page AL, Miller RH, Keeney DR (Eds.) Methods of soil analysis Part 2, 2nd edn. Agronomy Monograph 9. American Society of Agronomy, Madison, WI, pp 539–579

Nielsen JS, Joner EJ, Declerck S, Olsson S, Jacobsen I (2002) Phospho-imaging as a tool for visualisation and non-invasive measurement of P transport dynamics in arbuscular mycorrhiza. New Phytol 154:809–819

Nogueira MA et al (2006) Promising indicators for assessment of agroecosystems alteration among natural, reforested and agricultural land use in southern Brazil. Agric Ecosyst Environ 115:237–247

Northup BK, Brown JR, Holt JA (1999) Grazing impacts on the spatial distribution of soil microbial biomass around tussock grasses in a tropical grassland. Appl Soil Ecol 13:259–270

Oberson A, Joner EJ (2005) Microbial turnover of phosphorous in soil. In: Turner BL, Frossard E, Baldwin DS (eds) Organic phosphorous in the environment. CABI, Wallingford, pp 133–164

Oberson A, Friesen DK, Rao IM, Bühler S, Frossard E (2001) Phosphorus transformations in an Oxisol under contrasting land-use systems: the role of the soil microbial biomass. Plant Soil 237:197–210

Oehl F, Frossard E, Fließbach A, Dubois D, Oberson A (2004) Basal organic phosphorus mineralization in soils under different farming systems. Soil Biol Biochem 36:667–675

Olsson PA, Larsson L, Bago B, Wallander H, van Aarle IM (2003) Ergosterol and fatty acids for biomass estimation of mycorrhizal fungi. New Phytol 159:1–10

Powlson DS, Brookes PC, Christensen BT (1987) Measurement of soil microbial biomass provides an early indication of changes in total soil organic matter due to straw incorporation. Soil Biol Biochem 19:159–164

Prasad P, Basu S, Behera N (1995) A comparative account of the microbiological characteristics of soils under natural forest, grassland and cropfield from Eastern India. Plant Soil 175:85–91

Priess JA, Fölster H (2001) Microbial properties and soil respiration in submontane forests of Venezuelian Guyana: characteristics and response to fertilizer treatments. Soil Biol Biochem 33:503–509

Rabary B, Sall S, Letourmy P, Husson O, Ralambofetra E, Moussa N, Chotte JL (2008) Effects of living mulches or residue amendments on soil microbial properties in direct seeded cropping systems of Madagascar. Appl Soil Ecol 39:236–243

Rajashekhara Rao BK, Siddaramappa R (2008) Evaluation of soil quality parameters in a tropical paddy soil amended with rice residues and tree litters. Eur J Soil Biol 44:334–340

Rillig MC, Wright SF, Nichols KA, Schmidt WF, Torn MS (2001) Large contribution of arbuscular mycorrhizal fungi to soil carbon pools in tropical forest soils. Plant Soil 233:167–177

Sahani U, Behera N (2001) Impact of deforestation on soil physicochemical characteristics, microbial biomass and microbial activity of tropical soil. Land Degrad Dev 12:93–105

Sakamoto K, Oba Y (1994) Effect of fungal to bacterial biomass ratio on the relationship between CO_2 evolution and total soil microbial biomass. Biol Fertil Soils 17:39–44

Salamanca EF, Raubuch M, Joergensen RG (2002) Relationships between soil microbial indices in secondary tropical forest soils. Appl Soil Ecol 21:211–219

Salamanca EF, Raubuch M, Joergensen RG (2006) Microbial reaction of secondary tropical forest soils to the addition of leaf litter. Appl Soil Ecol 31:53–61

Salinas-García JR, de Velázquez-García J, Gallardo-Valdez M, Díaz-Mederos P, Caballero-Hernández F, Tapia-Vargas LM, Rosales-Robles E (2002) Tillage effects on microbial biomass and nutrient distribution in soils under rain-fed corn production in central-western Mexico. Soil Till Res 66:143–152

Sall SN, Masse D, Ndour NYB, Chotte JL (2006) Does cropping modify the decomposition function and the diversity of the soil microbial community of tropical fallow soil. Appl Soil Ecol 31:211–219

Saynes V, Hidalgo C, Etchevers JD, Campo JE (2005) Soil C and N dynamics in primary and secondary seasonally dry tropical forests in Mexico. Appl Soil Ecol 29:282–289

Scheller E, Joergensen RG (2008) Decomposition of wheat straw differing in N content in soils under conventional and organic farming management. J Plant Nutr Soil Sci 171:886–892

Shen H, Xu ZH, Yan XL (2001) Effect of fertilization on oxidizable carbon, microbial biomass carbon, and mineralizable carbon under different agroecosystems. Commun Soil Sci Plant Anal 32:1575–1588

Sierra J, Marbán L (2000) Nitrogen mineralization pattern of an oxisol of Guadeloupe, French West Indies. Soil Sci Soc Am J 64:2002–2010

Sierra J, Nygren P (2005) Role of root inputs from a dinitrogen-fixing tree in soil carbon and nitrogen sequestration in a tropical agroforestry system. Aust J Soil Res 43:667–675

Singh RS, Srivastava SC, Raghubanshi AS, Singh JS, Singh SP (1991) Microbial C, N and P in dry tropical savanna: effects of burning and grazing. J Appl Ecol 28:869–878

Srivastava SC, Lal SC (1994) Effects of crop growth and soil treatments on microbial C, N and P in dry tropical forest land. Biol Fertil Soils 17:109–117

Srivastava SC, Singh JS (1991) Microbial C, N and P in dry tropical forest soils: effects of alternate land-uses and nutrient flux. Soil Biol Biochem 23:117–124

Templer PH, Groffman PM, Flecker AS, Power AG (2005) Land use change and soil nutrient transformations in the Los Haitises region of the Dominican Republic. Soil Biol Biochem 37:215–225

Tian G (1998) Effect of soil degradation on leaf decomposition and nutrient release under humid tropical conditions. Soil Sci 163:897–906

Toledo-Aceves T, Garcia-Oliva F (2008) Effects of forest–pasture edge on C, N and P associated with *Caesalpinia eriostachys*, a dominant tree species in a tropical deciduous forest in Mexico. Ecol Res 23:271–280

Tornquist CG, Hons FM, Feagley SE, Haggar J (1999) Agroforestry system effects on soil characteristics of the Sarapiquí region of Costa Rica. Agric Ecosyst Environ 73:19–28

Torsvik V, Øvreås L (2007) Microbial phylogeny and diversity in soil. In: van Elsas JD, Jansson JK, Trevors JT (eds) Modern soil microbiology, 2nd edn. CRC Press, Boca Raton, FL, pp 23–54

Tripathi S, Kumari S, Chakraborty A, Gupta A, Chakrabarti K, Bandyapadhyay BK (2006) Microbial biomass and its activities in salt-affected coastal soils. Biol Fertil Soils 42:273–277

Turgay OC, Lumbanraja J, Yusnaini S, Nonaka M (2002) Effect of land degradation on soil microbial biomass in a hilly area of south Sumatra, Indonesia. Soil Sci Plant Nutr 48:769–774
Vance ED, Nadkarni NM (1990) Microbial biomass and activity in canopy organic matter and the forest floor of a tropical cloud forest. Soil Biol Biochem 22:677–684
Vance ED, Brookes PC, Jenkinson DS (1987) An extraction method for measuring soil microbial biomass C. Soil Biol Biochem 19:703–707
Villenave C, Bongers T, Ekschmitt K, Fernandes P, Oliver R (2003) Changes in nematode communities after manuring in millet fields in Senegal. Nematol 5:351–358
Villenave C, Ekschmitt K, Nazaret S, Bongers T (2004) Interactions between nematodes and microbial communities in a tropical soil following manipulation of the soil food web. Soil Biol Biochem 36:2033–2043
Vineela C, Wani SP, Srinivasarao C, Padmaja B, Vittal KPR (2008) Microbial properties of soils as affected by cropping and nutrient management practices in several long-term manurial experiments in the semi-arid tropics of India. Appl Soil Ecol 40:165–173
Waldrop MP, Balser TC, Firestone MK (2000) Linking microbial community composition to function in a tropical soil. Soil Biol Biochem 32:1837–1846
Wardle DA (1998) Controls of temporal variability of the soil microbial biomass: a global scale synthesis. Soil Biol Biochem 30:1627–1637
Wick B, Kühne RF, Vlek PLG (1998) Soil microbiological parameters as indicators of soil quality under improved fallow management systems in south-western Nigeria. Plant Soil 202:97–107
Winding A, Hund-Rinke K, Rutgers M (2005) The use of microorganisms in ecological soil classification and assessment concepts. Ecotoxicol Environ Safety 62:230–248
Witt C, Gaunt JL, Galicia CC, Ottow JCG, Neue HU (2000) A rapid chloroform-fumigation extraction method for measuring microbial biomass carbon and nitrogen in flooded rice soils. Biol Fertil Soils 30:510–519
Wu J, Joergensen RG, Pommerening B, Chaussod R, Brookes PC (1990) Measurement of soil microbial biomass C — an automated procedure. Soil Biol Biochem 22:1167–1169
Wu J, He ZL, Wei WX, O'Donnell AG, Syers JK (2000) Quantifying microbial biomass phosphorus in acid soils. Biol Fertil Soils 32:500–507
Xu X, Han L, Wang Y, Inubushi K (2007) Influence of vegetation types and soil properties on microbial biomass carbon and metabolic quotients in temperate volcanic and tropical forest soils. Soil Sci Plant Nutr 53:430–440
Yavitt JB, Wieder RK, Wright SJ (1993) Soil nutrient dynamics in response to irrigation of a Panamanian tropical moist forest. Biogeochemistry 19:1–25
Zhang H, Zhang GL (2003) Microbial biomass carbon and total organic carbon of soils as affected by rubber cultivation. Pedosphere 13:353–357
Zhang H, Zhang GL, Zhao YG, Zhao WJ, Qi ZP (2007) Chemical degradation of a Ferralsol (Oxisol) under intensive rubber (*Hevea brasiliensis*) farming in tropical China. Soil Till Res 93:109–116
Zunino H, Borie F, Aguielera S, Martin JP, Haider K (1982) Decomposition of ^{14}C-labeled glucose, plant and microbial products and phenols in volcanic ash-derived soils of Chile. Soil Biol Biochem 14:37–43

Chapter 3
Impacts of Forest Conversion to Agriculture on Microbial Communities and Microbial Function

Louis V. Verchot

3.1 Introduction

Over the past century, humans have dramatically altered the structure and function of ecosystems across the globe. The expansion of agriculture is one of the most significant of these changes. According to the Food and Agriculture Organization of the United Nations (FAO), agricultural land now occupies around 4.7 million km^2, which is about 40% of the Earth's land surface (Foley et al. 2005), and these cultivated lands are expanding. Most agricultural land is under pasture (~70%), and only a small percentage (<3%) is under permanent crops such as cacao, coffee and tea. The remaining 27% is under arable crops. For the last four decades, global agricultural production has been increasing steadily, at a rate averaging 2.3% per year (FAO 2007), and on average, 6 million ha of forest and grassland have been converted to agriculture annually. Increasing population, technological change, public policies, increasing food and oil prices, and economic growth drive this conversion. Given that projections for population growth indicate a global population of between 9 and 10 billion by 2050, and if the experiences of the Green Revolution are a guide for future expectations, agricultural land could expand to 10 million km^2 (Tilman et al. 2001).

Conversion of land to agriculture is largely a tropical phenomenon at the moment (Fig. 3.1), whereas the area occupied by agricultural land is decreasing in developed countries. Land under row crops and permanent crops has increased since 1965 in sub-Saharan Africa (37%), West Asia and North Africa (28%), East, South and Southeast Asia (23%), and Latin America and the Caribbean (48%). Recent trends suggest that land area for cropping is leveling off only in Latin America. Likewise, the area under meadow and pasture is increasing in West

L.V. Verchot
Center for International Forestry Research (CIFOR), P.O. Box 0113 BOCBD, Bogor 16000, Indonesia
e-mail: L.verchot@cgiar.org

P. Dion (ed.), *Soil Biology and Agriculture in the Tropics*, Soil Biology 21,
DOI 10.1007/978-3-642-05076-3_3,

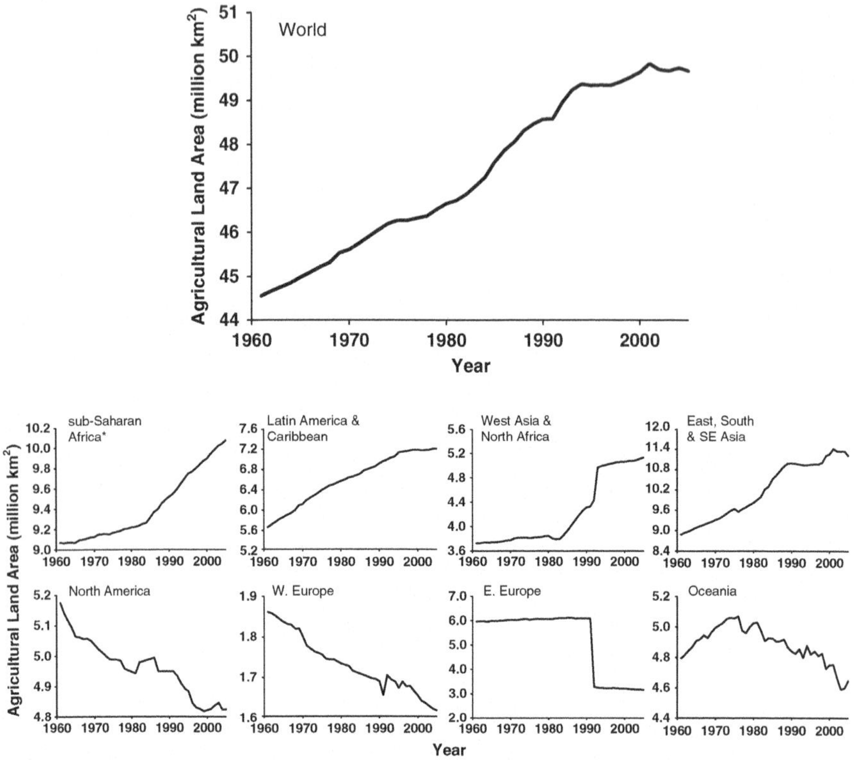

Fig. 3.1 Global and regional land-use change to agricultural land (cropland + pasture land). Ethiopia was not included in Africa panel as there were significant reporting discrepancies following the separation with Eritrea. (Source of data: http:// faostat.fao.org)
Note: different *Y* axis scales for each graph

Asia and North Africa (40%), East, South and Southeast Asia (24%), and Latin America and the Caribbean (48%). Short-term trends suggest that growth in pasture area may be leveling off in all regions, with the exception of sub-Saharan Africa.

A major focus of ecological research has been to understand how structural alterations such as land-use change affect the biological functioning of ecosystems. This concern is driven in part by the perception that land-use change results in degradation of natural ecosystems and particularly of the soil. Oldeman (1994) estimated that soil degradation is extensive, covering about 1,780 million ha across the world, and indicated that land-use change is a major driver of degradation. Soil biological processes respond directly to modifications of ecosystems and provide feedbacks that alter larger scale ecosystem processes, such as productivity, decomposition, and production and consumption of greenhouse gases (GHGs). In turn, these biotic changes alter the state of the world's ecosystems and the services they provide to humanity. For example, in many smallholder-farming systems, inorganic nutrient inputs are non-existent following conversion to agriculture, and only small amounts of organic nutrients are added as manure or green manures. Consequently, soil degradation is an important problem throughout the tropics and is an important driver in decreasing food security in Africa (Sanchez 2002).

Biological research in soils and land-use change has primarily focused on macrobiota. This work has shown how organisms such as earthworms and termites regulate carbon turnover and nutrient cycling (Black and Okwakol 1997; Fragoso et al. 1997; Jimenez et al. 1996; Brussaard et al. 1993). In this chapter, we will look at the current state of our understanding of the impacts of land-use change on microbial communities and how these impacts alter biogeochemical cycles. The literature on this topic is much more fragmentary than that concerning macrobiota. Nevertheless, a significant amount of work has been done in this area, and our knowledge is advancing, particularly with the advent and application of novel biotechnological approaches to understanding belowground biodiversity.

3.2 Soil Organic Matter and Microbial Biomass

Understanding the dynamics of soil organic matter (see Chap. 2) following land-use change is the starting point for understanding how microbial communities respond to the change. The simplest models of SOM dynamics distinguish between organic inputs and stabilized organic matter (van Noordwijk et al. 1997). Between 50 and 90% of the organic inputs are lost from the soil in the first year. By contrast, stabilized organic matter decomposes at rates of 1–5% per year (Nye and Greenland 1960; van Faasen and Smilde 1985; Young 1989). More recent models subdivide soil organic carbon (SOC) into functional pools with different turnover times (Parton et al. 1989; van Noordwijk and Lusiana 1999; Zimmermann et al. 2007). These models recognize that factors associated with the nature of the soil (Fig. 3.2)

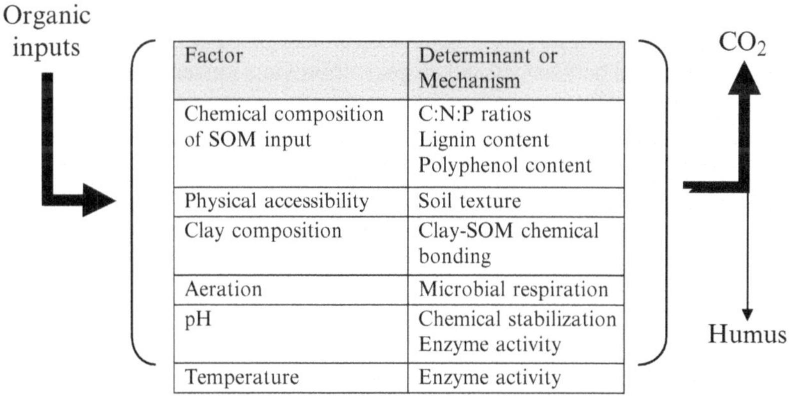

Factor	Determinant or Mechanism
Chemical composition of SOM input	C:N:P ratios Lignin content Polyphenol content
Physical accessibility	Soil texture
Clay composition	Clay-SOM chemical bonding
Aeration	Microbial respiration
pH	Chemical stabilization Enzyme activity
Temperature	Enzyme activity

Fig. 3.2 Factors and their determinants or mechanisms affecting decomposition of organic inputs and providing partial or temporary protection from decomposing organisms; SOM: soil organic matter

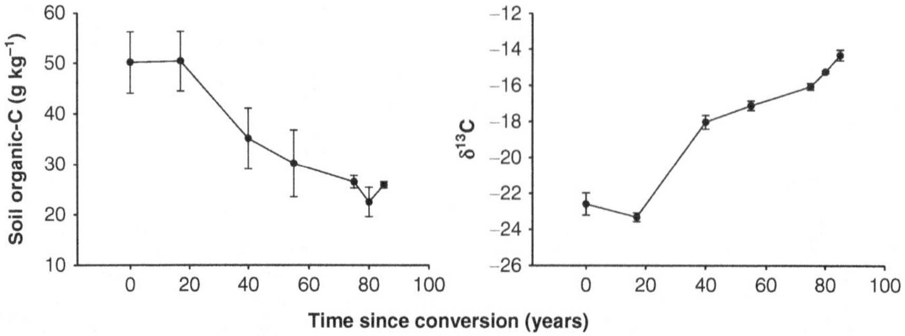

Fig. 3.3 Changes in soil C following land-use change in a landscape in Western Kenya (Verchot et al. unpublished) and change in $\delta^{13}C$ composition of the soil organic carbon (SOC). Time = 0 are forested conditions. Soil organic matter (SOM) remains relatively constant during an initial phase following conversion and decreases as soil is continuously cultivated. SOM initially maintains a high proportion of forest derived C as shown by the $\delta^{3}C$ signature of the SOM, but as carbon content decreases, the SOC becomes dominated by agricultural C (as shown by an increase in the $\delta^{13}C$ signature)

and management practices (e.g., cultivation intensity) play important roles in determining how fast carbon is lost from the system. For example, soil aggregation in medium- to fine-textured soils protects carbon from mineralization, and these soils maintain high carbon stocks for a number of years following conversion. Sandy soils provide little protection for organic matter decomposition, and lose carbon more quickly.

Following conversion from indigenous ecosystems to arable or permanent cropping systems, SOC decreases over time (Davidson and Ackerman 1993) and nutrient availability declines (Fig. 3.3). Carbon stocks are generally high in soils under forest vegetation. As carbon inputs decline following conversion, and tillage exposes protected carbon to mineralization, SOC declines. Forest-derived carbon persists, as the isotope data show, and even after 90 years following conversion, forest-derived carbon makes up a small percentage of SOM. A new equilibrium SOC level is reached between 30 and 80 years following conversion (Solomon et al. 2007; Awiti et al. 2008). Intensification of agriculture, particularly if it involves organic matter management, can stem the losses and stabilize SOC at higher levels (Mitchell et al. 2008); however, this is not the case in many parts of the tropics where subsistence agriculture is practiced. The story is different and perhaps less clear with regard to conversion of forests to pastures. A comprehensive literature review showed that conversion of forest to pasture can lead to either carbon stock increases or decreases (Murty et al. 2002). The authors reviewed 109 studies, and found that about half reported carbon gains and half reported losses.

As forestland is converted to pasture or cropland, microbial biomass follows a trend parallel to that of SOC. Basu and Behera (1993) characterized this relationship in the Eastern Ghat mountain range of Orissa. They used the $CHCl_3$ fumigation–incubation method to assess microbial biomass C, and they measured basal respiration in soils from forest, savanna and agricultural fields. Basal respiration is

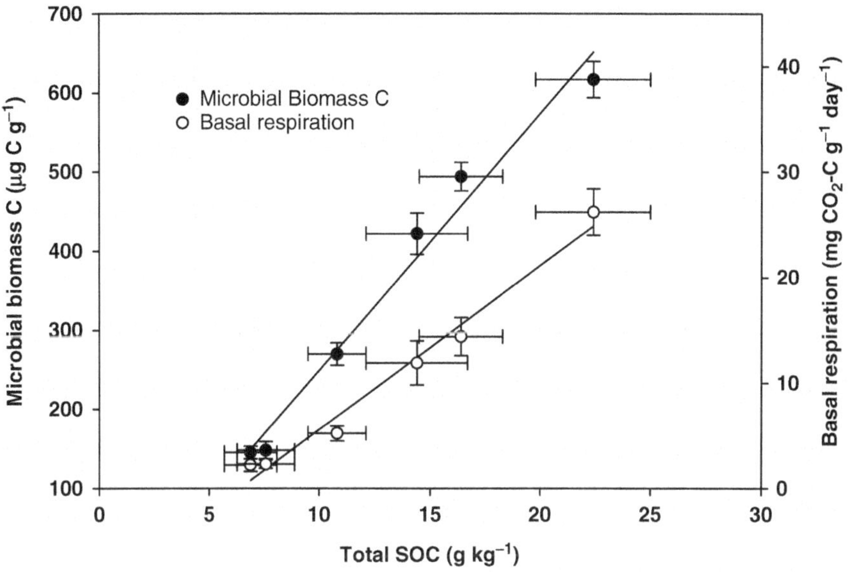

Fig. 3.4 Relationships between total soil organic carbon (SOC) and microbial biomass, and between SOC and basal respiration in an Inceptisol in Southern India. Microbial community size and activity show linear relationships across a gradient of indigenous ecosystems and permanent cropping systems with total SOC
Source: Dinesh et al. (2004)

an operationally defined parameter that integrates microbial activity and carbon availability in the laboratory and can be used to compare soil samples. As SOC declined by 40% in the land converted to agriculture or savanna, microbial biomass declined by 46–52% and basal respiration also declined. Dinesh et al. (2004) showed similar decreases when forests in southern India were replaced with permanent cropping systems (Fig. 3.4). Total SOC decreased by 69% with the conversion of moist deciduous forest to coconut plantation. Microbial biomass C decreased by 77% and basal respiration decreased by 91%. Linear relationships existed between SOC and the size of the microbial biomass, and between SOC and microbial activity measured in terms of basal respiration.

3.3 Microbial Communities

With the advent of novel techniques, it is now possible to study the effects of land-use change and management on entire microbial communities. There are several approaches based on the extraction of "signature" lipid biomarkers (Table 3.1) from the cell membrane and wall of micro-organisms (White and Ringelberg 1998). Short-chain phospholipids (C10–C20) derive from microbes, while longer-chain phospholipids derive from plants. Two approaches look at either fatty-acid

Table 3.1 Signature PLFA markers for different components of the soil microbial community

Community component	PLFA markers
General bacteria	15:0 anteiso, 17:0 anteiso, 17:0 iso, 14:0, 16:1 ω7c, 19:0 cyclo
Gram-positive bacteria	15:0 iso, 16:0 iso
Gram-negative bacteria	17:0 cyclo, 18:1ω7c
Actinomycetes	16:0 10 methyl, 17:0 10 methyl, tbsa 10me18:0
Fungi	18:2ω6c, 18:1ω9c
Arbuscular mycorrhizal fungi	16:1ω5c
Protozoa	20:2ω6,9c, 20:4ω6,9,12,15c

methyl esters (FAME) or phospholipid fatty acids (PLFA). The two measures differ in that FAME examines all phospholipids, while PLFA considers only the polar phospholipids. Because these fatty acids decompose rapidly in the soil following microbial death, FAME and PLFA analyses give a profile of the active microbial community. These methods offer opportunities to quantify the biomass of several groups such as the Gram-negative bacteria, the Gram-positive bacteria, actinomycetes, fungi, arbuscular mycorrhizal fungi (AMF), protozoa, and even some functional groups (e.g., methane oxidizers, sulfur reducers). Thus, these methods offer relatively easy means for quantitative assessment of the structure of microbial communities, although the presence of a large number of background unspecific fatty acids may mask differences in community structure (Marschner 2007). The major disadvantage of these methods is that they provide no information at the species level.

There are also several community "fingerprinting" techniques based on extraction and analysis of 16S ribosomal DNA (rDNA) fragments. Two analytical techniques, denaturing gel gradient electrophoresis (DGGE) and terminal restriction fragment length polymorphism (TRFLP), are popular for soil studies. These techniques involve extraction of rDNA from the soil, and specific amplification of this DNA through a polymerase chain reaction (PCR). There are some problems with these methods. Martin-Laurent et al. (2001) showed that the rDNA extraction method affected the results of observations on abundance and composition of phylotypes from bacterial communities of agricultural soils in France. They also showed that both the extraction method and the soil matrix introduce biases that affect the amplification efficiency of the PCR reaction, making comparisons across studies and sites difficult. There are also unknown biases in the PCR reaction, and the products of these reactions do not truly reflect community composition because of the selectivity of the primers (Marschner 2007). Thus, it is essential to consider these limitations when drawing conclusions from studies on the relative abundance of microbial phylotypes in soils.

One important limitation of the rDNA approaches is that we cannot look at the whole community, nor ascribe changes in community structure to particular functional groups. Primers are selective for either bacteria or fungi or for subgroups within these kingdoms, and we have only had adequate fungal primers for environmental samples since 2000 (Hawksworth 2001; Borneman and Hartin 2000). Thus, most of the rDNA work on microbial communities has focused on their bacterial components. However, we know that fungi play a major role in ecosystem processes, and fungal species remain largely undescribed (Hawksworth and Rossman 1997). Because of the selectivity of the primers for important genes, DNA-based techniques are particularly well suited for studies that seek to link the microbial population with biogeochemical or enzymatic functions in the soil.

Several authors have suggested that community structure and biochemical function should be related, but the experimental results to date are ambiguous (Lucas et al. 2007; Manefield et al. 2002; Torsvik and Øvreås 2002). Advanced molecular and biochemical ecological approaches have only been used to study micro-organisms in a limited number of tropical systems (Bossio et al. 2005; Gomes et al. 2003; Waldrop et al. 2000; Nusslein and Tiedje 1999; Borneman and Triplett 1997). Most of these studies focus on cataloging diversity of soil bacteria, profiling communities, and documenting how these communities are affected by disturbance or by some form of environmental change. However, most of this work has been site-specific.

The emerging picture is that different constituents of the microbial community respond differently to changes in the quality of organic matter inputs associated with land-use change. De Ruiter et al. (1995) showed that both top-down and bottom-up processes drive food webs in grassland soils. Organisms at high trophic levels were more subject to bottom-up drivers (e.g., resource quality and quantity), whereas top-down forces (e.g., consumers) regulated organisms at lower trophic levels. This work identified stabilizing feedbacks in these systems that maintain either states of low or high productivity (Wardle et al. 2004; Moore et al. 2003; De Ruiter et al. 1995). Plant species that are adapted to fertile conditions return high quality litter (characterized by low C:N ratios, low phenolics, low lignin and structural carbohydrates) and support soil food webs in which energy transfers are accomplished through bacterial channels. Low quality organic matter inputs from the plant community condition food webs on infertile soils, where fungal energy channels predominate. Thus, there is a tight linkage between the trajectories of the microbial community and the aboveground vegetation, which bears direct relevance to the effect of land-use change.

We have a very limited understanding of the factors that structure soil microbial communities, the effects of disturbance on structure, and how this disturbance plays out in larger-scale biogeochemical cycles. The results of the few studies that have looked at the effects of land-use change on the structure of microbial communities are consistent. Changes in vegetation in young Hawaiian soils led to dramatic changes in microbial community structure (Nusslein and Tiedje 1999), and conversion of forest to agriculture decreased microbial biomass and produced compositionally distinct microbial communities in Tahiti (Waldrop et al. 2000); Borneman and Triplett (1997) demonstrated significant differences between soil microbial populations in a mature forest and adjacent pasture in eastern Amazonia. Bossio et al. (2005) found similar results in eastern Kenya. In addition, they found that a regenerating secondary forest on one site was more similar to an indigenous forest at another site than it was to nearby agricultural sites. Many of these studies are anecdotal and are not based on replicated field trials. Therefore, little is known about how agricultural practices or land-use change affect microbial communities.

The legacy of land-use change has been appreciated from the point of view of biogeochemical function (Verchot et al. 2001), but the effects on microbial community structure is underappreciated in the literature. Fraterrigo et al. (2006) used PLFA and FAME analyses to determine that community composition varied significantly with past land use in the Appalachian region of western North Carolina, USA. They studied forest stands on sites that had been farmed in the 1930s and other sites that had been logged and regenerated in the 1950s. Microbial communities in forest stands that had previously been farmed had a higher relative abundance of markers for Gram-negative bacteria and a lower abundance of markers for fungi compared with previously logged and undisturbed stands. This lasting effect of land-use history on microbial community structure affected N mineralization rates, which were negatively correlated with fungal and Gram-negative bacteria markers. These results indicate that contrary to expectations of microbial community resilience, there is a persistent legacy of disturbance on microbial communities and the nutrient cycling processes they mediate.

3.4 Effects of Land-Use Change on Mycorrhizal Fungi

Mycorrhizal fungi warrant particular attention, as they are keystone species in the microbial community because of their importance in regulating the aboveground plant community (Wardle et al. 2004; Matson et al. 1997; Grime et al. 1987; see Chap. 9). There has been a considerable amount of effort in improving our understanding of the effects of land-use change and management on mycorrhizal fungi, because of their importance to ecosystem health. Because of the symbiotic relationships with plants in the ecosystem, where each plant benefits from a unique fungal isolate (Klironomos 2003), mycorrhizae can affect the composition of the aboveground plant community (Gange et al. 2003; Grime et al. 1987). This in turn can feed back to the microbial community through effects on quality and quantity of organic matter inputs to the soil from the plant community. Thus, changes to the composition of the mycorrhizal component of the soil microbial community may influence ecosystem productivity, standing biomass, nutrient cycling, carbon allocation, and relative abundance of plants (Stampe and Daehler 2003; van der Heijden et al. 1998). For example, increasing diversity of ectomycorrhizal fungi has been found to promote tree seedling productivity in low-fertility but not high-fertility substrates (Jonsson et al. 2001). These effects, in turn, can control the aboveground consumers associated with individual plant species (Vicari et al. 2002; Goverde et al. 2000; Gange and Nice 1997).

The effect of land-use change on mycorrhizae in general, and on AMF in particular, has been difficult to study because of the lack of adequate indicators of AMF activity. Determination of spore counts and infection rates are tedious and give poor indication of the biomass of mycorrhizae. However, these methods can help understand shifts in community structure. Studies employing these approaches generally show that land-use change results in decreased AMF abundance. Ling-Fei et al. (2007) measured AMF colonization and spore density in a landscape in southwest China and found that both parameters were lower in fallow and cropped agricultural land compared to native forest.

Mathimaran et al. (2007) showed that management and crop rotation have a significant effect on the AMF, particularly through alterations of the plant community present in the agroecosystem. The mycorrhizal communities in soils planted with a tree (Crotolaria) or a cereal crop (maize) were very different from the community in soils under sunflower or leeks. This specificity between mycorrhizae and the plant host sets up a positive feedback within the ecosystem (Wardle et al. 2004). AMF are associated with fertile soils, while intermediate- and low-fertility soils are associated with ectomycorrhizal and ericoid mycorrhizal fungi, respectively (Cornelissen et al. 2001). Because plants that are associated with AMF have high N content, and also low lignin and low phenol content, they return high quality organic materials to the soil, thus maintaining rapid nutrient cycling and high productivity. Plants associated with ectomycorrhizal and ericoid mycorrhizal fungi have lower N and higher phenol and lignin content. Thus, they return a lower quality litter to the soil, which reinforces low production on the site. We see that land-use change and management following conversion of forest to agriculture represents a phase change for the mycorrhizal community that alters the function of the ecosystem and has the potential to set up a new dynamic between the detrital and plant communities.

3.5 Microbial Community Activity

The microbial community regulates nutrient cycles by its effects on the decomposition process, and affects nutrient acquisition by plants through symbiotic relationships such as mycorrhizal or rhizobial associations with plant roots. The microbial community also represents a dynamic pool of organic matter that serves as a nutrient reservoir. Many studies (Verchot et al. 1999, 2000, 2006; Cleveland et al. 2003; Ishizuka et al. 2002; Erickson et al. 2001; Crill et al. 2000; Veldkamp et al. 1997; Neill et al. 1995) have looked at how land-use change affects biogeochemical processes. The general trend is that nutrient availability decreases as forests are converted to agriculture and pasture, the nitrogen economy changes from one dominated by NO_3^- to one dominated by NH_4^+, and extracellular enzyme activities decrease (Fig. 3.5).

Ecologists have been thinking about how the changes in microbial community structure relate to biogeochemical function (Marschner 2007; Chapin et al. 1997). Schimel and Gulledge (1998) suggested that land-use change alters microbial community activity in two principal ways (Fig. 3.6). First, change can alter the function of the existing assemblage of organisms. For example, the rates of many

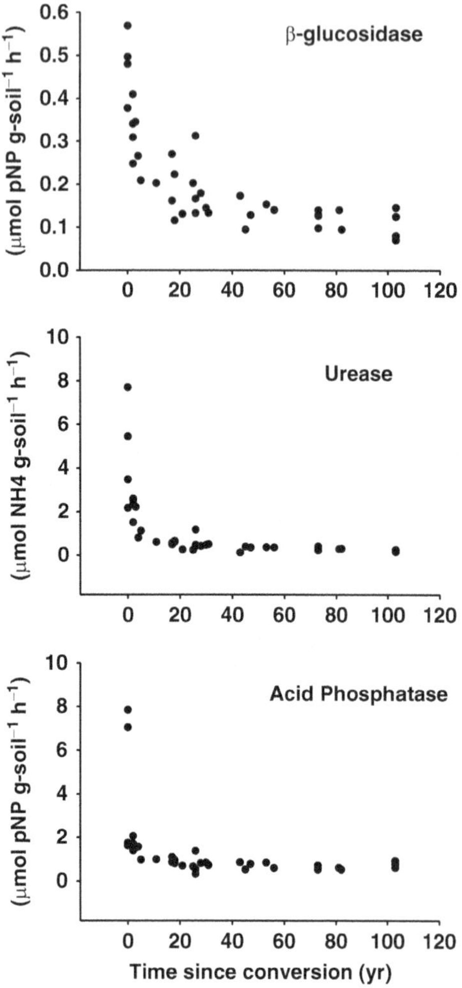

Fig. 3.5 Effects of conversion of forest to agriculture on extracellular enzyme activity in soils in Western Kenya (Verchot et al. unpublished). For the enzymes associated with narrow metabolic processes (acid phosphatase and urease), new equilibrium levels were established around 20 years after conversion, before SOC stabilized and when forest-derived SOC made up most of the total SOC. For the enzyme associated with the broader metabolic process (β-glucosidase), new stable levels of activity were established around 50 years following conversion, when forest-derived SOC made up only around 40% of the total SOC

biochemical processes increase with temperature up to an optimum and then decrease as enzymes denature. By provoking an increase in soil temperature, vegetation removal would alter the rates of decomposition, but not the nature of the physiological processes in the microbial community. Second, land-use change can alter the structure of the microbial community, which in turn would modify the physiological processes that drive biogeochemical cycling. For example, the temperature sensitivity of the decomposer community might shift, or the population of lignin decomposers may become dominated by N-insensitive fungi (Schimel and Gulledge 1998; Kaal et al. 1993). In an extreme case, key functions (e.g., denitrification, sulfate reduction, lignin decomposition) could be lost from the community. Thus, two categories of effects result from land-use change. In the first case, the rates of processes vary and this variation is reversible; in the second case the nature

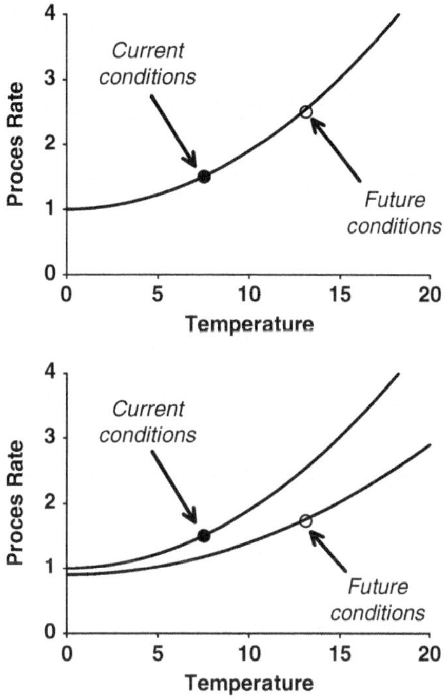

Fig. 3.6 Example adapted from Schimel and Gulledge (1998) of how land-use change could alter the relationship between a process rate and an environmental variable. *Top panel*: the community remains unchanged, and the process rate is altered by the changed environment. *Bottom panel*: the microbial community changes, altering the underlying drivers of the relationship between the environmental factor and the process

of the microbial community is modified in such a way that alterations in biogeochemical processes are not readily reversible, as in the case of the old farmed sites that we discussed at the end of Sect. 3.4.

The information presented in the preceding sections leads us to suggest that both reversible and irreversible phenomena are likely to result from land-use change. Very few studies examine quantitatively or qualitatively how land-use change alters the microbial community and how any modification in community structure, in turn, affects function. Limits imposed by the current biogeochemical and microbiological measurement methods make it difficult to collect appropriate datasets. For example, any relevant ecosystem process is the result of the activities of a subset of organisms in the microbial population. Differences captured in whole community profiles as defined by PLFA, or in profiles of major segments of a community, as revealed by 16S rDNA approaches, may not capture structural changes that are significant for the process in question. The difference in scale that exists between experimental designs of biogeochemists and environmental microbiologists is also an impediment to improving our understanding. Microbes accomplish their work at the scale of 10^{-6} m and many microbiologists are working at scales of the order of 10^{-4} m. By contrast, ecosystem processes are important at the scale of around 10^3–10^4 m^2 (Groffman et al. 2006). Thus, it has been difficult for ecologists to demonstrate that the composition and structure of the microbial community matter to ecosystem function.

There are good reasons to believe that microbial community structure is unimportant for understanding biogeochemical processes within the scale of a whole ecosystem. Indeed, various ecosystem models are successful over a wide range of temporal and spatial scales without taking variation in microbial community structure into account. This is the case of models of soil respiration, which often rely on temperature functions and organic carbon inputs (Davidson et al. 2006). Given the widely held assumption in microbiology that everything is everywhere and the environment selects (Hooper et al. 2008; Finlay 2002; Finlay and Clarke 1999; Baas-Becking 1934; see Chap. 12), it is unlikely that land-use change would eliminate whole groups of organisms responsible for the efflux of CO_2 from the soil. Thus, modifications of substrate availability, rather than alterations to microbial community structure, are likely to be the long-term driver of ecological alterations accompanying land-use change.

On the other hand, the importance of microbial community structure is suggested by site-specific differences in the functional relationship between microbially-mediated processes and environmental conditions. For example, the wide range of temperature response values for soil respiration obtained by Neff et al. (1996), in a standard incubation of soils from different sites, suggests that different microbial communities have different temperature responses. The incubations were short enough that consumption of the labile C during the incubations was unlikely. Organic matter quality may have been a factor, requiring different enzymes for decomposition.

Approaches using analyses of extracellular enzymes are shedding light on the importance for key processes of certain functional groups within microbial communities. To get a handle on the relationship between the structure of the microbial community and function, we will digress briefly from the theme of land-use change and look at several studies that have examined this relationship by manipulating the microbial community composition or by looking at different ecosystems within a landscape. This work generally focuses on metabolically "narrow" processes.

In one experiment, Lucas et al. (2007) showed that different parts of the microbial community responded differently to inputs of a high quality organic N source; in particular, ectomycorrhizal fungi and Gram-positive bacteria responded positively to N addition. No response was obtained with more complex substrates, despite similar levels of N addition to the system. Fungi play an important role in the breakdown of lignin and produce extracellular lignases. The authors manipulated several components of the microbial community and showed that changes in the fungal component had little effect on extracellular lignase activities. They looked at two lignases: phenol oxidase and peroxidase. Neither of the enzyme activities increased as a result of the increased importance of fungi in the microbial community structure.

Denitrification is not specific to one phylogenetic group; rather it can be found in about 50 genera, most of which belong to Proteobacteria (Zumft 1999). Rich and Myrold (2004) looked at the denitrifier component of the microbial community using the functional *nosZ* gene, which is responsible for producing N_2O reductase. They worked on four adjacent sites in a landscape in Willamette Valley in Oregon and compared different parameters of the denitrification process, namely denitrifying enzyme activity (DEA) and maximum potential N_2O reductase activity. The strongest correlations were between the nonmetric multidimensional scaling ordination results of *nosZ* and the DEA, but the ordination results between *nosZ* and the proportion of gas emitted as N_2O differed across habitats. The authors concluded that denitrifying community composition and activities were uncoupled across the ecosystems. However, in this study DEA was generally low at all sites, and hence the variation was low as well. In a study on sites with DEA activities an order of magnitude greater, Rich et al. (2003) found tighter coupling between denitrifying community composition and functioning in adjacent meadow and forest soils in Oregon. While Rich and Myrold (2004) concluded that relationships between denitrifying community structure and activities appeared to be ecosystem-specific, it is altogether possible that low variations may sometimes make this relationship undetectable.

It appears that there are only two studies in the literature that explore the effect of land-use change on microbial communities and how this translates into altered function. Waldrop et al. (2000) looked at the effects of conversion of tropical forest to pineapple plantations on microbial community structure and function in Tahiti. Bossio et al (2005) looked at different land uses in several landscapes and used a replicated experiment to explore these relationships in Kenya.

Using a chronosequence of newly established pineapple plantations, Waldrop et al. (2000) showed that SOC and nutrients decreased in the soil over time. This was associated with an increase in saturated fatty acids, which are generally indicative of bacteria, and a decline in branched fatty acids, which are indicative of Gram-positive bacteria. The markers for fungi and actinomycetes also increased in the pineapple plantations. These authors used extracellular enzyme activities to assess community function and showed that activity was largely driven by the size of the microbial community, which as we know is correlated with SOC. Specific activities (enzyme activity per unit of microbial biomass) of phenol oxidase, peroxidase, phosphatase, and sulphatase, all enzymes involved in relatively narrow metabolic processes, correlated significantly with site scores on the first principal component axis of the PLFA assessment of community composition. Specific activities of β-glucosidase and β-xylosidase, enzymes involved in broad metabolic processes, were not correlated with community composition.

In the second study, Bossio et al. (2005) used several methods to investigate microbial community structure and function in distinctly different soil types at five sites of western Kenya. In this study, 16S rDNA analysis by DGGE using universal bacterial primers showed that ecosystem and site were the primary determinants of total bacterial community composition. The 16S rDNA and PLFA profiles showed differences between forested and agricultural soils. Higher levels of Gram-negative bacteria in the forested soils accounted for the difference in PLFA profiles. Agricultural soils separated into two groups, one with higher relative abundances of branched fatty acids, and a second group with higher relative abundances of monounsaturated fatty acids. Extracellular enzyme activities and BIOLOG Gram-negative microtiter plates were used as functional indicators. These showed less specificity with respect to soil type, and greater variability than DNA- and PLFA-based measures. Thus, there appeared to be a high degree of functional redundancy in the microbial community. In replicated field experiments comparing traditional continuous maize cropping with an improved N-fixing tree fallow system in which both maize yields and microbial biomass C increased, 16S rDNA and PLFA analyses revealed differences in microbial communities between treatments, although these differences were not necessarily associated with increases in microbial diversity. Microbial biomass and enzyme activities were generally found to increase in soils with the N-fixing tree fallows; the relationship with soil type was not significant. The differences between the fallows and conventional agricultural practices were largely explained by increases in activities of enzymes associated with carbon cycling, and decreases in activities of those associated with P cycling. Thus, management practice such as the reintroduction of trees affected both microbial community composition and function.

In the studies on land-use change, the relative abundances of different microbial groups appear to affect function. To the degree that land-use change induces repeated stress on the soil microbial community (e.g., stress due to increased severity of drying/wetting cycles), substrate availability may be less important for rates of biogeochemical processes than the size and structure of the population, which would be controlled by the stress-induced death and regrowth cycles (Schimel and Gulledge 1998).

3.6 Conclusions

There is a rich body of knowledge about the effects of land-use change on carbon stocks and on biogeochemical processes. However, these studies often give conflicting results and we cannot always draw generalizations. Certainly, microbial activity depends on carbon availability, but the processes that are the result of microbial activity — soil respiration, N mineralization, trace gas production and consumption — defy generalizations. Our biogeochemical models are poor at predicting the magnitudes and sometimes even the direction of fluxes at new sites. Davidson and Janssens (2006) called for a more reductionist approach to assess the importance of kinetic properties of individual components of organic matter and the effect of in situ constraints on organic matter decomposition. Few studies have tried to develop an understanding of how alteration of environmental conditions produces physiological responses in the microbial community and alters community composition so as to modify processes at higher spatial scales. Thus, any reductionist approach to improving our biogeochemical models might also factor these elements into the research.

Difficulties in measurement remain a constraint to advance in this area. Even though we are now capable of measuring unculturable organisms, we cannot easily measure microbial diversity at meaningful scales. Assays that measure microbial function usually determine the overall rate of an entire metabolic process and do not break the process down into steps that can be associated with particular organisms. It also appears that diversity in itself will be a meaningless measure for understanding microbial function as there is significant redundancy within the community. Community composition may be more meaningful.

As Schimel et al. (2007) noted, plant ecology offers useful conceptual frameworks for advancing efforts to integrate microbial ecology into ecosystem ecology. In plant ecology, understanding how individual organisms respond to environmental stress has been the key to integrating population and ecosystem ecology. However, there are significant differences between plants and microbes that we need to consider as we design research to achieve the integration of microbial ecology into ecosystem ecology. In ecosystem ecology, we often differentiate between stress and disturbance. Recurrent stress factors such as drought, pollution or seasonal fire pose chronic challenges to organisms and engender physiological costs. Occasional disturbance (provoked or accidental fire, flooding, windstorms, land-use change) results from pulse events that involve physical disruption to the ecosystem; these induce direct mortality of organisms and alter, at least temporarily, the composition and structure of the ecosystem. For microbes, the distinction may be less clear. Microbes are likely to experience both stress and disturbance events through altered microclimate and resource availability that impose physiological costs. With the exception of tillage, which disrupts fungal hyphae, microbes will not experience physical disturbance as plants do.

Schimel et al. (2007) went on to propose that concepts related to life strategies, similar to those used to characterize plants (e.g., Grime 1977), may be useful for characterizing microbes. Looking at how microbes deal with stress — they may be inherently resistant or else physiological costs may be associated with acclimating to the stress — will provide a basis for understanding life history strategies of microbes. Thus, in addition to the processes that different suites of organisms undertake, integrating the nature of their response to stress will provide for more robust models of function.

Extrapolating future conditions based on observations of current responses to environmental change will be inadequate as environmental change accelerates. We are not yet capable of integrating the physiological ecology of microbes with population biology to explain ecosystem processes. However, the state of knowledge reviewed in this chapter suggests that we are at the point of being able to open the black box of microbial ecology and integrate this new knowledge into ecosystem ecology. This is likely to be essential for building more robust ecosystem models that can help us address the challenges posed by global environmental change.

References

Awiti AO, Walsh MG, Kinyamario J (2008) Dynamics of topsoil carbon and nitrogen along a tropical forest–cropland chronosequence: Evidence from stable isotope analysis and spectroscopy. Agric Ecosyst Environ 127:265–272

Baas-Becking L (1934) Geobiologie of inleiding tot de milieukunde. Van Stockkum and Zoon, The Hague
Basu S, Behera N (1993) The effect of tropical forest conversion on soil microbial biomass. Biol Fertil Soils 16:302–304
Black H, Okwakol M (1997) Agricultural intensification, soil biodiversity and agroecosystems function: the role of termites. Appl Soil Ecol 6:37–53
Borneman J, Hartin RJ (2000) PCR primers that amplify fungal rRNA genes from environmental samples. Appl Environ Microbiol 66:4356–4360
Borneman J, Triplett EW (1997) Molecular microbial diversity in soils from Eastern Amazonia: evidence for unusual microorganisms and microbial population shifts associated with deforestation. Appl Environ Microbiol 63:2647–2653
Bossio DA, Girvan MS, Verchot LV, Bullimore J, Borelli T, Albrecht A, Scow KM, Ball AS, Pretty JN, Osborn MA (2005) Soil microbial community response to land use change in an agricultural landscape in Western Kenya. Microb Ecol 49:50–62
Brussaard L, Hauser S, Tian G (1993) Soil faunal activity in relation to the sustainability of agricultural systems in the humid tropics. In: Soil organic matter dynamics and the sustainability of tropical agriculture: Proceedings of an international symposium. Leuven, pp 241–256
Chapin FS III, Walker BH, Hobbs RJ, Hooper DU, Lawton JH, Sala OE, Tilman D (1997) Biotic control over the functioning of ecosystems. Science 277:500–504
Cleveland CC, Townsend AR, Schmidt SK, Constance BC (2003) Microbial dynamics and biogeochemistry in tropical forests and pastures, southwestern Costa Rica. Ecol Appl 13:314–326
Cornelissen JHC, Aerts R, Cerabolini B, Werger MJA, van der Heijden MGA (2001) Carbon cycling traits of plant species are linked with mycorrhizal strategy. Oecologia 129:611–619
Crill PM, Keller M, Weitz A, Grauel B, Veldkamp E (2000) Intensive field measurements of nitrous oxide emissions from a tropical agricultural soil. Global Biogeochem Cycles 14:85–95
Davidson EA, Ackerman IL (1993) Changes in soil carbon inventories following cultivation of previously untilled soils. Biogeochemistry 20:161–193
Davidson EA, Janssens IA (2006) Temperature sensitivity of soil carbon decomposition and feedbacks to climate change. Nature 440:165–173
Davidson EA, Janssens IA, Luo Y (2006) On the variability of respiration in terrestrial ecosystems: moving beyond Q_{10}. Global Change Biol 12:154–164
de Ruiter PC, Neutel AM, Moore JC (1995) Energetics, patterns of interaction strengths, and stability in real ecosystems. Science 269:1257–1260
Dinesh RS, Chaudhuri G, Sheeja TE (2004) Soil biochemical and microbial indices in wet tropical forests: effects of deforestation and cultivation. J Plant Nutr Soil Sci 167:24–32
Erickson H, Keller M, Davidson EA (2001) Nitrogen oxide fluxes and nitrogen cycling during postagricultural succession and forest fertilization in the humid tropics. Ecosystems 4:67–84
FAO (2007) The State of Food and Agriculture. Food and Agriculture Organization of the United Nations. Rome, p 222
Finlay BJ (2002) Global dispersal of free-living microbial eukaryote species. Science 296: 1061–1063
Finlay BJ, Clarke KJ (1999) Ubiquitous dispersal of microbial species. Nature 400:828
Foley JA et al (2005) Global consequences of land use. Science 309:570–573
Fragoso C, Brown G, Parton C, Blanchartg E, Lavelle P, Pashanasi B, Senapati B (1997) Agricultural intensification, soil biodiversity and agroecosystems function: The role of earthworms. Appl Soil Ecol 6:17–35
Fraterrigo JM, Balser TC, Turner MG (2006) Microbial community variation and its relationship with nitrogen mineralization in historically altered forests. Ecology 87:570–579
Gange A, Nice HE (1997) Performance of the thistle gall fly, *Urophora cardui,* in relation to host plant nitrogen and mycorrhizal colonization. New Phytol 137:335–343
Gange C, Brown VK, Alphin DM (2003) Multitrophic links between arbuscular mycorrhizal fungi and insect parasitoids. Ecol Lett 6:1051–1055

Gomes NCM, Fagbola O, Costa R, Rumjanek NG, Buchner A, Mendona-Hagler L, Smalla K (2003) Dynamics of fungal communities in bulk and maize rhizosphere soil in the tropics. Appl Environ Microbiol 69:3758–3766

Goverde M, Van der Heijden MGA, Wiemken A, Sanders IR, Erhardt A (2000) Arbuscular mycorrhizal fungi influence life history traits of a lepidopteran herbivore. Œcologia 125:362–369

Grime JP (1977) Evidence for the existence of three primary strategies in plants and its relevance to ecological and evolutionary theory. Am Nat 111:1169–1194

Grime JP, Mackey JML, Hillier SH, Read DJ (1987) Floristic diversity in a model system using experimental microcosms. Nature 328:420–422

Groffman PM, Venterea RT, Verchot LV, Potter CS (2006) Landscape and regional scale studies of nitrogen gas fluxes. In: Wu J, Jones KB, Li H, Loucks OL (eds) Scaling and uncertainty analysis in ecology: Methods and applications. Springer, Dordrecht, pp 191–203

Hawksworth DL (2001) The magnitude of fungal diversity: the 1.5 million species estimate revisited. Mycol Res 105:1422–1432

Hawksworth DL, Rossman AY (1997) Where are all the undescribed fungi? Phytopathology 87:888–891

Hooper SD, Raes J, Foerstner KU, Harrington ED, Dalevi D, Bork P (2008) A molecular study of microbe transfer between distant environments. PLoS ONE 3:e2607. doi:10.1371/journal.pone.0002607

Jimenez JJ, Moreno AG, Lavelle P, Decaens T, Angel AF, Thomas RJ (1996) Ecology and biology of a new tropical earthworm *Martiodrilus carimaguensis* (Oligochaeta, Glossoscolecidae). In: Swift M (ed) The biology and fertility of tropical soils. Report of the tropical soil biology and fertility programme. TSBF, Nairobi, p 15

Jonsson L, Nilsson M-C, Wardle DA, Zackrisson O (2001) Context dependent effects of ectomycorrhizal species richness on tree seedling productivity. Oikos 93:353–364

Kaal EEJ, De Jong E, Field JA (1993) Stimulation of ligninolytic peroxidase activity by nitrogen nutrients in the white rot fungus *Bjerkandera* sp. strain BOS55. Appl Environ Microbiol 59:4031–4036

Klironomos J (2003) Variation in plant response to native and exotic arbuscular mycorrhizal fungi. Ecology 84:2292–2301

Ling-Fei L, Zhang Y, Zhi-Wel Z (2007) Arbuscular mycorrhizal colonization and spore density across different land-use types in a hot and arid ecosystem, Southwest China. J Plant Nutr Soil Sci 170:419–425

Lucas RW, Casper BB, Jackson JK, Balser TC (2007) Soil microbial communities and extracellular enzyme activity in the New Jersey Pinelands. Soil Biol Biochem 39:2508–2519

Manefield M, Whiteley AS, Griffiths RI, Bailey MJ (2002) RNA stable isotope probing, a novel means of linking microbial community function to phylogeny. Appl Environ Microbiol 68:5367–5373

Marschner P (2007) Soil microbial community structure and function assessed by FAME, PLFA and DGGE – advantages and limitations. In: Varma A, Oelmüller R (eds) Advanced techniques in soil microbiology. Springer, Berlin, pp 182–200

Martin-Laurent F, Philippot L, Hallet S, Chaussod R, Germon JC, Soulas G, Catroux G (2001) DNA extraction from soils: old bias for new microbial diversity analysis methods. Appl Environ Microbiol 67:2354–2359

Mathimaran N, Ruh R, Jama B, Verchot L, Frossard E, Jansa J (2007) Impact of agricultural management on arbuscular mycorrhizal fungal communities in Kenyan ferralsol. Agric Ecosyst Environ 119:22–32

Matson PA, Parton WJ, Power AG, Swift MJ (1997) Agricultural intensification and ecosystem properties. Science 277:504–509

Mitchell CC, Delaney DP, Balkcom KS (2008) A historical summary of Alabama's old rotation (circa 1896): the world's oldest, continuous cotton experiment. Agron J 100:1493–1498

Moore JC, McCann K, Setälä H, DeRuiter PC (2003) Top-down is bottom-up: does predation in the rhizosphere regulate aboveground dynamics? Ecology 84:846–857
Murty D, Kirschbaum MUF, McMurtrie RE, McGilvray H (2002) Does conversion of forest to agricultural land change soil carbon and nitrogen? A review of the literature. Global Change Biol 8:105–123
Neff JC, Holland EA, Townsend AR (1996) Variability in the temperature response of decomposition in sub-tropical and tropical soils. Bull Ecol Soc Am 77:322
Neill C, Piccolo MC, Steudler PA, Melillo JM, Feigl BJ, Cerri CC (1995) Nitrogen dynamics in soils of forests and active pastures in the western Brazilian Amazon Basin. Soil Biol Biochem 27:1167–1175
Nusslein K, Tiedje JM (1999) Soil bacterial community shift correlated with change from forest to pasture vegetation in a tropical soil. Appl Environ Microbiol 65:3622–3626
Nye PH, Greenland DJ (1960) The soil under shifting cultivation. Technical communication no. 51. Commonwealth Bureau of Soils, Harpenden
Oldeman LR (1994) The global extent of soil degradation. In: Greenland DJ, Szabolcs I (eds) Soil resilience and sustainable land use. CAB International, Wallingford, pp 99–118
Parton WJ, Sanford RL, Sanchez PA, Stewart JWB (1989) Modeling soil organic matter in tropical soils. In: Coleman DC, Oades JM, Uehara G (eds) Dynamics of soil organic matter in tropical ecosystems. University of Hawaii Press, Honolulu, pp 153–171
Rich JJ, Myrold JJ (2004) Community composition and activities of denitrifying bacteria from adjacent agricultural soil, riparian soil, and creek sediment in Oregon, USA. Soil Biol Biochem 36:1431–1441
Rich JJ, Heichen RS, Bottomley PJ, Cromack K Jr, Myrold DD (2003) Community composition and functioning of denitrifying bacteria from adjacent meadow and forest soils. Appl Environ Microbiol 69:5974–5982
Sanchez PA (2002) Soil fertility and hunger in Africa. Science 295:2019–2020
Schimel JP, Gulledge J (1998) Microbial community structure and global trace gases. Global Change Biol 4:745–758
Schimel J, Balser TC, Wallenstein M (2007) Microbial stress-response physiology and its implications for ecosystem function. Ecology 88:1386–1394
Solomon D, Lehmann J, Kinyangi J, Amelung W, Lobe I, Ngoze S, Riha S, Pell A, Verchot L, Mbugua D, Skjemstad J, Schäfer T (2007) Long-term impacts of anthropogenic perturbations on dynamics and speciation of organic carbon in tropical forest and subtropical grassland ecosystems. Global Change Biol 13:511–530
Stampe EC, Daehler C (2003) Mycorrhizal species identity affects plant community structure and invasion: A microcosm study. Oikos 100:362–372
Tilman D et al (2001) Forecasting agriculturally driven global environmental change. Science 292:281–284
Torsvik V, Øvreås L (2002) Microbial diversity and function in soil: From genes to ecosystems. Curr Opin Microbiol 5:240–245
van der Heijden MGA, Klironomos JN, Ursic M (1998) Mycorrhizal fungal diversity determines plant biodiversity, ecosystem variability and productivity. Nature 396:69–72
van Faasen HG, Smilde KW (1985) Organic matter and nitrogen turnover in soils. In: Kang BT, van der Heide J (eds) Nitrogen management and farming systems in humid and sub-humid tropics. Institute for Soil Fertility Research, Haren, pp 39–55
van Noordwijk M, Lusiana B (1999) WaNuLCAS, a model of water, nutrient and light capture in agroforestry systems. Agroforest Syst 45:131–158
van Noordwijk M, Cerri CC, Woomer PL, Nugroho K, Bernoux M (1997) Soil carbon dynamics in the humid tropical forest zone. Geoderma 79:187–225
Veldkamp E, Keller M, Nuñez M (1997) Effects of pasture management on N_2O and NO emissions from soils in the humid tropics of Costa Rica. Global Biogeochem Cycles 12:71–79

Verchot LV, Davidson EA, Cattânio JH, Ackerman IL, Erickson HE, Keller M (1999) Land-use change and biogeochemical controls of nitrogen oxide emissions from soils in eastern Amazonia. Global Biogeochem Cycles 13:31–46

Verchot LV, Davidson EA, Cattânio JH, Ackerman IL (2000) Land-use change and biogeochemical controls of methane fluxes in soils of eastern Amazonia. Ecosystems 3:41–56

Verchot LV, Holmes Z, Mulon L, Groffman PM, Lovett GM (2001) Gross versus net rates of N mineralization and nitrification as indicators of functional differences between forest types. Soil Biol Biochem 33:1889–1901

Verchot LV, Hutabarat L, Hairiah K, van Noordwijk M (2006) Nitrogen availability and soil N_2O emissions following conversion of forests to coffee in southern Sumatra. Global Biogeochem Cycles 20:GB4008. doi:10.1029/2005GB002469

Vicari MP, Hatcher E, Ayres PG (2002) Combined effect of foliar and mycorrhizal endophytes on an insect herbivore. Ecology 83:2452–2464

Waldrop MP, Blaser TC, Firestone MK (2000) Linking microbial community composition to function in a tropical soil. Soil Biol Biochem 32:1837–1846

Wardle DA, Bardgett RD, Klironomos JN, Setälä H, van der Putten WH, Wall DH (2004) Ecological linkages between aboveground and belowground biota. Science 304:1629–1633

White DC, Ringelberg DB (1998) Signature lipid biomarker analysis. In: Burlage RS, Atlas R, Stahl D, Geesey G, Sayler G (eds) Techniques in microbial ecology. Oxford University Press, New York, pp 255–272

Young A (1989) Agroforestry for soil conservation. CAB International, Wallingford

Zimmermann M, Leifeld J, Schmidt MWI, Smith P, Fuhrer J (2007) Measured soil organic matter fractions can be related to pools in the RothC model. Eur J Soil Sci 58:658–667

Zumft WG (1999) The denitrifying prokaryotes. In: Dworkin M (ed) The prokaryotes: An evolving electronic resource for the microbiological community, Springer, New York, www.prokaryotes.com

Chapter 4
The Structural and Functional Biodiversity of Soil: An Interdisciplinary Vision for Conservation Agriculture in Brazil

Gustavo Ribeiro Xavier, Maria Elizabeth Fernandes Correia, Adriana Maria de Aquino, Jerri Édson Zilli, and Norma Gouvêa Rumjanek

4.1 Introduction

Brazil is a country of continental dimensions with more than 8.5 million km^2, and for this reason it has a great climatic and soil diversity, which culminates in the formation of major biomes of national and international interest. Although some states have a sub-temperate climate, Brazil is essentially a tropical country, with highly weathered soils of low natural fertility. All biomes harbor a great plant and animal biodiversity, part of it still unknown.

Although the conservation and preservation of the Brazilian biomes is a permanent concern, several of them are being threatened by deforestation or expansion of agricultural frontiers. In the Amazon biome, the area deforested over the period ranging from August 2001 to August 2007 reached 118,542 km^2. From August 2006 to August 2007, an estimated 11,532 km^2 was deforested (INPE 2008), which is an indication of a decreasing annual deforestation rate, but nonetheless suggests that understanding and preserving indigenous biodiversity remains a big challenge.

An even more critical situation is observed in the Atlantic Coastal Rainforest (SOS Mata Atlântica 2008) and the Brazilian Cerrado (Machado et al. 2004), where more than 90% and 60% respectively of the native vegetation has been destroyed. These biomes present not only most of the agricultural activities in the country, but also the highest population density.

In parallel, it is estimated that the planted area increased from 1.2% to 2.7% for the 2008/2009 grain harvest in relation to the previous harvest year. This means that

G.R. Xavier (✉), M.E.F. Correia, A.M. de Aquino, and N.G. Rumjanek
Embrapa Agrobiologia, BR 465, km 7, Seropédica, Cep 23.890-000, Rio de Janeiro, Brazil
e-mail: gustavo@cnpab.embrapa.br

J.É. Zilli
Embrapa Roraima, BR 174, km 8, Boa Vista, Cep 69301-970, Roraima, Brazil

P. Dion (ed.), *Soil Biology and Agriculture in the Tropics*, Soil Biology 21,
DOI 10.1007/978 3 642-05076-3_4, © Springer-Verlag Berlin Heidelberg 2010

14 grain crops cover an area of about 48 million ha, of which soybean [*Glycine max* (L.) Merrill] accounts for slightly less that half of the total (MAPA 2008). Agribusiness has a strategic participation in the Brazilian trade balance, accounting for about 25% of the GDP over the years 2000–2007.

Thus, Brazil is experiencing the dichotomy of expanding the agricultural sector, which is important for the economy, while preserving its biomes. In addition, the combined promotion of agricultural production and environmental conservation has often led to a clash between the agricultural and the environmental sectors.

Research in agricultural production systems has been focused on conservation, in the search for new paradigms both for agricultural production and for environmental conservation. This strategy is based on the notion that it is feasible for agricultural management to incorporate ecological principles related both to restoration and maintenance of environmental services such as those provided by soil and water, and to the promotion of biodiversity conservation. In addition, conservation agriculture systems should introduce designs that allow recovery and preservation of threatened natural systems, such as those which subsist in highly fragmented areas.

This chapter intends to illustrate the involvement of agrobiological processes in conservation agriculture, and will place an emphasis on the soil biota community and its importance for soil function. It will also examine how biodiversity and biological activity relate to the productive capacity of soils. Because the soil biological component must be considered in the design of resource management strategies (Lavelle et al. 1997), it is necessary to know the soil community, to assess its functions, and finally to optimize its activities through proper practices.

Serious environmental problems have arisen with effects that go beyond the agricultural ecosystem, impacting natural and even urban environments. For example, the erosion of land and contamination due to chemical inputs such as fertilizers and pesticides are observed throughout the Brazilian territories. The loss of biodiversity, which is currently occurring all over terrestrial landscapes, disrupts food chains, exposes pathogens to new hosts, and diminishes nutrient availability at landscape scale.

4.2 The Unknown Soil Biodiversity

Soil microbial biodiversity is essential for maintenance of ecological processes such as organic matter decomposition, nutrient cycling, soil aggregation and control of pathogens (Kennedy 1999). Acting intensively in food chains and in several ecological processes, soil micro-organisms reflect the past history of the environment. It is therefore essential to understand the relationship between organisms and their environment through an investigation of structural and functional diversity of microbial communities and of their response to various natural and anthropogenic disturbances (Ranjard et al. 2000).

The most obvious benefit of biodiversity is to ensure that a multiplicity of functions can be performed by soil organisms. These biodiversity issues have also led to extensive discussions on functional redundancy (Giller et al. 1997). Thus, agricultural management should be concerned with the various groups of organisms that perform a given function, rather than with the abundance or distribution of a particular species.

The soil harbors more biodiversity than any other ecosystem on Earth. One reason for this great diversity of micro-organisms and invertebrates is the soil horizontal and vertical heterogeneity. In addition, the combined action of biotic and abiotic factors results in the formation of functional domains such as the rhizosphere. These domains are formed by the action of regulators such as plants, soil fauna, and edaphoclimatic conditions, among others.

4.3 Agrobiological Processes for Conservation Systems

4.3.1 Processes Mediated by Nitrogen-Fixing Bacteria

The energy efficiency and economic benefits of agricultural conservation systems are key determinants of the sustainability of these systems and of their potential for further development. In Brazil, little attention has been given to the distribution of energy flows in agricultural systems. Commonly, systems have been developed based on the intensive use of fertilizers obtained as oil by-products, and especially of nitrogen fertilizers (Urquiaga and Zapata 2000).

Agrobiological processes based on the soil biota may contribute to the generation of biological inputs for agriculture (see Chap. 11), and hence to the development of soil conservation practices. For example, the sustainability of food crops, forages, and green manure legumes is mainly associated with their ability to establish symbiotic associations with stem- and root-nodulating N_2-fixing rhizobia.

In the Brazilian agribusiness context, the impact of biological nitrogen fixation (BNF) in soybean, which is grown without the use of nitrogen fertilizers, brings to the economy an estimated US$ 6 billion per harvest (US$ 147/barrel), contributing to the competitiveness and success of the soybean agribusiness and boosting the Brazilian trade balance.

In addition to soybeans, other legumes in Brazil stand out for their economic and social importance. Cowpea (*Vigna unguiculata*) is quite suited to semi-arid regions, due to its high nutritional value, rusticity, adaptability to low soil fertility, and tolerance to drought, high temperatures and salinity. Cowpea was traditionally grown in northeast Brazil, a region with recognized edaphoclimatic limitations that affect the productivity of most crops. The sustainable activity of farmers in this environment requires technological innovation aiming at grain yield increase (Rumjanek et al. 2005).

For years, it was considered that the promiscuous nodulation ability of cowpea with native rhizobia from the group "cowpea miscellaneous" hindered the selection and exploitation of rhizobia for this culture (Rumjanek et al. 2005). However, based on efficiency studies (Martins et al. 1997), rhizobial ecology and characteristics such as tolerance to antibiotics (Xavier et al. 1998), salinity, and temperature (Xavier et al. 2007), it was possible to select a strain capable of increasing productivity by up to 30% (Martins et al. 2003) (Fig. 4.1). These data indicate that it is possible to select for competitive strains useful for cowpea inoculation in Brazil, so as to increase the productivity of this culture and reduce poverty among farmers.

Nowadays, the cowpea cultivation area is expanding to the North and Middle West regions and gradually the practice of rhizobial inoculation is following the same trend.

Menna et al. (2006) analyzed the molecular phylogeny based on the 16S rRNA gene of 68 elite rhizobial strains used in Brazil as commercial legume inoculants, and found that some differed from the type strains, which suggests that they may represent new species of the *Methylobacterium* and *Burkholderia* genera. The great diversity observed emphasizes that the tropics are an important reservoir of unknown N_2-fixation species and genes. The optimization of BNF in these conditions requires a greater knowledge of the dynamics of populations established in

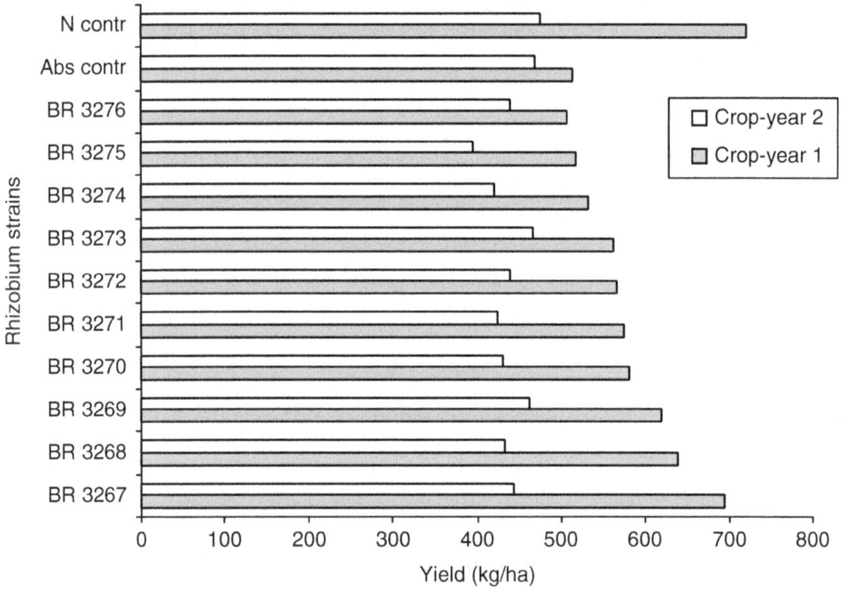

Fig. 4.1 Cowpea grain productivity during two consecutive crops. Data are means of the results obtained with three inoculation schedules differing according to the time of inoculant application, compared by LSD at the 5% level for each crop. Data include grain yields obtained following inoculation with various rhizobial strains and for uninoculated controls with (N contr) or without (Abs contr) nitrogen fertilization

tropical regions as well as of the specificity level and N-fixing efficiency of the symbiotic associations (Santos et al. 2008).

In spite of their potential for increasing nitrogen fixation, the new elite strains are of controversial use, because they belong to genera also containing human and plant pathogenic species. Additional research will be required to resolve this issue.

Other N-fixing bacteria such as those associated with grasses (*Poaceae*) present a potential for improving biological inputs in tropical agriculture. Brazil is the world's largest sugarcane producer with more than 6.7 million ha of planted area. This culture is currently strategic for the Brazilian bioenergy supply. The capacity of diazotrophic plant growth-promoting bacteria to improve sugarcane performance has been demonstrated under both greenhouse and field conditions. A BNF contribution of around 30% of the total nitrogen accumulated was observed in micropropagated plants inoculated with a bacterial consortium, suggesting that the use of a mixed inoculum is a promising strategy for improving biological nitrogen fixation in sugarcane crops (Oliveira et al. 2004, 2006).

Despite the positive results reported in the international literature on the use of inoculants in corn, wheat, and other cereals, such inoculants remain poorly used in Brazil, particularly because of the difficulties associated with evaluation, registration, and quality control of commercial products. Since 2007, a bacterial consortium (*Gluconacetobacter diazotrophicus* BR 11281, *Herbaspirillum seropedicae* BR 11335, *Herbaspirillum rubrisubalbicans* BR 11504, *Azospirillum amazonense* BR 11145 and *Burkholderia tropica* BR 11366) is recommended as an inoculant for sugarcane by Embrapa Agrobiologia. A partnership with the commercial inoculum-producing sector is currently envisioned to promote the widescale use of this mixed inoculant.

In an attempt to improve the inoculant quality in Brazil, different carrier blends together with cowpea rhizobium cells were evaluated at room temperature, and their performance as inoculant was compared to a peat-based inoculant carrier. Rhizobial cells were maintained best in blends containing 50–60 wt% carboxymethyl cellulose (CMC), and the cowpea nodulation of polymer blends was similar statistically to that of peat-based inoculants. CMC/starch polymer blends result in an efficient carrier for rhizobial inoculants, show competitive advantages such as being biodegradable, non-toxic and water soluble, and enable pre-inoculation of seed and maintenance of rhizobium numbers at room temperature comparable to those of traditional peat inoculants (Oliveira et al. 2007; Fernandes Júnior et al. 2009).

4.3.2 Processes Mediated by Plant Growth-Promoting Rhizobacteria

Various bacteria belonging to the *Pseudomonas* genus deserve attention since they are plant growth-promoting rhizobacteria (PGPR). Strains of *Pseudomonas fluorescens* and *Pseudomonas putida* have the ability to colonize the rhizosphere and rhizoplane of cultivated plants. These bacteria promote the growth of plants through a series of mechanisms such as: (1) production of siderophores, which chelate iron and inhibit the growth of certain microbiota components, (2) antibiosis through the excretion of antibiotics (Thomashow 1996), (3) solubilization of phosphates (see Sect. 10.4.2.5), and (4) production of compounds similar to auxins and cytokinins (Antoun and Prévost 2005).

Thomashow and Weller (1996) identified the production of antibiotics by *Pseudomonas* spp. as an efficient mechanism of disease suppression and/or plant growth stimulation. In suppressive soils, there is a direct relationship between the prevalence of soil micro-organisms and levels of their metabolites involved in the biological control process or induction of plant resistance. Moreover, continuous culture of the plant host may lead to the selection of bacterial antagonists to plant pathogens, as was reported for wheat and the fungal agent of take-all, *Gaeumannomyces graminis* var. *tritici* (Mazzola 2002).

Pseudomonas spp. isolates were obtained from soil, rhizosphere, rhizoplane, and inner part of root tissue of lettuce (*Lactuca sativa*) and carrot (*Daucus carota*) grown under organic management, and these bacteria were examined for cultural and molecular characters. The bacterial communities associated with lettuce and carrot differed in composition (Zago et al. 2000). In addition, the populations of bacteria isolated from the rhizoplane and from root tissues differed in composition from those isolated from soil samples. These data suggest the existence of specificity in the plant–bacteria association.

Xavier et al. (2004) also compared the bacteria diversity associated with lettuce and carrot roots, this time examining single strand conformation polymorphisms. The plants were cultivated as single culture (monocrop) or as intercrop in two soil types: Yellow Red Argissol (pH 7.0, Al 0.0 cmol dm^{-3}, Ca 3.3 cmol dm^{-3}, Mg 1.7 cmol dm^{-3}, P 104.5 cmol dm^{-3}, K 240 cmol dm^{-3}, organic carbon 0.755%) and Planosol (pH 7.0, Al 0.0 cmol dm^{-3}, Ca 1.25 cmol dm^{-3}, Mg 1 cmol dm^{-3}, P 16 cmol dm^{-3}, K 71 cmol dm^{-3}, organic carbon 0.355%). It was possible to detect specificities in bacterial communities depending on crop, soil type, and plant growth stages. On the other hand, differences related to intercropping and monoculture management were only detected when group-specific primers were used, especially primers for *Alphaproteobacteria*.

The production gains observed when intercropped cultural systems are used may be related to changes in the taxonomic and functional diversity of micro-organisms associated with the cultures; however, knowledge about these interactions is still very limited. It is possible that intercropping leads to an enrichment of microbial groups with plant growth-promoting activity.

Dias et al. (2008) tested in vitro the antagonism of PGPR strains against *Rhizoctonia solani* and *Sclerotium rolfsii*. A total of 94 bacterial strains isolated from the rhizosphere of four vegetable species under organic cultivation were evaluated. Twenty-two isolates, identified as *P. fluorescens*, were predominant in lettuce and rudbeckia rhizospheres, while in kale and parsley rhizospheres *P. putida* strains prevailed. Sixty percent of the evaluated strains showed antagonistic potential and, among those, 24 isolates expressed antagonism to both target fungi, with *P. fluorescens* being the most representative bacterial species. These data suggest that antibiosis promoted by soil micro-organisms may represent a promising field for the development of new agrobiological inputs.

4.3.3 Processes Mediated by Soil Fauna

Although soil fauna participates actively in several ecosystem processes, little emphasis has been given to a possible management of its functionality, as has historically been done for micro-organisms. Most studies on soil fauna in Brazil are aimed at the effect of different agricultural practices on diversity and structure of these communities. However, some of the benefits of the diversity of invertebrates that colonize the soil and their relationship with conservation farming practices have already been reported for different Brazilian biomes.

One of the effects of soil fauna activity is the production of biogenic structures, i.e., soil aggregates and organic matter formed by the action of soil biota. These products of soil fauna determine in large part the soil structure and, as a consequence, porosity and moisture retention, which are essential to plant growth (Jouquet et al. 2006). Contrary to what is commonly believed, the fauna action on soil structure can be of great magnitude even in the short term. In an experiment involving the transposition of soil monoliths between an area of degraded pasture and a forest area in Central Amazonia region, Barros et al. (2001) found that the forest soil fauna activity was able to restore the structure of compacted soil from the pasture. The time required for this recovery of the soil physical quality was only 1 year.

Another effect of soil fauna is the action of saprophagous species on the litter in the case of natural systems or on the straws or mulches in agricultural systems, particularly those based on conservation. Various invertebrates belong to the guild of the "litter transformers" (Lavelle et al. 1997), and, although they do not promote a profound biochemical transformation in these plant residues, their action on physical transformation and microbial stimulation triggers further steps in the decomposition process (Swift et al. 1979).

There are few estimates of litter consumption potential by saprophagous invertebrates in Brazilian ecosystems. However, it is known that in a native semi-deciduous forest in southern Brazil, two particular woodlice species, *Alantoscia floridana* and *Balloniscus glaber*, together were able to process 860 kg of leaves per hectare per year, which can represent 16% of the annual intake of leaves (Quadros and Araujo 2008). In another estimate for arboreal leguminous plantations in southeastern Brazil, Correia (2003) showed that a single exotic species of millipede, *Leptogoniulus sorornus*, was able to consume in only 1 month 185 kg ha^{-1} of leaf material deposited on the soil.

Among the factors that affect the litter consumption rate by saprophagous fauna, the C/N ratio and the concentration of polyphenols are probably the most important. Bianchi and Correia (2007) evaluated the feeding activity of the millipede *Trigoniulus corallinus* on the litter of two arboreous species with potential for use in agroforestry systems in southeastern Brazil, namely the leguminous species *Mimosa caesalpinifolia* and the *Myrtaceae* species, *Syzygium cumini*. While the daily consumption rate of the leguminous species was 74.2 mg per individual, in the case of *S. cumini* it was only 2.57 mg per individual. This difference in consumption rate correlated with a difference in the C/N ratio of the leaf material from both arboreous species, which was 15 for *M. caesalpinifolia* and 33 for *S. cumini*. Thus, in conservation systems, to manage the C/N ratio of consortia and soil coverage is an indirect way of managing the saprophagous fauna of soil and of maximizing the benefits of this group of organisms in the decomposition and cycling of nutrients.

4.4 Conservation Systems

4.4.1 No-tillage

The Brazilian no-tillage initiative, which has become an example for tropical countries around the world, has triggered a change in behavior of farmers and technicians in the pursuit of sustainable agriculture. The no-tillage system expanded from about 1 million ha planted with annual crops at the beginning of the 1990s, to over 25 million in 2005/2006. It is now used in all perennial crops, in sugarcane, in the recovery of pastures through rotation between crops and pastures, in reforestation, and in horticulture. Its attractiveness lies in the fact that it decreases the need for manual operations and for animal, tractor, or aerial traction (Plataforma de Plantio Direto 2009; FEBRAPD 2009).

Moreover, this system relies on an integrated vision involving the combination of practices such as: green manuring for the formation of soil coverage; the maintenance of crop residues on the soil surface; the adoption of integrated weed control methods through the use of soil coverage and herbicides; and the avoidance of soil disturbance, except in the sowing furrow. In this system, the soil fauna in a Latosol proved to be more diverse than that of a conventional system, under the climatic conditions of both the Brazilian Cerrado (Silva et al. 2006) and the Atlantic Forest (Rodrigues 2006).

In a study carried out to genetically characterize 30 fast-growing rhizobial strains isolated from the nodules of field-grown Asian and modern soybean genotypes which had previously been inoculated, Hungria et al (2006) observed a putatively new rhizobial species that was present only in undisturbed soils, showing high relatedness to *Rhizobium* OR 191, and another strain resembling *Agrobacterium*. Three species, *Rhizobium tropici, Bradyrhizobium japonicum* and *Bradyrhizobium elkanii*, were found under the no-tillage sustainable management system, while the only species isolated from soils under conventional tillage was *R. tropici*. These results suggested that rhizobial diversity becomes drastically reduced when a conventional soil management system is adopted, as compared to that associated with a no-tillage system.

More recently, soybean nodules were collected from 12 sites from the State of Mato Grosso, in the Brazilian Cerrado, where both soybean and bradyrhizobial strains have been introduced over the past 18 years. Diversity was higher under a no-tillage system than under conventional tillage management, highlighting the importance of maintaining crop residues at the surface of tropical soils. Understanding the ecology of exotic rhizobia after their introduction to new cropping areas represents the first step towards the design of better inoculation strategies, which in turn may result in sustainability and higher plant yields (Loureiro et al. 2007).

In a no-tillage system, the soil biota structure varies according to the crop rotation system and the quality of plant residues used as mulch. Earthworms and other fauna groups become more abundant when mulches with a high nitrogen content, such as soybean and turnip, are used. Upon complete degradation of crop residues, as occurs in the Cerrado, the bare and exposed soil becomes an unsuitable habitat for soil organisms (Aquino et al. 2008). On the other hand, termites are dominant in soils amended with lower quality straw mulching such as that of oats and corn. The distribution of fauna individuals in the soil profile is variable depending on management, time of assessment, and plant cover species used (Silva et al. 2007). Generally, as the amount of residues on the soil surface diminishes due to the decomposition process, there is a tendency for fauna components to concentrate in the lower layers of the soil (Silva et al. 2007).

Aquino et al (2006) observed that several taxonomic groups, such as Isopoda, Diplopoda, Diplura, Gastropoda, Blattodea and Dermaptera, disappeared in the Parana region (Brazil), because of injuries caused by agricultural implements. Members of the first three groups are living in decomposing organic matter and are vital to its fragmentation (Swift et al. 1979; Hashimoto et al. 2004; Tuck and Hassall 2005). Following the reduction in resources and shelters occurring in soils having undergone agricultural transformation, some fauna groups may occupy the remaining available niches, establish effectively and dominate the community. Among these, social insects, especially ants (Formicidae) and termites (Isoptera) stand out in pastures of the Brazilian Cerrado (Benito et al. 2004; Silva et al. 2006). Whereas social insects naturally occur in forest areas, deforestation results in the maintenance of a more specialized population of these insects (Constantino and Acioli 2005). In several regions of Brazil and particularly in the Cerrado, termites modify the landscape in degraded pastures, leaving them covered by *murundus* (earthmounds) (Aquino et al. 2008b).

4.4.2 Agroforestry Systems

In almost all Brazilian biomes, the tree component is dominant and it is no accident that many such biomes receive the designation of forest, such as the Amazon Rainforest and the Atlantic Forest. The forest element has been historically seen in Brazil either as an obstacle to the establishment of intensive agricultural systems such as pastures and grain monocultures, or as a source of wood for different purposes. In both situations, the land use eventually led to the eradication of the forest ecosystem. This type of predatory use of forests has brought 93% of the Atlantic Forest, in the past covering 15% of Brazil's territory, to devastation (Ceccon and Miramontes 2008). Currently, due to greater awareness of environmental problems, there is a tendency to valorize the environmental products and services that the forests can sustainably provide.

In this context, it is not the forest that should make room for agriculture but rather the agricultural activity that should be incorporated into the forest environment. The principle of agroforestry systems is to carry out agricultural production in an environment with a forest structure, allowing the restoration of important ecological functions such as the cycling of nutrients and soil protection (Macdicken and Vergara 1990; see Chap. 9). In the past 20 years, the search for different models and designs of agroforestry systems tailored to specific purposes and environmental conditions has increased significantly. The agroforestry activity in Brazil provides an economic stimulus to forest recovery, leading to the incorporation of the tree component in rural settings, and especially in those that are family-based (Rodrigues et al. 2007).

At least in the case of soil macrofauna, biodiversity loss is minimized upon replacement of the natural ecosystem with another which has a similar structure (Decaëns et al. 2004). For example, when a savanna is transformed into a pasture and a forest into an agroforestry system, a large fraction of the soil fauna community is retained, together with the processes involved.

One of the oldest agroforestry production models in Brazil is cacao production (*Theobroma cacao*), which is established in the state of Bahia, in the Atlantic Forest domain of northeastern Brazil. In this region, cacao is grown in large part either under the canopy of the forest, or in a consortium with the leguminous tree *Erythrina* sp. In these shaded cacao systems, especially those of greater diversity, many ant species are preserved, including some considered rare. Even in the system of cacao with *Erythrina* sp, which harbors less tree diversity, 192 ant species were found, composing a community similar to that of native environments (Delabie et al. 2007).

Soil macrofauna has been investigated in some agroforestry systems with low plant diversity of the Amazon region, in the states of Acre and Rondonia. Compared with that associated with other land uses such as fallow land, annual crops, and pastures of *Brachiaria humidicola*, the soil macrofaunal community in agroforestry systems was more abundant and diverse, with a clear dominance of ants and termites (Barros et al. 2002).

Some agroforestry models in Brazil are based on the functional diversity of plant species, defined according to the successional stage at which these species occur. These systems involve the high density cultivation of a variety of plants with particular architectures, such as herbs, bushes and trees. Some plants are used to promote the recovery of soil fertility, such as N-fixing legumes, whereas others are important to create a shaded habitat or to attract pollinators (Götsch 1995; Vaz 2001). An evaluation of the benefits of this agroforestry model for soil quality was done in an Atlantic Forest area in southeastern Brazil. The agroforestry system examined comprised a total of 20 annual crop and tree species. The stocks of nutrients, organic matter, microbial activity, and composition of the soil fauna were similar in the soils from the agroforestry system and the forest, while in a cassava monoculture (*Manihot esculenta*) in the same area, all these indicators were lower (Silva 2006).

The use of agroforestry systems can also represent an alternative to slash-and-burn systems used by small farmers throughout Brazil (see Chap. 5 for a description of slash-and-burn agriculture). Soil quality was assessed in a transition environment between the Cerrado and Caatinga biomes. The environments compared were agroforestry systems with 6 and 10 years of implantation, traditional slash-and-burn agriculture systems, and the secondary forest in the state of Piaui, in mid-northern Brazil. The largest stocks of nutrients, organic C and N, as well as the greatest diversity of soil macrofauna, were found in the agroforestry system with 10 years of implantation (Lima 2008).

Whereas agroforestry systems promote biodiversity conservation and the provision of environmental services, a forest cover may be incompatible with some farming systems. One example was the simulation of the sugarcane cultivation with rubber (*Hevea brasiliensis*) or eucalyptus (*Eucalyptus grandis*) in alley cropping (Pinto et al. 2005). In the case of grasslands, however, several designs of forestry–agricultural systems with a sparse tree cover have been proposed, with environmental and economic gains (Dias 2008). The introduction of trees, particularly legumes that fix atmospheric N, improves the quality of the environment under the tree canopy, which is reflected in an increase in abundance and diversity of the soil macrofauna. In particular, the abundance of saprophagous and hygrophilous soil fauna groups such as earthworms and woodlice was increased. These benefits of the tree component were related to both the input of organic matter with a lower C/N ratio, and to the change in the microclimatic conditions under the tree canopy (Dias et al. 2006, 2007; see also Chap. 6 for a description of similar tree effects in desert settings).

4.4.3 Agroecological Systems

A new approach to conservation agriculture is based on agroecological systems. The components of these systems are economic (potential for income and employment, market access), environmental (maintenance or improvement of the quality of natural resources and of ecological relationships), social (alleviation of poverty and improvement of food security), cultural (as relates to use of traditional crops), political (organizations involved in changes, participatory decision processes), and ethical (transcendent moral values) (Embrapa 2006). These systems are supported mainly by biological processes aiming to ensure a constant organic matter supply for fertility buildup.

An experimental station dealing with research, extension, and training of human resources in conservation agriculture has been established since 1993 in Seropédica, State of Rio de Janeiro (Brazil). This is known as the "Agroecological Production Integrated System", or as "Fazendinha Agroecológica Km 47." The management adopted in the "Fazendinha" seeks to optimize the recycling of nutrients, and aims at the integration of animal and plant production activities. Additional goals are: a nitrogen self-sufficiency through the systematic use of crop rotation and diversification; the minimization of nutrient loss occurring by percolation and erosion; the maintenance of the plant nutritional balance; the avoidance of stress situations, to insure that plant defense mechanisms be expressed to their full potential; the maintenance of phytoparasites and weed populations to tolerable levels, while avoiding the use of techniques with negative eco-toxicological impacts; the deployment and tracking of agroforestry booths; the development of alternative practices for the management of dairy cattle and laying hens; and finally, the scientific treatment of the system components through a multidisciplinary approach. It is expected that this type of agricultural management involving biota preservation in soil-plant systems will increase the microbial diversity-based antagonism among the rhizobacteria and lead to a natural soilborne plant pathogen suppression (Dias 2006).

Organic management organized along those principles favors populations of different species of Collembola and Acarina when compared with forest and pasture (Badejo et al. 1998). In turn, collembolans change the community of fungi and contribute to the elimination of pathogens such as *R. solani*, *Fusarium oxysporum* and others (Bettiol et al. 2002).

In another study conducted in the same experimental station, Zilli et al. (1999) compared the diversity of rhizobia in cowpea cultivated in soils of a secondary forest, a pasture, and an area under organic production crop. More diversity was observed in the forest, followed by the organic cultivation area and then the pasture.

4.5 Conclusions

The principal functions in soil agroecosystems are those involved with biogeochemical cycles and maintenance of the soil structure: these functions strictly depend on the effective participation of soil biota components. Each of these processes involves the participation, not of a single species, but rather of several species that may be very different and interrelated. The soil biota can be harnessed so as to improve the sustainability of farming systems. More efficient agrobiological processes pave the way for biotechnological applications compatible with the novel soil conservation systems.

Despite the growing interest aroused by the soil conservation systems, there is still little information about the soil microbial community structure and soil biota composition. The study of soil biota and research on soil biodiversity are crucial to the understanding, optimization and control of the mechanisms involved in the maintenance of soil functions in agricultural production systems.

The suggestion is made here that such studies on soil biological processes should be coupled with efforts aimed at the genetic improvement of plants, and that the integration of soil biology and plant genetics has the potential to transform tropical regions into major zones of sustainable food production.

Yet another crucial element in this transformation is the cooperative action of all the different actors in the food production chain. This will be achieved through participatory research and the collective evaluation of the economic, social, and environmental impact of new technologies.

References

Antoun H, Prévost D (2005) Ecology of plant growth promoting rhizobacteria. In: Siddiqui ZA (ed) PGPR: Biocontrol and biofertilization. Springer, Dordrecht, pp 1–38

Aquino AM, Silva VVE, Guerra JGM (2006) Macrofauna edáfica em relação ao pré-cultivo de adubos verdes na produção orgânica de brócolos em sistema de plantio direto. In: Anais da Reunião Brasileira de Fertilidade do Solo e Nutrição de Plantas, 27, Reunião Brasileira Sobre Micorrizas, 11, Simpósio Brasileiro de Microbiologia do Solo, 9, Reunião Brasileira de Biologia do Solo, 6, SBCS/SBM/Embrapa, Bonito (MS), CD-ROM

Aquino AM, Silva RF, Mercante FM, Correia MEF, Guimarães MF, Lavelle P (2008a) Invertebrate soil macrofauna under different ground cover plants in the no-till system in the Cerrado. Eur J Soil Biol 44:191–197

Aquino AM, Correia MEF, Alves MV (2008b) Biodiversidade da macrofauna edáfica no Brasil. In: Moreira FMS; Siqueira JO; Brussaard L (eds). Biodiversidade do Solo em Ecossistemas Brasileiros. Lavras, UFLA, pp 147–175

Badejo MA, Aquino AM, Almeida DL, Correira MEF (1998) Preliminary investigations on microarthropods in different agroecosystems under organic management. In: REUNIÃO BRASILEIRA DE FERTILIDADE DO SOLO E NUTRIÇÃO DE PLANTAS, 23, REUNIÃO BRASILEIRA SOBRE MICORRIZAS, 7, SIMPÓSIO BRASILEIRO DE MICROBIOLOGIA DO SOLO, 5, REUNIÃO BRASILEIRA DE BIOLOGIA DO SOLO, 2, 1998, Lavras/MG. Resumos... Fertbio 98. Lavras/MG: SBCS

Barros E, Curmi P, Hallaire V, Chauvel A, Lavelle P (2001) The role of macrofauna in the transformation and reversibility of soil structure of an oxisol in the process of forest to pasture conversion. Geoderma 100:193–213

Barros E, Pashanasi B, Constantino R, Lavelle P (2002) Effects of land-use system on the soil macrofauna in western Brazilian Amazonia. Biol Fertil Soils 35:338–347

Benito NP, Brossard M, Pasini A, Guimarães MF, Bobillier B (2004) Transformations of soil macroinvertebrate populations after native vegetation conversion to pasture cultivation (Brazilian Cerrado). Eur J Soil Biol 40:147–154

Bettiol W, Ghini R, Galvão JAH, Ligo MAV, Mineiro LC (2002) Soil organisms in organic and conventional cropping systems. Sci Agric 59:565–572

Bianchi MO, Correia MEF (2007) Mensuração do consumo de material vegetal depositado sobre o solo por diplópodes. Circulatr Técnica, p 4

Ceccon E, Miramontes O (2008) Reversing deforestation? Bioenergy and society in two Brazilian models. Ecol Econ 67:311–317

Constantino R, Acioli ANS (2005) Termite diversity in Brazil (Insecta: Isoptera). In: Moreira FMS, Siqueira JO, Brussard L (eds) Soil biodiversity in Amazonian and other Brazilian ecosystems. CAB International, UK, pp 117–128

Correia MEF (2003) Distribuição, preferência alimentar e transformação de serapilheira por diplópodes em sistemas florestais. 100 f. Tese (Doutorado em agronomia- Ciência do Solo) – Instituto de Agronomia, Universidade Federal Rural do Rio de Janeiro, Seropédica

Decaëns T, Jiménez JJ, Barros E, Chauvel A, Blanchart E, Fragoso C, Lavelle P (2004) Soil macrofauna communities in permanent pastures derived from tropical forest or savanna. Agricult Ecosys Environ 103:301–312

Delabie JHC, Jahyny B, Nascimento IC, Mariano CSF, Lacau S, Campiolo S, Philpott SM, Leponce M (2007) Contribution of cocoa plantations to the conservation of native ants (Insecta: Hymenoptera: Formicidae) with a special emphasis on the Atlantic Forest fauna of southern Bahia. Brazil Biodivers Conserv 16:2359–2384

Dias A (2008) Rizobactérias associadas a hortaliças sob cultivo orgânico: caracterização e seleção para o biocontrole de fungos fitopatogênicos habitantes do solo. Dissertação (Mestrado em Biotecnologia Vegetal) – Universidade Federal do Rio de Janeiro, Instituto de Ciências da Saúde, Rio de Janeiro, p 93

Dias PF, Souto SM, Correia MEF, Rocha GP, Moreira JF, Rodrigues KM, Franco AA (2006) Árvores fixadoras de nitrogênio e macrofauna do solo em pastagem de híbrido de Digitaria. Pesq Agropec Bras Brasília 41:1015–1021

Dias PF, Souto SM, Correia MEF, Rodrigues KM, Franco AA (2007) Efeito de leguminosas arbóreas sobre a macrofauna do solo em pastagem de *Brachiaria brizantha* cv Marandu. Pesq Agropec Trop 37:38–44

Dias PF, Souto SM, Azevedo BC, Vieira MS, Colombari AA, Dias J, Franco AA (2008b) Estabelecimento de leguminosas arbóreas em pastos de capim-marandu e tanzânia. Pesq Agropec Bras 43:1413–1419

Embrapa (2008) Marco referencial em agroecologia. Empresa Brasileira de Pesquisa Agropecuária. Embrapa Informação Tecnológica, Brasília DF

FEBRAPD (2009) Evolução da área cultivada no sistema de plantio directo na palha – Brasil. http://www.febrapdp.org.br/arquivos/EvolucaoAreaPDBr72A06.pdf. Accessed 05 Feb 2009

Fernandes Júnior PI, Rohr TG, Oliveira PJ, Xavier GR, Rumjanek NG (2009) Polymers as carriers for rhizobial inoculant formulations. Pesq Agropec Bras Brasíia 44:1184–1190

Giller KL, Beare MH, Lavelle P, Izac AMN, Swift MJ (1997) Agricultural intensification, soil biodiversity and agroecosystem function. Appl Soil Ecol 6:3–16

Götsch E (1995) Break-through in agriculture. AS-PTA, Rio de Janeiro, p 22

Hashimoto M, Kaneko N, Ito MT, Toyota A (2004) Exploitation of litter and soil by the train millipede (Diplopoda: Xystodesmidae) in larch plantation forest in Japan. Pedobiologia 48:71–81

Hungria M, Chueire LMO, Megías M, Lamrabet Y, Probanza A, Guttierrez-Mañero FJ, Campo RJ (2006) Genetic diversity of indigenous tropical fast-growing rhizobia isolated from soybean nodules. Plant Soil 288:343–356

INPE (2008) Desmatamentos no bioma Cerrado: Uma análise temporal (2001-2005) com base nos dados MODIS – MOD13Q1. http://marte.dpi.inpe.br/col/dpi.inpe.br/sbsr%4080/2006/11.15.21.21/doc/3877-3883.pdf. Accessed 10 Dec 2008

Jouquet P, Dauber J, Lagerlöf J, Lavelle P, Lepage M (2006) Soil invertebrates as ecosystem engineers: intended and accidental effects on soil and feedback loops. Appl Soil Ecol 32:153–164

Kennedy AC (1999) Bacterial diversity in agroecosystems. Agricult Ecosys Environ 74:65–76

Lavelle P, Bignell D, Lepage M, Wolters V, Roger P, Ineson P, Heal OW, Dhillion S (1997) Soil function in a changing world: the role of invertebrate ecosystem engineers. Eur J Soil Biol 33:159–193

Lima SS (2008) Impacto do manejo agroflorestal sobre a dinâmica de nutrientes e a macrofauna invertebrada nos compartimentos serapilheira-solo em área de transição no norte do Piauí. 2008. 76 f. Thesis (Master in Environment and Development) Universidade Federal do Piauí, Teresina, PI, Brazil

Loureiro MF, Kaschuk G, Alberton O, Hungria M (2007) Soybean [*Glycine max* (L.) Merrill] rhizobial diversity in Brazilian oxisols under various soil, cropping, and inoculation managements. Biol Fertil Soils 43:665–674

Macdicken KG, Vergara NT (1990) Introduction to agroforestry. In: Macdicken KG, Vergara NT (eds) Agroforestry: Classification and management. Wiley, New York, pp 1–30

Machado RB, Ramos Neto MB, Pereira PGP, Caldas EF, Gonçalves DA, Santos NS, Tabor K, Steininger M (2004) Estimativas de perdas da área do Cerrado brasileiro. Conservação Internacional, Brasília, DF. http://www.conservation.org.br/arquivos/RelatDesmatamCerrado.pdf. Accessed 10 Dec 2008

MAPA (2008) Acompanhamento da safra brasileira: Grãos safra 2008-2009, Intenção de plantio, p 39

Martins LMV, Neves MCP, Rumjanek NG (1997) Growth characteristics and symbiotic efficiency of rizobia isolated from cowpea nodules of the north-east of Brazil. Soil Biol Biochem 29:1005–1010

Martins LMV, Rangel FW, Xavier GR, Ribeiro JRA, Morgado LB, Neves MCP, Rumjanek NG (2003) Contribution of biological nitrogen fixation to cowpea: a strategy for improving grain yield in the semi-arid region of Brazil. Biol Fertil Soils 38:333–339

Mazzola M (2002) Mechanisms of natural soil suppressiveness to soilborne diseases. Antonie Van Leeuwenhoek 81:557–564

Menna P, Hungria M, Barcellos FG, Bangel EV, Hess PN, Martínez-Romero E (2006) Molecular phylogeny based on the 16S rRNA gene of elite rhizobial strains used in Brazilian commercial inoculants. Syst Appl Microbiol 29:315–332

Oliveira ALM, Urquiaga S, Döbereiner J, Baldani JI (2004) The effect of inoculating endophytic N_2-fixing bacteria on micropropagated sugarcane plants. Plant Soil 242:205–215

Oliveira ALM, Canuto EL, Urquiaga S, Reis VM, Baldani JI (2006) Yield of micropropagated sugarcane varieties in different soil types following inoculation with diazotrophic bacteria. Plant Soil 284:23–32

Oliveira PJ, Fernandes PI Jr, Xavier GR, Rumjanek NG (2007) Estudo do comportamento reológica da mistura carboximetilcelulose/amido e seu uso como veículo de inoculação rizobiano. In: 9th Congresso Brasileiro de polímeros, vol 1. CBPOL, Campina Grande

Pinto LFG, Bernardes MS, Van Noordwijk M, Pereira AR, Lusiana B, Mulia R (2005) Simulation of agroforestry systems with sugarcane in Piracicaba, Brazil. Agricult Syst 86:275–292

Plataforma de Plantio Direto (2009) Sistema planto direto. http://www22.sede.embrapa.br/plantiodireto/IntroducaoHistorico/sistemaPlantioDireto.htm. Cited 05 Feb 2009

Quadros AF, Araujo PB (2008) An assemblage of terrestrial isopods (Crustacea) in southern Brazil and its contribution to leaf litter processing. Rev Bras Zool 25:58–66

Ranjard L, Poly F, Nazaret S (2000) Monitoring complex bacterial communities using culture-independent molecular techniques: application to soil environment. Res Microbiol 51:167–177

Rodrigues ER, Cullen JRL, Beltrame TP, Moscogliato AV, Silva IC (2007) Avaliação econômica de sistemas agroflorestais implantados para recuperação de reserva legal no pontal do Paranapanena, São Paulo. Árvore 31:941–948

Rumjanek NG, Martins LMV, Xavier GR, Neves MCP (2005) Contribuição da FBN em caupi. In: Freire Filho FR, Lima JAA, Ribeiro VQ (eds). Feijão caupi: avanços tecnológicos. Embrapa Informação Tecnológica, Brasília, DF, pp 281–335

Santos CERS, Freitas ADS, Vieira IMMB, Colaço W (2008) Fixação simbiótica do N_2 em leguminosas tropicais. In: Fiqueired MVB, Burity HA, Stamford NP, Santos CERS (eds) Microorganismos y agro biodiversidad: un nuevo desafio para la agricultura. Agrolivros, Guaíba, pp 17–68

Silva MSC (2006) Indicadores de qualidade do solo em sistemas agroflorestais em Paraty RJ. Thesis (Master in Soil Science), Universidade Federal Rural do Rio de Janeiro, Seropédica, Brazil

Silva RF, Aquino AM, Mercante FM, Guimarães MDEF (2006) Macrofauna invertebrada do solo sob diferentes sistemas de produção em Latossolo da região do Cerrado. Pesq Agropec Bras 41:697–704

Silva RF, Tomazi M, Pezarico CR, Aquino AM, Mercante FM (2007) Macrofauna invertebrada edáfica em cultivo de mandioca sob sistemas de cobertura do solo. Pesq Agropec Bras 42:865–871

SOS MATA ATLÂNTICA (2008) Mata Atlântica. http://www.sosmatatlantica.org.br/index.php?section=info&action=mata. Cited 10 Dec 2008

Swift MJ, Heal OW, Anderson JM (1979) Decomposition in terrestrial ecosystems. Blackwell, Oxford

Thomashow LS (1996) Biological control of plant root pathogens. Curr Opin Biotechnol 7:343–347

Thomashow LS, Weller DM (1996) Current concepts in the use of introduced bacteria for biological disease control: mechanisms and antifungal metabolites. In: Stacey G, Keen NT (eds) Plant-microbe interactions, vol 1. Chapman and Hall, New York, pp 187–235

Tuck JM, Hassall M (2005) Locating food in a spatially heterogeneous environment: implications for fitness of the macrodecomposer *Armadillidium vulgare* (Isopoda: Oniscoidea). Behav Ecol Sociobiol 58:545–551

Urquiaga S, Zapata F (2000) Manejo eficiente de la fertilización nitrogenada de cultivos anuales en América Latina y el Caribe. Genesis, Porto Alegre

Vaz P (2001) Agroforestería en Brasil: una experiencia de regeneración análoga. Boletin de ILEIA 16:5–7

Xavier GR, Martins LMV, Neves MCP, Rumjanek NG (1998) Edaphic factors as determinants for the distribution of intrinsic antibiotic resistance in a cowpea rhizobia population. Biol Fert Soils 27:386–392

Xavier GR, Tebbe CC, Goncalves LSA, De-Polli H, Guerra JGM, Martins CM, Rangel FW, Rumjanek NG (2004) Molecular profiles of non-culturable bacteria associated to lettuce and carrot roots: a strategy to understand intercropping in an organic farm system. In: 10th International symposium on microbial ecology, Cancun, p 56

Xavier GR, Martins LMV, Ribeiro JRA, Rumjanek NG (2007) Especificidade simbiótica entre rizóbios e acessos de feijão-caupi de diferentes nacionalidades. Caatinga 1:25–33

Zago VCP, Almeida DL, Rumjanek NG (2000) Estudo das populações de *Pseudomonas fluorescens* spp. associadas a olerícolas sob sistema de produção orgânica. 7th Encontro Nacional de Microbiologia Ambiental, Recife, Fundação Antônio dos Santos Abranches – FASA, Resumos, p 115

Zilli JE, Ferreira EPB, Neves MCP, Rumjanek NG (1999) Efficiency of fast-growing rhizobia capable of nodulating cowpea. An Acad Bras Cienc 71:553–560

Chapter 5
The Potential of Soil Beneficial Micro-Organisms for Slash-and-Burn Agriculture in the Humid Forest Zone of Sub-Saharan Africa

Dieudonné Nwaga, Jan Jansa, Monique Abossolo Angue, and Emmanuel Frossard

Abbreviations

AABNF	African Association for Biological Nitrogen Fixation
AGRA	Alliance for a Green Revolution in Africa
AfNet-TSBF	African Network for Soil Biology and Fertility Institute
Africa NUANCES	Africa Nutrient Use in Animal and Cropping Systems: Efficiencies and Scales
AM	Arbuscular mycorrhiza
AUF	Agence Universitaire de la Francophonie
ASB	Alternatives to Slash-and-Burn Program
BIOVEG	Biotechnologie végétale: amélioration des plantes et sécurité alimentaire
BCA	Bio-control agents
BM	Beneficial micro-organisms
CARBAP	Centre Africain de Recherche sur Bananiers et Plantains
CFU	Colony forming units
CIAT	International center for Tropical Agriculture
CSM-BGBD	Conservation and sustainable management of below ground biodiversity

D. Nwaga (✉)
Laboratory of Soil Microbiology, Faculty of Science/Biotechnology Centre, P.O. Box 812, University of Yaoundé I, Yaoundé, Cameroon, Africa
e-mail: dnwaga@yahoo.fr

J. Jansa and E. Frossard
ETH Zurich, Institute of Plant Sciences, CH-8315, Lindau (ZH), Switzerland

M. Abossolo Angue
Department of Earth Sciences, University of Yaoundé I, Yaoundé, Cameroon, Africa

P. Dion (ed.), *Soil Biology and Agriculture in the Tropics*, Soil Biology 21,
DOI 10.1007/978-3-642-05076-3_5, © Springer-Verlag Berlin Heidelberg 2010

EM	Ectomycorrhiza
EMBRAPA	Empresa Brasileira de Pesquisa Agropecuária
ETH	Swiss Federal Institute of Technology
FAO	United Nations Food and Agriculture Organisation
ICRAF	International Centre for Research in Agroforestry
IFDC	International Fertilizer Development Center
IITA	International Institute for Tropical Agriculture
IRD	Institut Français de Recherches pour le Développement
IRAD	Institute for Research on Agriculture for Development
ISFM	Integrated soil fertility management
LNB	Legume-nodulating bacteria
LSTM	Laboratoire des Symbioses Tropicales et Méditerranéennes
MINESUP	Ministry of Higher Education
MINEP	Ministry of Environment and Protection of Nature
MF	Mycorrhizal fungi
MIRCEN	Microbial Resources Centre
NARS	National Agricultural Research System
PGPR	Plant growth-promoting rhizobacteria
PSM	Phosphorus-solubilizing micro-organisms
SB	Slash-and-burn
SOM	Soil organic matter
SS	Sub-Saharan
TSBF	Tropical Soil Biology and Fertility Institute
UNESCO-BAC	UNESCO Biotechnology Action Council

5.1 Introduction

In sub-Saharan (SS) Africa, about a third of the population lives on less than 1 US dollar a day, and food production per capita declined between 1980–2000 (Kates and Dasgupta 2007; Mahendra et al. 2008). According to the Food and Agriculture Organization of the United Nations (FAO), about 30% of the SS African population is at risk of food crisis. Thus, reducing poverty and food insecurity remains the main challenge for SS Africa, which is the only zone in the world where food production per capita has remained stagnant over the past 40 years (Sanchez 2002; FAO 2001). Food scarcity increases susceptibility to malaria, AIDS, and tuberculosis. Although agriculture involves about 70% of the population of SS Africa, more than 25,000 human beings are dying each day from malnutrition, and food aid is estimated to be 2.8 million tons per year. Because of the actual context of world crisis, food production in SS Africa must be increased. Climate change poses an additional threat, in particular to poor farmers (Mahendra et al. 2008). Forest clearing and burning is done by logging companies and by farmers alike for food subsistence agriculture. Biodiversity loss, nutrient losses and resulting soil infertility, continuous soil degradation, reduction in crop productivity, mixed field crops, as well as long fallows are some typical characteristics of slash-and-burn agriculture (SB) in humid forest zones (Kotto-Same et al. 2000; Gockowski et al. 2005). This is because SB agriculture cannot sustainably feed the current human populations of the SS African countries. Alternatives to SB agriculture require the management of local biological resources in order to sustain food production in SS Africa. Research results obtained in humid forest zones strongly support the idea of implementing soil biotechnology for a sustainable agriculture and forestry.

Poor soil fertility, drought, and land degradation have been described as the most important constraints to food security in SS Africa (Bationo et al. 2004; Swift and Shepherd 2007). The main processes regulating soil fertility in SS Africa are the rate of litter fall and soil organic matter (SOM) decomposition, leaching and recycling, and soil acidification. The level of annual soil mining in African agroecosystems is 4.4 million tons N against 0.8 million tons applied, 0.5 million tons P against 0.3 million tons applied, and 3 million tons K against 0.2 million tons applied (FAO 1995; Sanchez et al. 1997). The World Institute of Phosphate has classified most tropical soils to be severely deficient in P (Koala et al. 1988), and most P-fixing soils exist in Africa, mainly in humid forest zones (Sanchez et al. 1997). A huge amount of P is stored in the soil, only a very small part of which, for example representing 0.25 and 0.82% of total P respectively in a nonfertilized and a fertilized Oxisol from Kenya, is available to plants (Oberson et al. 2006).

5.1.1 Slash-and-Burn Agriculture in the Humid Forest Zones of SS Africa

Slash-and-burn agriculture is a form of agriculture in which an area of forest is cleared by cutting and burning and is then planted, usually for several seasons, before being abandoned. From 300 to 500 million people practice this system worldwide on about 500 million hectare of arable land, of which 45% is located in tropical areas. SB agriculture is only sustainable when the human population density is low and about 60% of Africa's food comes from SB agriculture (Morgan 2005).

This system is characterized by its heterogeneity, complexity, and diversity in terms of management practices: cropping seasons, crop species, fallow duration and management, plot size, soil fertility, land-use systems, available labor, credit, and market all vary. The benefits of burning are: a flush of nutrients from plant biomass to the soil in the form of ash, an interruption of the life cycle of insect pests, a destruction of weed seeds and pathogens, and a pH increase; however, these are short-term effects. Furthermore, tropical forests are also habitats for extremely different organisms, containing large numbers of endemic and endangered species which can be threatened by SB actions.

Shifting cultivation, fallow systems, plantations, mixed farming, or permanent farming are the main farming systems in humid forest zones. The cropping phase is usually short, only 1 or 2 years, and a fallow period of at least 15 years follows. In less populated areas, such as in the southern humid forest zones of Cameroon, SB agriculture is still the dominant farming practice (Kotto-Same et al. 2000; Nounamo and Yemefack 2001). In Africa, 70% of deforestation is attributed to SB agriculture, compared to 50% in Asia and 30% in Latin America (ICRAF 2000). According to Gockowski et al. (2005), the annual rate of deforestation in Cameroon is estimated at 0.6%, with about 108,000 ha of closed-canopy forest being lost annually; about half of this clearing is for agricultural purposes, and the remainder is largely for logging.

Clearing and burning affects soil chemical properties, nutrient availability and soil organisms. The use of fire may raise soil temperature in the top few centimeters by 50°C–80°C above the temperature before burning (Guichuru et al. 2003). Burning at medium temperature may have beneficial effects on N cycling, available nutrients, SOM content, and cation exchange capacity. In southern Cameroon, after burning, increased availability of nutrients such as S (2%–60%), P (50%–300%), Ca (50%–200%), Mg (15%–45%) and K (6%–80%) was reported by Nounamo and Yemefack (2001). Whereas these are short-term effects, soil heating could have a much more sustained and less desirable influence on soil P and N availability besides inputs of ash. Indeed, most organic matter, including N-containing compounds and organic acids, are combusted at temperatures of 200°C–300°C for 20–30 min. Long-term adverse effects of SB agriculture are: water and air pollution, deforestation, soil degradation, biodiversity destruction, and climate-change effects. In addition, burning induces loss to the atmosphere of C, N, and S. Nutrients such as P could become fixed onto soil particles while other ones from ash are washed out before the crop can absorb them, or transported away by the wind. Soil exposure to high temperature affects properties of soil colloids and disturbs living soil organisms. Because of rain leaching and risks of erosion, crops have to be planted quickly to take up soluble nutrients (Nandwa 2003).

5.1.2 Alternatives to Slash-and-Burn Agriculture: Integrated Soil Fertility Management

Integrated Soil Fertility Management (ISFM) takes advantage of nutrient cycles, belowground biodiversity, and enhanced ecosystem services for sustainable agricultural production. The African Network (AfNet) of the Tropical Soil Biology and Fertility (TSBF) Institute of the International Center for Tropical Agriculture (CIAT) has conducted numerous studies on ISFM in SS Africa over the past 20 years. In particular, their work has shown that nutrient release must be synchronized with crop needs and that the management of SOM and the soil biodiversity should be integrated with local resources and indigenous knowledge (Brown et al. 1994; TSBF 1999). In order to boost agricultural production by African farmers, several soil fertility management strategies have been proposed (TSBF 1999; FAO 2001; Bationo et al. 2004). More recently, the Soil Health Project of Alliance for a Green Revolution in Africa (AGRA) was launched in 2008 as a 5-year program aiming to restore fertility of African soils (AGRA 2008; Nyobe and Nwaga 2008). Through collaborative action with 4.1 million farmers, the program intends to regenerate 6.3 million ha of farm land by means of a balanced approach to improved soil management.

A strategy involving integrated pest management, belowground biodiversity (BGBD) inventory, soil-based ecosystem services, and capacity building is recommended for a sustainable empowerment of farmers in SS Africa (Bationo et al. 2004; Bignell et al. 2005). A holistic approach involving a biophysical, chemical, and socio-economic approach is envisaged for tropical soils. Although improved crop varieties were adopted to a similar extent in Asia, Latin America and SS Africa during past decades, these varieties permitted yield increases of the order of 66%–88% in Asia and Latin America, but of only 28% in Africa (Sanchez 2002). In SS Africa, the cost of inorganic fertilizers is four to eight times higher than that in Europe, North America or Asia. Options that are available for increasing soil fertility include: (1) use of on-farm-generated inputs, (2) replacement of nutrients via incorporation of plant biomass (leguminous cover crops, improved fallows, green manures, *Chromolaena* or *Tithonia* biomass transfer), (3) increase of nutrient use efficiency (through the use of improved legume microbial inoculants), and (4) mineral fertilizer application. A minimum NPK nutrient application rate of 50 kg ha^{-1} has been recommended in the Abuja summit on fertilizers in 2006; by contrast, the actual average rate is only 9 kg ha^{-1} in African countries.

5.1.3 Soil Beneficial Micro-Organisms to Sustain Crop Productivity and Development

Few data are available on the importance of SS African BGBD for enhancing ecosystem services and plant productivity. The main functions of soil biota include: (1) nutrient acquisition, storage and cycling, (2) biological control of soil-borne pests and diseases, (3) synthesis and decomposition of SOM, (4) regulation of carbon sequestration and greenhouse gas emission, (5) modification of soil structure, (6) bioremediation of environment by detoxification, and (7) bioturbation through the action of earthworms and other "soil ecosystem engineers." Total bacterial biomass in natural soil ecosystems ranges between 0.3–3.0 t ha^{-1} live weight; the fungal soil biomass has been estimated to range between 1 and 20 t ha^{-1} and amounts to 10–100 m of active fungal hyphae per gram of soil (FAO 2001; Danso 2004; Moreira et al. 2006; Swift et al. 2008). It is suspected that a high microbial genetic diversity exists in low-input tropical soils, which could be exploited through simple agricultural practices or soil biotechnology. Possible management options are direct introduction of selected microbes through inoculation or indirect management of native soil microbes through soil and cropping system manipulation. One objective of these actions would be to replace or supplement synthetic fertilizer inputs in SS Africa. According to Swift (1998), agricultural intensification reduces BGBD, which should then be restored to insure an optimal biological intervention.

The ecology and functioning of tropical soil organisms and their interaction with plant roots, litter and soil under SB agriculture are not well explored. The described species of soil organisms in SS Africa could represent less than 0.1% of total microbial diversity, as compared to some 1%–5% at a global scale (Hawksworth 1991; UNEP 1995). Soil biology studies applied to SB agriculture should consider the following major functional groups of soil micro-organisms: legume-nodulating bacteria (LNB), phosphorus-solubilizing micro-organisms (PSM), plant growth-promoting rhizobacteria (PGPR), biological control agents (BCA) including plant, insect and microbial pathogens, (endo- and ecto-) mycorrhizal fungi (MF), and saprophytes. Most of these are present in tropical African soils (Table 5.1).

The consequence of traditional management of soil fertility in a context of rapid population growth in SS Africa is an increase of food insecurity and poverty. By 2050, the African population could reach 2 billion. Strategies to feed this growing population must take into account the major biophysical constraints for agricultural production in SS Africa, which are: inappropriate cultural practices, lack of availability of improved seed cultivars and their poor adoption, soil degradation, soil fertility and soil biota losses, soil toxicity (due to acidity, Al and Mn toxicities), crop pests, and diseases.

5.1.4 Objectives of this Chapter

The main objective of the present chapter is to provide a synthesis on the potential of beneficial micro-organisms (BM) in SB agriculture for sustainable food production in the humid forest zone of SS Africa. To do so, emphasis will be placed on results from local experimentations which aimed to assess the function and effect of BM such as decomposers, MF, LNB, phosphorus-solubilizing bacteria, and BCA. The chapter will attempt to take into account the diversity of tropical ecosystems and cultural practices. Some key questions which have been considered are:

- What are the characteristics of SB agriculture in the humid forest?
- What is the real impact of SB on soil BM related to nutrient cycling?
- Can we use BM to improve nutrient use and nutrient capture efficiency of plants?
- To what extent do BM influence agricultural production and how can they be managed?
- To what extent can BM and ISFM replace or supplement inorganic fertilizers and pesticides?
- Why integrate soil health and fertility management in agricultural practice?
- What are the socio-economic implications of microbial diversity management?
- What are the institutional considerations related to BM?

Table 5.1 The main groups of beneficial micro-organisms tested in SS African tropical soils

Microbial group[a]	Benefits to agriculture	State of technology	Main process	Action on:
Arbuscular mycorrhizal fungi (*Glomus*, *Gigaspora*)	Increased rooting, improved P nutrition, stress tolerance, yield and soil aggregation, reduced nutrient losses and production costs	Selection and inoculation technologies available, problems with competition from indigenous strains	Mine soil inorganic P reserves, low host plant specificity, favors other soil beneficial organisms	Most crops and fruit trees (e.g., banana, cassava)
Ectomycorrhizal fungi (*Boletus*, *Scleroderma*)	Increased plant growth, protection and stress tolerance, reduced nutrient losses, some are edible mushrooms	Production on culture media is slow, lack of inoculation technology	Decompose organic matter to mineral N and P; high host plant specificity	Specific forest crops or trees [e.g., *Gnetum* (liana), *Acacia* (legume)]
Legume-nodulating bacteria (*Rhizobium*, *Bradyrhizobium*)	Improved N nutrition and yield, reduced N fertilizer costs	Basic technology available	Fix N_2 from atmosphere	Legumes (e.g., groundnuts, *Stylosanthes*)
Phosphate-solubilizing microorganisms (*Bacillus*, *Pseudomonas*)	Increased P uptake and yield	Basic technology available	Mobilize inorganic and organic P	Most crops and trees (e.g., cowpea, maize)
Disease control agents (*Pseudomonas*, *Trichoderma*)	Reduced disease severity	Basic technology available	Antagonism, competition, antibiosis	Bacterial and fungal diseases [e.g., tomato wilt (*Ralstonia*), root rot or wilt (*Fusarium*)]
Entomopathogenic fungi (*Metarhizium*)	Control of insect pests, increased flowering	Basic technology available	Pathogenicity	Crop pests [e.g., Cowpea flower thrips (*Megalurothrips*)]
Plant growth-promoting rhizobacteria (*Pseudomonas*)	Plant growth promotion	Basic technology available	Variety of mechanisms, including phytohormone production	Most crops (e.g., cowpea, maize, Citrus)

[a]Examples of organisms belonging to a particular group are provided in parentheses

5.2 Methodological Approach and Strategy

5.2.1 Description of Climate, Soils, Vegetation and Land Use Systems

The humid forest (HF) zone is characterized by an equatorial climate classified as humid tropical, with bimodal rainy seasons and a short and a long dry season. The bimodal rainfall pattern determines two potential cropping seasons of a combined duration of at least 270 days; there are no more than 4 months per year with precipitation below 200 mm. The total annual rainfall in the HF zone is 1,500–2,500 mm and the temperatures are generally constant over the year, ranging within monthly averages of 22°C–26°C. The dominant soils are classified as Ferralsols and Acrisols, with poor fertility, low pH (3.5–6.0), low cation exchange capacity (16–24 cmol kg^{-1}), low base saturation, relatively high organic matter content, and risk of Al and Mn toxicity (Ambassa-Kiki and Tiki Manga 1993; Nandwa 2003). These soils are suitable for a wide range of uses, despite their low activity clays and their low water and nutrient holding capacities. Native vegetation is evergreen rainforest with little seasonal variation, but the secondary forest is currently an important component. During the dry season, lasting for 2–3 months, water can become a limiting factor for crop production. Agricultural systems in this zone are complex and heterogeneous (Kotto-Same et al. 2000; Guichuru et al. 2003; Gockowski et al. 2005). They comprise agroforestry, home gardens, managed fallows, shifting cultivation, forest plantations, and permanent cropping. Land-use systems are characterized by high diversification, with cacao (*Theobroma cacao*) and coffee (*Coffea arabica* and *Coffea robusta*) plantations being the primary source of income. Other very important sources of income include non-wood forest products, such as a leafy legume called "Okok" or "Eru" (*Gnetum* spp.), fruit vegetables (*Cucumeropsis manii*, *Citrullus* spp.), native trees such as oil palm (*Elaeis guineensis*), bush mango (*Irvingia gabonensis*), Kola (*Cola* spp), bitter kola (*Garcinia kola*), planted tree crops such as mango (*Manguifera indica*), African plum (*Dacryodes edulis*), avocado (*Persea americana*), and citrus fruits. Progression of cropping from year to year involves an annual crop such as a cereal (maize) for the first year; root crops [cassava (*Manihot esculenta*), cocoyam (*Xanthosoma sagitifolia*), yam (*Dioscorea spp.*), and sweet potatoes (*Ipomoea batatas*)] for the second and third years; and finally banana/plantain (*Musa* spp.) and fruit trees for several years. Mixed cropping is the rule, and groundnuts (*Arachis hypogaea*) play an important role in this land-use system (called "afub owondo" in southern Cameroon). P fixation, Al toxicity, and micronutrients (Mo) deficiency may constrain agricultural production. Continuous cultivation is not possible without appropriate management such as crop rotation and fallowing. Soils from the humid forest zone in southern Cameroon have mostly a negative nutrient balance at farm level: losses were estimated at −21 to −73 kg N, −3 kg P and −13 to −23 kg K ha^{-1} $year^{-1}$ and were due to exportation by crops, burning, leaching, and absence of recycling of farm residues (Kanmegne 2004). Only the cacao farms had a positive nutrient balance (+9.6 kg N, +1.4 kg P, +7.6 kg K ha^{-1} $year^{-1}$).

5.2.2 Isolating, Selecting and Testing Beneficial Microbes

Diverse BM have been isolated from Central African soils or root samples in various land use systems of Cameroon, selected for crop improvement, and tested under nursery and farm conditions using cereals, legumes, vegetables, root and tuber crops, and fruit trees (Nwaga 1997; Nwaga et al. 2000, 2004, 2007a; Ngonkeu 2003; Fankem et al. 2006; Ngakou et al. 2008; Ngo Nkot et al. 2008). Table 5.1 presents the main groups of BM residing and tested in African tropical soils.

The BM from our resource bank were screened for specific activities or functions (microbial biomass C, symbiotic activity, plant dry weight increase, effect on plant nutrient uptake, or acidity tolerance) under controlled conditions before being tested under farm conditions in diverse agro-ecological zones and soil types. Standard protocols were used for the different types of BM. For inoculation of a single plant with a selected MF strain, a mixture with 50–1,000 arbuscular MF spores and colonized roots was applied under a seed or on the roots, the exact number of applied spores depending on crop type (fewer spores for micro-propagated plants and more spores for direct seeding). For LNB, 100 g solid inoculum with more than 10^8 cells g^{-1} was applied for inoculating 5–10 kg legume seeds before sowing. Between 10^5 and 10^6 cells per plant were applied for PSM. For BCA such as *Metarhizium*, 50–70 g spores were applied on an area of 500 m^2. Microbial quality control is important to ensure a high density of the desired micro-organism and to avoid contaminants or pathogenic agents. Hence, these microbes were produced and controlled by a recognized microbiology laboratory. The potential of these microbial preparations as an alternative to traditional SB agriculture was assessed using classical experimental designs and laboratory analysis. These include randomized complete block design on farm or simple randomization with replicates in nursery, followed by rigorous statistical evaluation through analysis of variance and comparative tests for significance between treatments.

5.3 Agricultural Uses of Beneficial Micro-Organisms in the Humid Forest Zone

Soil organisms contribute to a wide range of essential services and functions in all ecosystems. Their biodiversity represents a key resource both for the functioning of natural ecosystems and also for sustainable agricultural production in the world (FAO 2001).

5.3.1 Beneficial Micro-Organisms for Plant Productivity Through Nutrient Improvement

5.3.1.1 Soil Microbial Activity, Land Use Systems and Intensification

Microbial diversity is recognized as important for sustainable food production in the world by FAO (2007). Belowground changes occur across diverse land use systems of clayey Ferralsols in Cameroon humid forests (Nwaga et al. 2003). For example, soil microbial biomass C was higher in primary forest (up to 1,223 $\mu g\ g^{-1}$ soil) as compared to forest fallow and old mixed farm (350–399 $\mu g\ g^{-1}$ soil) or cacao plantation (632 $\mu g\ g^{-1}$ soil) (Nwaga et al. 2003; Nounamo and Yemefack 2001). A study of sandy soils of Senegal showed that microbial biomass C doubled after manure application and increased with tillage (Niane-Badiane et al. 1998). Similarly, studies of a Kenyan Nitosol showed a 140% increase in microbial biomass C during 8 weeks after a 60 kg ha^{-1} green manure input (Baaru et al. 2007). Forest soils with a higher C content and acidity could favor microbial activity by promoting development of SOM-decomposing fungi. It should be stressed that traditional low-input agriculture practices, such as shifting cultivation and fallow-based management systems, rely heavily on biological activity for soil fertility maintenance. In contrast, in high-input agriculture, biological impacts are less obvious because of the effects of pesticides, fertilizers, and tillage.

Some sites in the tropics show particularly high arbuscular MF diversity as compared to temperate sites (Mangan et al. 2004; Opik et al. 2006), and 5–10 MF species may be detected on a single tropical field site (Jansa et al. 2002; Nwaga et al. 2003; Oehl et al. 2004; Hijri et al. 2006; Mathimaran et al. 2007; Tchabi et al. 2008). Agricultural management practices such as tillage and fertilization may affect community composition of the native MF in cropped fields, but these interventions do not necessarily diminish overall mycorrhizal diversity (Jansa et al. 2003; Mathimaran et al. 2005, 2007). For its part, shifting cultivation tends to increase the activity of arbuscular MF, but to decrease that of ectomycorrhizal fungi. Land use changes and burning may affect arbuscular mycorrhizal diversity and occurrence in the humid forest of southern Cameroon, where 21 arbuscular mycorrhizal species were found (Nwaga et al. 2003).

After slashing and burning, forest soil parameters such as root phosphatase activity, arbuscular MF infectivity, spore numbers, cation exchange capacity, available P level, organic matter content, and soil pH increased respectively by 118%, 89%, 40%, 80%, 62%, 18% and 5% (Table 5.2). It has been reported that root colonization by arbuscular MF may respond to fire to a greater extent upon infrequent, as opposed to regular, burning (Hartnett et al. 2004). Thus, burning may affect belowground functioning of humid forest soils.

Table 5.2 Effect of burning in farmers' plots on belowground functioning of humid forest soils in southern Cameroon

Parameter[a]	Unburned soil	Burned soil	Effect (%)
pH (water)	3.83 (3.55–4.21)	4.02 (3.84–4.18)	–
Organic matter (%)	1.58 (1.44–1.82)	1.86 (1.44–2.77)	+18
Available P (mg kg^{-1})[b]	5.36 (4.74–6.59)	8.67 (4.74–14.36)	+62
Cation exchange capacity (cmol kg^{-1})	1.69 (0.57–3.33)	3.03 (2.02–4.43)	+80
Root phosphatases ($\times 10^3$ µmol min^{-1} mg^{-1})	10.0 (5–14)	21.8 (16–25)	+118
Number of AMF species after trapping[c]	8.8 (8–10)	8.3 (7–10)	−6
AMF spores number (per gram soil)	6.2 (5.0–7.5)	8.6 (7.0–9.5)	+40
AMF infective units (per gram soil)[d]	209 (57–423)	396 (142–737)	+89

[a]Average from four composite soils (primary forest, secondary forest and 10- and 12-year-old fallow). Samples were obtained by mixing 1 kg of five random sub-samples (0–20 cm depth) from lightly burned and unburned 5 × 40 m transects
[b]Available P by Murphy and Riley (1962)
[c]AMF: Arbuscular mycorrhizal fungi
[d]Using the most probable number method
(*Source*: Nwaga et al. unpublished data)

5.3.1.2 Mycorrhizal Symbiosis

The main functions of arbuscular MF are to improve plant uptake of immobilized nutrients such as P, stimulate NH_4^+ assimilation, and increase tolerance to drought and soil acidity, as well as to pests, diseases and weeds (Table 5.1). They can also stimulate photosynthesis and symbiotic nitrogen fixation on legumes. Effectively creating a mycorrhizosphere, arbuscular MF may interact with other micro-organisms to accelerate SOM decomposition and N release, build up networks between plant roots (see Chap. 9), increase exploration of a greater soil volume, and improve soil aggregation and stability. The responsiveness of plants to arbuscular MF colonization varies, and certain legumes (cowpea, groundnuts), tuber crops (cassava, sweet potato, yam), fruit trees (cacao, coffee, citrus, palm oil), vegetables (onion, leek), and cereals (maize, sorghum) are examples of very responsive tropical crops and trees. The responsiveness also varies according to soil type and plant age. Banana was more dependent on mycorrhizal colonization in an Oxisol than in an Andosol (Nwaga et al. 2009). Arbuscular MF activity in rhizosphere soil was greater in old, as compared to young plantations (Nwaga et al. unpublished data).

Following the isolation and screening of more than 200 local isolates of arbuscular MF including *Glomus*, *Gigaspora*, *Acaulospora*, *Scutellospora*, and *Paraglomus*, 11 isolates were tested for stimulation of P uptake by millet, cowpea and leek under controlled conditions. Marked differences in fungal strain efficiency were observed as well as some preference of certain fungal strains towards certain crop species (Fig. 5.1). As compared to the non-mycorrhized control, P uptake per plant was 3–10.5, 1.4–28 and 2.5–35 times higher respectively for mycorrhized cowpea, leek, and millet.

In subsequent trials conducted under farm conditions in diverse sites of Cameroon, plant inoculation with mixtures of two or three selected MF strains

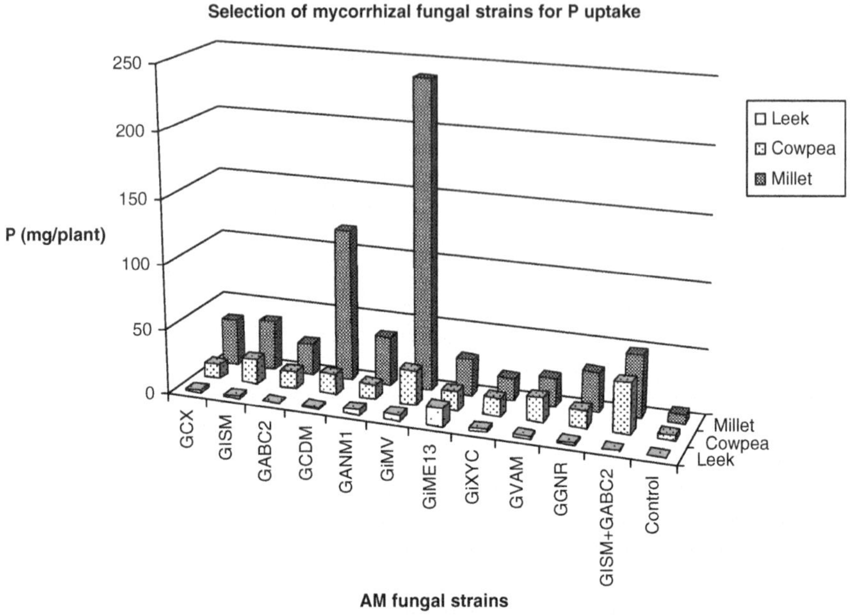

Fig. 5.1 Selecting strains of arbuscular mycorrhizal fungi from Cameroonian soils using P uptake (expressed in mg per plant) by different crops grown on sterilized substrate. Twenty kilograms per pot of ferralitic substrate (soil/sand: 3/1) was twice sterilized (120°C, 1 h) after sieving. Substrate characteristics were as follows: clay, 31%; pH, 5.3; C content, 2.1%; total N content, 0.69%; available P content (Olsen), 16 ppm. Nutrient uptake (mg P per plant) and plant growth (dry weight) were assessed under nursery conditions. The identity of the strains was: *Glomus clarum* (GCHX), *Glomus intraradices* (GIYM); *Glomus albidum* (GABC2); *Glomus clarum* (GCDM); *Glomus aggregatum* (GANV) *Gigaspora margarita+Glomus aggregatum* (GiMN+GANV); *Gigaspora margarita* (GiME) *Gigaspora* sp. (GiXXC); *Glomus versiforme+Glomus macrocarpus* (GVAC)+GMAC); *Glomus geosporum* (GGRN); *Glomus intraradices+Glomus albidum* (GIYM +GABC2). These strains were from the resource bank of the soil microbiology laboratory of University of Yaoundé I. About 50 g of inoculum (10 spores g^{-1} and 50% root colonization) was mixed with 500 g of sterilized sand for inoculating one test plant. A randomized 12 × 2 factorial design repeated five times with 11 treatments and an uninoculated control was used. The uninoculated control received 50 g of a sterilized inoculum. Rorison nutrient solution containing most macro and micronutrients and only 0.23 g per liter of soluble P was used to fertilise the plants weekly. After 45 and 60 days for cowpea and 90 days for millet and leek, aboveground plant parts were harvested and analyzed (Ngonkeu 2003; Nwaga et al. 2004)

resulted in increases of grain yield ranging from 48% to 478% (Table 5.3). The response observed varied with the mycorrhizal dependency of the different crops and was lowest for cereals and highest for tubers, legumes, vegetables, and fruit trees. N and P content and crop taste were also improved by MF. MF inoculation increased maize yield by 52%–59%, as compared to a 30%–35% increase resulting from inorganic fertilizer application of NPK (37–27–14). Inoculation increased maize tolerance to acidity and to toxic Al levels encountered in the tested Oxisol, where the pH was 3.9 and Al content 2.6 cmol kg^{-1} (Nwaga et al. 2004). Six varieties of hot pepper were inoculated in an Oxisol and fruit yield increases varied

Table 5.3 Response of various crops to inoculation by selected arbuscular mycorrhizal fungi under farm conditions in diverse sites in Cameroon

Crop[a]	Locality	Yield (t/ha)[b]		Yield increase (%)	Soil type
		Uninoculated	Inoculated		
Banana[c]	Yaoundé	2.50	11.25	438	Ultisol
Cowpea	Yaoundé	0.12	0.35	200	Oxisol
Groundnuts	Yaoundé	0.83	1.98	137	Oxisol/Ultisol
Leek	Nkolbisson	0.85	4.58	438	Oxisol
Maize	Ebolowa	2.62	4.10	58	Oxisol
Mucuna[d]	Minkoameyos	4.68	7.41	58	Oxisol/Ultisol
Oil palm[c]	Mbankomo	4.75	11.82	149	Ultisol
Pepper	Nkolbisson	1.98	3.62	86	Oxisol
Sorghum	Maroua	1.54	2.28	48	Vertisol
Soybean	Ngaoundéré	0.27	0.41	50	Oxisol/Ultisol
Tomato	Ebolowa	3.24	7.00	116	Oxisol
Yam[e]	Ekona	10.23	59.10	478	Andosol

[a]Complete block design using four replicates and at least 80 plant samples per treatment (except banana, oil palm and yam). Plants were inoculated with a mixed inoculum containing two or three MF strains belonging to the species *Glomus clarum*, *Glomus intraradices* and *Gigaspora margarita*
[b]Unless specified otherwise
[c]Measured yield parameter: bunch weight per plant
[d]Measured yield parameter: aboveground biomass at flowering
[e]Tuber weight under nursery conditions (gram per plant)
(*Source*: adapted from Nwaga 2001)

from 34% to 175%. A tomato yield increase of 116% resulted from MF inoculation, which was equivalent to the results of inorganic NPKMg (12–14–19–5) fertilizer application. In addition, MF inoculation reduced the incidence of bacterial wilt of tomato, from 12% in the uninoculated control, to 0.8%. A reduction of 8–12 months in oil palm precocity for fructification and a yield increase of 149% in bunch production were observed as a result of inoculation. For banana, nursery mycorrhizal plants showed greater precocity for fruit maturation and a bunch yield increase of 438%, as compared to the uninoculated control. Fungal inoculation of yam under nursery conditions showed an improvement of P use efficiency and an increase of 478% in mini tuber production. This is of particular interest in the context of promoting tubers as seedlings for yam planting materials. In all the tested legumes, inoculation of MF increased nodulation; for example, in the legume cover crop *Mucuna*, arbuscular MF increased N accumulation from 198 kg in the control to 283 kg N ha^{-1}, and also improved P assimilation efficiency (Jemo et al. 2007).

5.3.1.3 Symbiosis with Legume-Nodulating Bacteria

The assessment of nodulation status of African legumes has been launched but needs to be further elaborated (Sprent 2001; Diabate et al. 2005). Various grain, forage, cover crop, and tree legumes associate with different types of LNB. Of these, grain legumes are of prime interest to farmers.

In developing countries, fertilizers account for more than 70% of the energy used in agriculture and most are N fertilizers. Whereas the main functions of LNB is N_2 fixation, other effects such as hormone production, phosphate solubilization, and possible suppression of pathogens may be of relevance. Grain legumes derive between 50% and 98% of their N supply from N_2 fixation, which is equivalent to 15–200 kg N ha^{-1} (Mafongoya et al. 2004; Dakora and Keya 1997). However, N fixation by native tree legumes may reach 584 kg N ha^{-1} $year^{-1}$ (Elkan 1992). Tropical grain legumes provide a residual effect of 0–205 kg ha^{-1} in fertilizer equivalent for the next crop and in this respect cowpea (*Vigna unguiculata*) offers the best performance amongst tropical grain legumes (Giller 2001). In Bambara groundnuts (*Vigna subterranea*), BNF may provide 70% of the total nitrogen assimilated by the plant (Gueye 1992).

Some major problems limiting N_2 fixation by LNB in the tropics include soil acidity, Al toxicity, Ca and P deficiency, drought, and temperature stress (Obaton

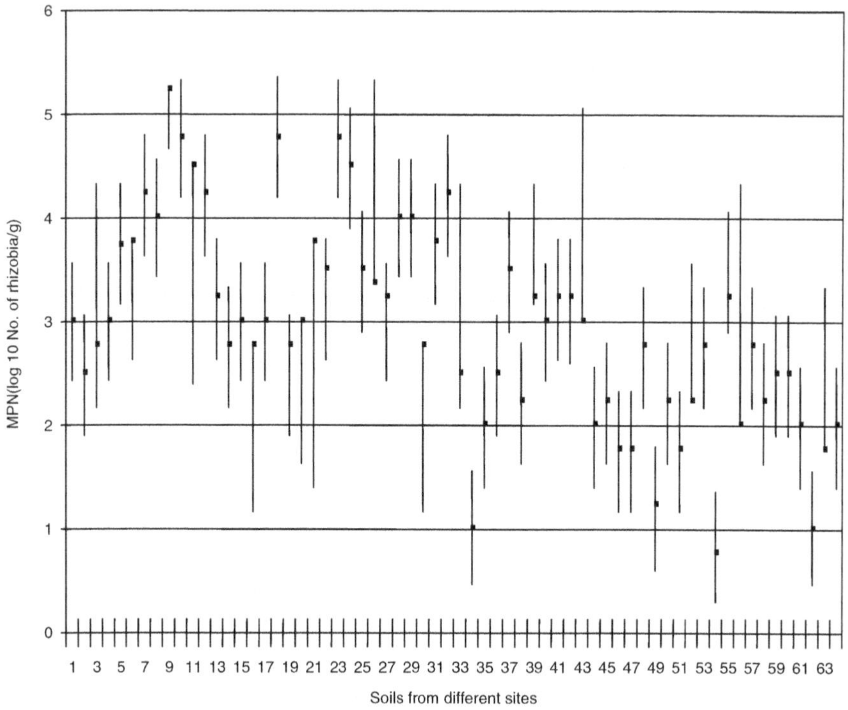

Fig. 5.2 Density of rhizobia, estimated as the $\log_{10}$ most probable number of cells g^{-1}, in sites representing diverse land use systems in humid forests of Cameroon. Data are means and standard deviations obtained from four replicates of six dilutions using *Macroptilium atropurpureum* host plants. Sites 1–16: mixed crop farms; sites 17–32: fallows; sites 33–48: plantations; sites 49–64: forests (Ngo Nkot 2009)

1992; Kimou and Zengbet 1994; Date 2000). Molybdenum and other micronutrient deficiencies and low soil densities of LNB in the soils are also major limiting factors (Giller 2001). Acid-tolerant LNB strains such as TAL1145 have been selected, but in the field these were not more effective than acid-sensitive strains. The average density of LNB from diverse land use systems in the humid forest of Cameroon varied from 31 to 76,000 cells g^{-1} soil (Fig. 5.2). LNB populations in some individual soils were as low as 6 cells g^{-1} soil (for forest soil) or as high as 180,000 cells g^{-1} soil (for mixed farm soil).

The respective effectiveness and competitiveness of inoculant and indigenous LNB influence the legume inoculation response under field conditions (Mafongoya et al. 2004). Spontaneous nodulation of some varieties of soybean (*Glycine max*) with promiscuous indigenous LNB has become exploited by many African farmers in Nigeria, South Africa, Tanzania, Zambia, and Zimbabwe (Giller 2001; Sanginga et al. 2001). Cowpea (*V. unguiculata*), a very promiscuous legume, has rarely been found to respond to inoculation unless the soil conditions are not suitable for LNB survival (Giller 2001).

To study the response of tropical promiscuous legumes to inoculation with LNB, a composite inoculum was assembled, consisting of three to five local strains previously selected for N_2 fixation efficiency and acidity tolerance (Nwaga 1997; Nwaga and Ngo Nkot 1998). Following inoculation, various grain legumes grown in Ferralsols from Cameroon exhibited increases in grain yield ranging from 74% to 556%, depending on legume species and site (Table 5.4). For example, common bean showed yield increases of 162% following inoculation and of 98% after 154 kg ha^{-1} N urea application, as compared to an uninoculated and unfertilized control. Seed protein content was improved by 50% following inoculation, and by 16% as a result of urea application. The cover crop and forage legumes *Mucuna*, *Stylosanthes*, and *Crotalaria* showed increases in biomass production of 66%, 82% and 107% respectively. For *Mucuna*, N accumulation without and with inoculation was 206 and 314 kg ha^{-1} respectively (Jemo et al. 2007). Results showed that

Table 5.4 Response of legumes to inoculation by selected legume-nodulating bacteria under farm conditions in diverse sites in Cameroon

Crop[a]	Locality	Yield (t ha^{-1})		Increase (%)	Soil type
		Control	Inoculated		
Common bean	Mbouda	0.98	2.48	162	Oxisol/Ultisol
Cowpea	Yaoundé	0.23	1.51	558	Ultisol
Crotalaria[b]	Nkolfoulou	5.60	11.60	107	Oxisol
Groundnut	Yaoundé	0.60	1.28	113	Oxisol
Groundnut	Yaoundé	0.22	0.41	82	Oxisol
Mucuna[b]	Minkoameyos	4.68	7.79	66	Oxisol/Ultisol
Soybean	Ngaoundéré	0.27	0.47	74	Oxisol
Stylosanthes	Ngaoundéré	0.46	0.76	64	Ultisols

[a]Complete block design using four replicates and at least 80 plant samples per treatment
[b]Above ground biomass at flowering
(*Source*: adapted from Nwaga 2001)

cowpea could also significantly respond positively to inoculation, despite its promiscuous association with indigenous LNB.

Groundnut (*Arachis hypogaea*) is one of the most important legumes cultivated in the humid forest zone of Cameroon. However, the diversity of groundnut-nodulating bacteria remains poorly understood. Forty-two strains from *Arachis hypogaea* cultivated on forest, fallow, plantation, and mixed-farm regimes were examined by restriction fragment length polymorphism (RFLP) analysis of their rDNA. A considerable level of genetic diversity was evidenced among those groundnut isolates, which were grouped into eight distinct genotypes. Populations of groundnut isolates were most diverse in cacao farms and fallows, and lowest in a groundnut mixed farms (Ngo Nkot et al. 2008). Inoculation of groundnuts by LNB provided contrasting responses on different sites in Cameroon, with yield increases of 169% on a relatively fertile clay soil from Yaoundé and of only 5% on a mixed farm low-fertility sandy soil from Bokito (Mandou et al. 2002). A synergistic effect of inoculation and molybdenum seed treatment was also noticed, since this dual treatment resulted in a 288% yield increase in the Yaoundé site and a 21% increase for the Bokito site, as compared to the untreated control (Fig. 5.3). The contrasting response to inoculation in the two study sites may be related to density of indigenous LNB populations, since the Bokito soil contained 17,000 cells g^{-1} and the Yaoundé soil only 170. An inoculation response is extremely unlikely when native LNB soil bacteria density is more than 1,000 cells g^{-1} (Thies et al. 1991; Mafongoya et al. 2004). LNB density is generally lower in plantations and forests soils than in fallows or mixed farms of southern Cameroon (Ngo Nkot 2009). Inoculation experiments done on 12 sites in Zambia, Zimbabwe, Tanzania, Kenya, and Uganda

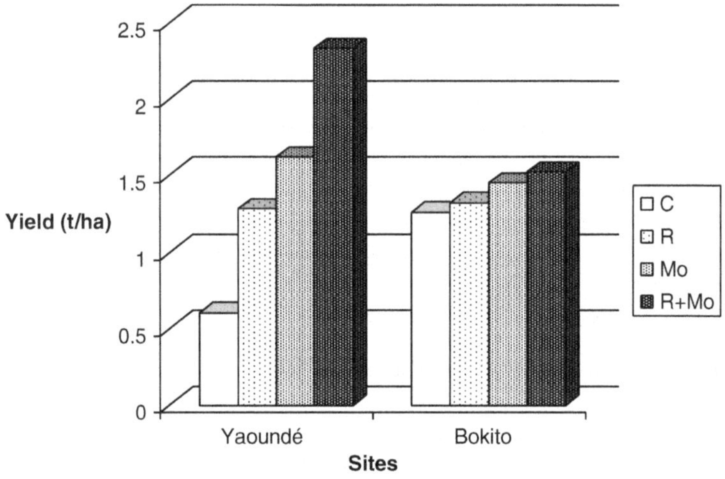

Fig. 5.3 Effect of inoculation with selected legume nodulating bacteria (R) and application of 25 g ha^{-1} molybdenum (Mo) or both treatments (R+Mo) on groundnut (*Arachis hypogaea*) yield under field conditions in two contrasting Ferralsols in Cameroon. C, untreated control (Mandou et al. 2002)

resulted in average yield increases of 125% for soybean, 43% for common bean, 40% for groundnut and 35% for cowpea (Woomer and Karanja 1996). However, another experiment with six legumes species grown on a moderately fertile Kenyan soil yielded little response to inoculation in the case of five tested legumes. In this latter case, the density of indigenous LNB was only 310–900 cells g^{-1} (Chemining'wa et al. 2007), but it may be that the bacteria used as inoculants were poorly adapted to the local soil conditions. Occurrence of nodulation was assessed in 156 tree legume species growing in six natural forest areas in Guinea. Inoculation with LN bacteria improved their growth (Diabate et al. 2005). Thus we conclude that it may be important to continue efforts to select and apply adapted LN strains to both crop and tree species, especially those trees involved in reforestation programmes.

5.3.1.4 Phosphorus-Solubilizing Micro-Organisms

About 95%–99% of soil P is present in an insoluble form and hence cannot be utilized by plants. As much as 42% of the world reserves of rock P is in Africa, most of it being exported (van Straaten 2007). PSM are soil fungi or bacteria with the capacity to mobilize P and make this element available for plants (Oberson et al. 2006). P solubilizers produce chelating agents, organic acids, and phosphatases (Table 5.1; see Sect. 10.4.2). In the rhizosphere of oil palm grown in a humid forest soil of Cameroon, the density of P-solubilizing bacteria was $0.18–3.70 \times 10^4$ colony-forming units (CFU) per gram of soil, which represented 3%–39% of the total soil cultivable bacteria (Fankem 2007). Some 277 isolates were studied and found to solubilize phosphates mainly through the secretion of carboxylic acids (Fankem et al. 2006). Preliminary work was done on the occurrence and selection of PSM from oil palm (*Elaeis guineensis*) rhizosphere in humid forest zone in Cameroon (Fankem 2007), and their potential for improving plant P acquisition and growth. When inoculated to maize, five out of ten isolates significantly increased plant yield and three increased the shoot and grain P contents. Three of the ten isolates examined in the study were later identified using molecular tools as *Pseudomonas fluorescens* (Fankem et al. 2008). On-farm studies are needed on diverse crops to confirm the benefits of the selected PSM strains.

5.3.2 *Action of Beneficial Micro-Organisms for Plant Health Improvement*

While plant pests and diseases are responsible for more than 25% of food loss worldwide, in the tropics losses of more than 50% occur for many crops (Nwaga 1988). Most plant diseases are caused by fungal pathogens. The soil-borne fungal pathogen *Fusarium oxysporum* causes a severe oil palm heart disease with damages reaching 40%–70% in some plots in Côte d'Ivoire (Allou et al. 2003). *Phytophthora megakarya* causes black pod disease of cacao and up to 50%–70% losses in

Cameroon. Other important fungal pathogens are *Phytophthora cinnamomi*, which attacks many trees and crops, *Pythium* spp. causing root rots and damping off in a variety of host plants, and *Pythium myriotylum*, which is responsible for a severe root rot of cocoyam and empty pods of groundnuts (Nwaga 1988). Bacterial and viral diseases are also very important because there are no efficient curative control methods. *Ralstonia solanacearum* causes a severe bacterial wilt in *Solanaceae* crops (tomato, potato). African cassava mosaic virus and rosette virus disease of groundnut are other examples of major constraints for crop productivity in SS Africa which can cause severe plant growth reduction and losses; these are disseminated by insect vectors. Important animal pathogens include nematodes such as *Meloidogyne* spp. causing root-knot of vegetables, and *Radopholus similis*, which is the agent of a root rot of banana/plantain. *Striga* spp. is a very important parasitic weed in some cereals such as sorghum and legumes such as cowpea. Insect pests are also responsible for numerous crop losses, one example being the bean flower thrips *Megalurothrips sjostedti*, a key pest of cowpea in the tropics.

The incidence of soil-borne diseases may be reduced by specific microbial antagonists such as fluorescent pseudomonads and *Trichoderma* sp. Thrips and various other insect pests may be controlled by the entomopathogenic fungi *Beauveria* sp. and *Metarhizium* sp. Integrated pest and disease management strategies involving BM inoculation, use of tolerant crop varieties, adapted cultural practices, and reduced applications of synthetic pesticides should be promoted. AMF inoculation of groundnut increased mycorrhizal colonization and nodulation, while it reduced rosette virus and cercosporiosis disease severity by about 40% (Table 5.5). Some of the improvement observed with diseased plants might be due to an enhanced precocity of flowering and fruiting in the mycorrhizal groundnuts, which would then change the dynamics of the rosette insect vector. A *Pseudomonas* biological control agent reduced the disease index of *Pythium aphanidermatum* root rot of cowpea from 3.44 to 1.06. By combining the *Pseudomonas* biological control agent with the arbuscular MF *Glomus deserticola*, a further reduction of the disease index to 0.13 was obtained (Nwaga et al. 2007b). *Metarhizium*, a thrips-controlling fungus, provided good control of this pest in four agro-ecological zones of

Table 5.5 Impact of selected arbuscular mycorrhizal fungi inoculation and urea treatment on diseases and yield of groundnut (*Arachis hypogaea*) grown on Ferralsol in Yaoundé, Cameroon

Treatment	Mycorrhizal root colonization (%)	Nodule no per plant	Cercosporiosis incidence (%)	Rosette incidence (%)	Yield (t ha^{-1})	
					Diseased plants[1]	Healthy plants[2]
Control	31[c]	222[c]	70[a]	34[d]	0.00[c]	0.44[d]
Urea	33[c]	242[c]	72[a]	43[a]	0.00[c]	0.58[c]
Mycorrhiza	93[a]	459[a]	44[b]	21[c]	0.18[a]	3.18[a]
Mycorrhiza + urea	86[b]	337[b]	30[c]	28[b]	0.05[b]	2.90[b]

[1]Rosette-diseased plants
[2]Non rosette-diseased plants
Data followed by different letters differ according to the Student-Fisher test at $p < 0.05$
(*Source*: Nwaga et al. unpublished data)

Cameroon (Ngakou 2007; Ngakou et al. 2008). Under farm conditions, a combined inoculation with two arbuscular MF, *Glomus clarum* and *Gigaspora margarita*, reduced the severity of bacterial wilt of tomato due to *Ralstonia solanacearum* (Nwaga unpublished data).

5.4 Discussion

5.4.1 Managing Beneficial Micro-Organisms in Agricultural Systems in the Humid Forest Zone of SS Africa

Sustainable agricultural production will be achieved through an integration of several management interventions. Microbial biomass in tropical forest soils may greatly influence organic matter turnover and nutrient supply. For example, arbuscular mycorrhizal inoculation and the use of a drought-tolerant variety of banana may be combined to improve root development and water use efficiency (Nwaga et al. 2009) or pest and disease tolerance (Jefwa et al. 2008). In a sandy loam soil of Senegal, inoculation of onion with arbuscular MF resulted in bulb yields of 39–42 t ha^{-1}, while the yields were only 27 t ha^{-1} following synthetic fertilizer (NPK: 23–7–23) application and 13 t ha^{-1} for the untreated control (Sow et al. 2008). Combining BM with resources such as organic matter inputs from crop residues, cover crops, forage legumes, or agro-industrial wastes stimulates biomass production by the forage legume *Stylosanthes*, and vegetable yield in tomato and other *Solanum* species. BM may be useful in a variety of systems, including fallow, fruit tree plantations, and agroforestry systems. Mycotrophic crops such as banana, cassava, cocoyam, cowpea, fruit trees, groundnut, sorghum, sweet potato, yam, and forages such as *Andropogon*, *Axonopus*, *Brachiaria*, *Crotalaria*, *Dolichos*, *Mucuna*, *Panicum*, *Pennisetum*, and *Stylosanthes* increase populations of arbuscular MF, while non-mycotrophic plants and absence of soil cover reduce their populations. Hence, planting mycotrophic plants stimulates the multi-functional action of these fungi (Finlay 2004). Low-input, multi-cropping systems may also positively affect the activity of MF and thus their impact (Plenchette et al. 2005). Leguminous tree fallows and herbaceous cover crops are potential sources of rich organic inputs for the succeeding crops and contribute to soil fertility replenishment (Sanchez et al. 1997; Carsky et al. 2003). Some fluorescent *Pseudomonas* species are beneficial to mycorrhiza, since they stimulate root colonization and solubilize inorganic P. Combining diverse microbes such as LNB, arbuscular MF, and *Metarhizium* may sustain cowpea production in SS Africa (Ngakou et al. 2007). More generally, crop farming could change the dynamics of forest MF community and, consequently, their functioning and potential benefits for food production (Oehl et al. 2004; Mathimaran et al. 2007). Hence, additional research is needed to understand BM effects and their relationship with land-use systems. However, it is unlikely that management of microbial communities could completely replace the need for external fertilizer inputs.

Training of farmers is required for efficient use of microbial fertilizers. Beneficial micro-organisms play a key role in crop productivity, soil conservation, and biological control of pest and diseases, but, as is shown here, cultural practices and environmental stress may significantly influence their activity and efficiency. Adequate management options should be adopted to enhance microbial beneficial effect on crops and trees (Bethlenfalvay and Linderman 1992; Nwaga 2001; Swift et al. 2008).

A global approach to soil fertility restoration is being developed in SS Africa by some international scientific research institutions (CIAT, FAO, ICRAF, IFDC, TSBF Institute of CIAT, IITA), scientific (CSM-BGBD, ETH, AGRA) and socio-economic (AfricaNUANCES, ASB, Soil Health-AGRA,) projects, networks (Bio-veg network of AUF, AfNet-TSBF Institute-CIAT), associations such as AABNF, and national partners (Universities and NARS). Working with farmers, these institutions recommend the participatory valorization of African natural resources and readily available organic waste, together with the increased and integrated use of: (1) LNB and MF in tree-based fallow or mixed cropping, (2) local sources of rock P in P-deficient soils, and (3) leaf biomass transfer (Sanchez 2002). Specifically, farmers should receive training on the following subjects: forestry and agroforestry, tree regeneration to counter desertification, soil fertility restoration, fruit tree planting, crop production, seed production, feed for livestock, and biodiversity conservation.

5.4.2 *Socio-Economic Considerations on Beneficial Micro-Organisms*

Increased exploitation of BM is recommended by FAO for farmers in developing countries. To the agricultural systems of the world, N_2 fixation and biological control provide benefits of over US$ 90 and 100 billion, respectively (Pimentel et al. 1997). Residual N left in the soil by tropical grain legumes may range from 0 to 205 kg N ha^{-1} (Giller 2001), which represents a significant added value of the legume crops for the subsequent use of the land. It has been shown that cowpea yields could be improved by 25%–150% through the use of suitable inoculation technology with local strains of LNB in Nigeria (Oloke and Odeyemi 1988). In Congo, inoculation of soybean with efficient LNB strains resulted in N fertilizer savings of 84–94 kg N ha^{-1} (Mandimba 1997). In India, LNB inoculation of groundnut resulted in net income farmer increases of 56% (Reddy et al. 2005). In Zimbabwe and Zambia farms, the cost of LNB inoculation is usually less than US$ 5 ha^{-1} and the estimated return on investment is 1:100 and 1:70 respectively, in comparison with 1:3 for synthetic N fertilizer. While the estimated annual economic value of N_2 fixation in soybean, especially in the promiscuous varieties, is of US$ 200 million in the 19 African countries (Anonymous 2007), the most significant example of economic benefits accrued through the use of LNB remains the case of soybean in Brazil (see Chap. 4).

Arbuscular MF are of great economic value to farmers. Their inoculation results in a 50% decrease in P fertilizer-related expenses and in yield increases, for example ranging between 25% and 46% over the uninoculated control in the case of tomato (Anonymous 2006), and 223% for onion (Sow et al. 2008). In Cameroon, where a small production unit of biofertilizers exists, the estimated cost of MF inoculation varies from less than US$ 0.01–0.02 per plant for nursery plants (vegetables and trees), to US$ 0.02–0.20 per plant for direct seeding plants (cereals) (Nwaga unpublished data).

Phosphorus-solubilizing bacteria could also be sold at a low cost, and might help in reducing the cost of mycorrhizal inoculation and increasing their efficiency and benefits for farmers.

One of the major concerns about the relevance of agricultural research in SS Africa is the adoption of new technologies by farmers. "The poor in Africa do not eat conferences, processes or technologies" (Adesina 2007). For improving soil productivity in SS Africa, holistic, integrated, and participatory approaches are recommended (FAO 2001). Thus, policy makers, traditional rulers, and scientists should make all reasonable efforts to promote knowledge about BM and sustainable production technologies among smallholder farmers.

5.5 Conclusions

Whether or not BM such as MF, N_2 fixers, P solubilizers, and BCA might some day replace synthetic fertilizers and pesticides is not the most important question. More urgently, it should be asked: what is the impact of these organisms on food and feed production and on environmental management in the humid tropic zones? Current research results evidence the potential of nitrogen fixers, phosphorus solubilizers, mycorrhiza, and BCA as useful inputs for African agriculture. Soil microbes accelerate recycling, but do not compensate for nutrient losses, except in the case of N. Although biological control will not prevent the development of all pests and diseases, microbial pesticides are safe to non-target organisms, self-replicating, and largely harmless to the environment. In addition to improving the nutrient status of soil and plants and protecting against pests and diseases, BM improve soil structure and C storage capacity. They also increase plant nutrient and water use efficiency, and tolerance to abiotic stresses. In SS Africa and other regions where high-input agriculture is out of the reach of most farmers, the use of these microbes should be promoted because of their ability to sustain agricultural and livestock production, provide environmental services, and improve the livelihood of farmers. Developing an African mycorrhizal network, microbial resource banks, and production units such as MIRCENs, which already exist in Senegal, Kenya, and Egypt, would be instrumental in achieving food security for this region.

With the development of biofuel demand and food insecurity increase in SS Africa, an eco-agriculture approach should be promoted in order to protect the environment and mitigate climate change adverse effects. Research on ISFM in SS Africa should fully integrate all the components of the system: organic resources, inorganic fertilizers, biological processes, and socio-economic aspects. Most African farmers are not aware of the economic benefits of integrating BM with organic and inorganic inputs, neither are they aware of the impact of climate and soil fertility changes on sustainability of agricultural production. How can we integrate indigenous knowledge and deal with the resistance to change which is observed in rural communities? We need holistic and participatory programs that will take into account the priorities of smallholder farmers, as well as the objectives of researchers, public extension services, international aid donors, and traditional village rulers. Given the increasing population density of SS Africa, SB agriculture will have to adhere to integrated management principles, and eventually to evolve into a more productive system, relying on increased fertilizer inputs, improved seed varieties, adapted cropping systems, and rational pest and disease control, and combining livestock production with cropping to improve nutrient cycling. The empowerment of farmers depends not only on technical issues, but also on socio-economical aspects and integration of indigenous knowledge (see Chap. 14). Policy, funding, and training issues dealing with soil biodiversity are critical for sustaining and alleviating poverty in SS Africa.

Acknowledgments We thank AfNet-TSBF Institute of CIAT (Nairobi, Kenya), AfricaNUANCES Project (Wageningen, Netherlands), AUF "Bioveg" Network (Paris, France), AGRA Soil Health Project (Accra, Ghana), students and colleagues in Cameroon (CARBAP, ICRAF, IITA-Nkolbisson, IRAD, IRD-Yaoundé, MINEP, MINESUP, Universities of Yaoundé I, Ngaoundéré, Buea, Douala, Dschang) and IITA (Ibadan, Nigeria and Cotonou, Benin), University of Cheik Anta Diop (Dakar, Senegal), EMBRAPA (Seropedica, Brazil), ETH Institute of Plant Science (Zurich, Switzerland), LSTM IRD (Montpellier, France), Universities of Dundee and York (Plant Biology Dept, UK), Institute of Soil Science and Plant Nutrition, Martin-Luther University (Halle-Wittenberg, Germany) for their collaboration, work and networking. We also thank UNESCO Biotechnology Action Plan for a memorable training scholarship in Brazil (EMBRAPA-RJ).

References

Adesina A (2007) Building sustainable rural input markets for poor farmer in Africa: approaches and lessons for the Africa Green revolution. In: International symposium on innovations as key to Green Revolution in Africa: exploring the scientific facts. AfNet-TSBF-CIAT/SOFECSA, Arusha, p 13

AGRA (2008) Alliance for a green revolution in Africa. AGRA Update 1:1–8

Allou K, Ahoussou N, Ake S, Diabate S, Franqueville D (2003) Comportement de clones de palmiers à huile au champ en zones de haute densité de *Fusarium oxysporum* f. sp. *Elaeidis* en Côte d'Ivoire. Agron Afr 15:29–38

Ambassa-Kiki R, Tiki Manga T (1993) Biophysical and socioeconomic characterisation of the humid forest zone of Cameroon: alternatives to slash-and-burn agriculture (ASB) project. ASB, Yaoundé

Anonymous (2006) Improved soil organic matter management, crop management and composting technologies for enhanced soil microbial function. http://www.tsbfsarnet.org/CSM-BGBD/CSM-BGBD-main-links/PhaseII/ex-ante%20analysis-uas.pdf. Accessed 12 Aug 2009

Anonymous (2007) Economic evaluation of the contribution of below ground biodiversity: Case study of biological nitrogen fixation by rhizobia. http://www.ciat.cgiar.org/tsbf_institute/pdf/report_2007/outcome.pdf. Accessed 12 Aug 2009

Baaru MW, Munguendi DN, Bationo A, Verchot L, Waceke W (2007) Soil microbial biomass carbon and nitrogen as influenced by organic and inorganic inputs at Kabete, Kenya. In: Bationo A, Waswa B, Kihara J, Kimetu J (eds) Advances in integrated soil fertility management in sub Saharan Africa: Challenges and opportunities. Springer, Berlin, pp 827–832

Bationo A et al (2004) The African network for soil biology and fertility: new challenges and opportunities. In: Bationo A (ed) Managing nutrient cycles to sustain soil fertility in sub-Saharan Africa. Academy Science Publications, Nairobi, pp 1–23

Bethlenfalvay GJ, Linderman RG (1992) Mycorrhiza in sustainable agriculture. American Society of Agronomy, Madison

Bignell DE et al (2005) Belowground biodiversity assessment: developing a key functional group approach in best-bet alternatives to slash and burn. In: Palm C, Vosti SA, Sanchez PA, Ericksen PJ (eds) Slash-and-burn agriculture, the search for alternatives. Columbia University Press, New York, pp 119–142

Brown S, Anderson JM, Woomer PL, Swift MJ, Barrios E (1994) Soil biological processes in tropical ecosystems. In: Woomer PL, Swift MJ (eds) The biological management of tropical soil fertility. Wiley, Chichester, pp 15–46

Carsky RJ, Sanginga N, Schulz S, Vanlauwe B (2003) Promising practices for sustainable intensified systems in the Savannah zone of West Africa. In: Jamin JY, Seiny Boukar L, Floret C (eds) Savanes africaines: des espaces en mutation, des acteurs face à de nouveaux défis. Actes de colloque, CIRAD, Montpellier, France

Chemining'wa GN, Muthomi JW, Theuri SWM (2007) Effect of rhizobia inoculation and starter-N on nodulation, shoot biomass and yield of grain legumes. Asian J Plant Sci 6:1113–1118

Dakora F, Keya SO (1997) Contribution of legume nitrogen fixation to sustainable agriculture in sub-Saharan Africa. Soil Biol Biochem 29:809–817

Danso SKA (2004) Managing soil genetic resources for enhanced biodiversity. International symposium on soil biology and fertility, AfNet-TSBF-CIAT, 17–21 May, Yaoundé

Date RA (2000) Inoculated legumes in cropping systems of the tropics. Field Crops Res 65:123–136

Diabate M, Munive A, de Faria SM, Ba A, Dreyfus B, Galiana A (2005) Occurrence of nodulation in unexplored leguminous trees native to the West African tropical forest and inoculation response of native species useful in reforestation. New Phytol 166:231–239

Elkan GH (1992) Biological nitrogen fixation systems in tropical ecosystems: an overview. In: Mulongoy K, Gueye M, Spencer DSC (eds) Biological nitrogen fixation and sustainability of tropical agriculture. Wiley, Chichester, pp 27–40

FAO (1995) FAO fertilizer yearbook 1994, vol 44. Food and Agriculture Organization of the United Nations, Rome

FAO (2001) Soil fertility management in support of food security in sub-Saharan Africa. Food and Agriculture Organization of the United Nations, Rome

FAO (2007) Biodiversity of micro-organisms and insects for food and agriculture: status and needs. Food and Agriculture Organization of the United Nations, Rome

Fankem H (2007) Diversity and potentials of phosphate solubilizing microorganisms associated with palm-tree (*Elaeis guineensis* Jacq.) rhizosphere in Cameroon. PhD thesis, University of Yaounde I, Cameroon

Fankem H, Nwaga D, Deubel A, Dieng L, Merbach W, Etoa F-X (2006) Occurrence and functioning of phosphate solubilizing microorganisms from oil palm tree (*Elaeis guineensis*) rhizosphere in Cameroon. Afr J Biotechnol 5:2450–2460

Fankem H, Ngo Nkot L, Deubel A, Quinn J, Merbach W, Etoa F-X, Nwaga D (2008) Solubilization of inorganic phosphates and plant growth promotion by strains of *Pseudomonas fluorescens* isolated from acidic soils of Cameroon. Afr J Microbiol Res 2:171–178

Finlay RD (2004) Mycorrhizal fungi and their multifunctional roles. Mycologist 18:91–96

Giller KE (2001) Nitrogen fixation in tropical cropping systems. CABI, Wallingford

Gockowski J et al (2005) The forest margins of Cameroon. In: Palm C, Vosti SA, Sanchez PA, Ericksen PJ (eds) Slash-and-burn agriculture, the search for alternatives. Columbia University Press, New York, pp 305–331

Gueye M (1992) Effect of *Rhizobium* and *Rhizobium-Glomus* inoculation on nitrogen fixation in Bambara groundnut. In: Mulongoy K, Gueye M, Spencer DSC (eds) Biological nitrogen fixation and sustainability of tropical agriculture. Wiley, Chichester, pp 283–287

Guichuru MP et al (2003) Soil fertility management in the lowland humid forest zone. In: Gichuru MP et al (eds) Soil fertility management in Africa: a regional perspective. Academy Science Publishers, Nairobi, pp 147–186

Hartnett DC, Potgieter AF, Wilson GWT (2004) Fire effects on mycorrhizal symbiosis and root system architecture in southern African savanna grasses. Afr J Ecol 42:328–337

Hawksworth DL (1991) The fungal dimension of biodiversity: magnitude, significance, and conservation. Mycol Res 6:641–655

Hijri I, Sykorova Z, Oehl F, Ineichen K, Mader P, Wiemken A, Redecker D (2006) Communities of arbuscular mycorrhizal fungi in arable soils are not necessarily low in diversity. Molec Ecol 15:2277–2289

ICRAF (2000) Alternatives to slash and burn programme. ICRAF, Nairobi

Jansa J, Mozafar A, Anken T, Ruh R, Sanders IR, Frossard E (2002) Diversity and structure of AMF communities as affected by tillage in a temperate soil. Mycorrhiza 12:225–234

Jansa J, Mozafar A, Kuhn G, Anken T, Ruh R, Sanders IR, Frossard E (2003) Soil tillage affects the community structure of mycorrhizal fungi in maize roots. Ecol Appl 13:1164–1176

Jefwa JM, Coyne D, Gaidasova S, Elsen A, van Asten P, Vanlauwe B (2008) Benefits and potential use of arbuscular mycorrhizal fungi in banana and plantain systems in Africa. http://www.banana2008.com/com/details/Production.pdf. Accessed 12 Aug 2009

Jemo M, Nolte C, Nwaga D (2007) Biomass production, N and P uptake of Mucuna after bradyrhizobia and arbuscular mycorrhizal fungi inoculation, and P-application on acid soil of southern Cameroon. In: Bationo A, Waswa B, Kihara J, Kimetu J (eds) Advances in integrated soil fertility management in Sub Saharan Africa: challenges and opportunities. Springer, Berlin, pp 277–281

Kanmegne J (2004) Slash and burn agriculture in the humid forest zone of southern Cameroon: soil quality dynamics, improved fallow management and farmers' perceptions. PhD Thesis, Wageningen University, The Netherlands

Kates RW, Dasgupta P (2007) African poverty: a grand challenge for sustainability science. Proc Natl Acad Sci U S A 104:16747–16750

Kimou A, Zengbet M (1994) Facteurs limitant la fixation de N_2 chez deux espèces de légumineuses à graines (*Glycine max* et *Vigna unguiculata*) en zone tropicale humide de Côte d'Ivoire. In: Sadiki M, Hilalé A (eds) Recent developments in biological nitrogen fixation research in Africa. IAV, Rabat, pp 381–386

Koala S, Sims JR, El-Actar H, El-Halfawi M (1988) Phosphorus deficiency in the semi-arid tropics and implications for grain legume production. IRDC, Kluwer, Dordrecht

Kotto-Same J et al (2000) Alternatives to Slash-and-Burn: summary report and synthesis of phase II in Cameroon. ASB-ICRAF, Nairobi

Mafongoya PL, Giller KE, Odee D, Gathumbi S, Ndufa SK, Sitompul SM (2004) Benefiting from N_2-fixation and managing rhizobia. In: van Noordwijk M, Cadish G, Ong CK (eds) Below-ground interactions in tropical agrosystems. CAB International, Wallingford, pp 227–242

Mahendra MS, Günther F, van Velthuizen H (2008) Food security and sustainable agriculture: the challenge of climate change in sub-Saharan Africa. International Institute for Applied System Analysis, Laxenburg

Mandimba GR (1997) L'inoculation du soja: une arme contre la faim pour les pays en développement. 6e JS du réseau biotechnologies végétales AUPELF/UREF, Orsay, p 180

Mandou MMS, Banlock DR, Nwaga D, Foko J (2002) Effet du molybdène et des rhizobia sur la nutrition azotée et le rendement de l'arachide (*Arachis hypogaea* L.) en champ: étude préliminaire. Biosci Proc 9:83–89

Mangan SA, Eom AH, Adler GH, Yavitt JB, Herre EA (2004) Diversity of arbuscular mycorrhizal fungi across a fragmented forest in Panama: insular spore communities differ from mainland communities. Oecologia 141:687–700

Mathimaran N, Ruh R, Vullioud P, Frossard E, Jansa J (2005) *Glomus intraradices* dominates arbuscular mycorrhizal communities in a heavy textured agricultural soil. Mycorrhiza 16:61–66

Mathimaran N, Ruh R, Jama B, Verchot L, Frossard E, Jansa J (2007) Impact of agricultural management on arbuscular mycorrhizal fungal communities in Kenyan Ferralsol. Agric Ecosyst Environ 119:22–32

Moreira FMS, Siqueira JO, Brussaard L (2006) Soil organisms in tropical ecosystems: a key role for Brazil in the global quest for the conservation and sustainable use of biodiversity. In: Moreira FMS, Siqueira JO, Brussard L (eds) Soil biodiversity in Amazonian and other Brazilian ecosystems. CAB International, Wallingford, pp 1–12

Morgan S (2005) Shifting cultivation. http://www.clas.ufl.edu/users/morgans/Lecture_March.29.pdf. Accessed 12 Aug 2009

Murphy J, Riley JP (1962) A modified single solution method for determination of phosphate in natural waters. Anal Chimica Acta 27:31–36

Nandwa SM (2003) Perspectives on soil fertility in Africa. In: Gichuru MP et al (eds) Soil fertility management in Africa: a regional perspective. Academy Science Publishers, Nairobi, pp 1–50

Ngakou A (2007) Potentials of rhizobia, arbuscular mycorrhizal fungi and *Metarhizium anisopliae* in managing *Megalothrips sjostedti* and improving cowpea production in Cameroon. PhD thesis, University of Buea, Cameroon

Ngakou A, Nwaga D, Ntonifor NN, Tamò M, Nebane CLN, Parh IA (2007) Contribution of arbuscular mycorrhizal fungi (AMF), rhizobia and *Metarhizium anisopliae* to cowpea production in Cameroon. Int J Agric Res 2:754–764

Ngakou A, Tamò M, Parh IA, Nwaga D, Ntonifor NN, Korie S, Nebane CLN (2008) Management of cowpea flower thrips, *Megalurothrips sjostedti* (Thysanoptera, Thripidae), in Cameroon. Crop Prot 27:481–488

Ngonkeu MEL (2003) Biodiversité et potentiel des champignons mycorhiziens à arbuscules de quelques zones agro-écologiques du Cameroun. Thèse de Doctorat 3e Cycle, Université of Yaoundé I, Cameroon

Ngo Nkot (2009) L'acidité du sol comme facteur limitant la fixation de l'azote atmosphérique chez l'arachide (*Arachis hypogaea* L.) en zone de forêt dense humide. PhD thesis, University of Yaoundé I, Cameroon

Ngo Nkot L, Krasova-Wade T, Etoa FX, Sylla SN, Nwaga D (2008) Genetic diversity of rhizobia nodulating *Arachis hypogaea* L. in diverse land use systems of humid forest zone in Cameroon. Appl Soil Ecol 40:411–416

Niane-Badiane A, Ganry F, Jacquin F (1998) Assessment of microbial biomass in a ferrugineous tropical soil of the central northern part of Senegal. In: Congrès mondial des sciences du sol, vol 16, CIRAD, Montpellier, pp 1–7

Nounamo L, Yemefack M (2001) Farming systems in the evergreen forest of southern Cameroon: shifting cultivation and soil degradation. Tropenbos-Cameroon Documents 8. Tropenbos-Cameroon Programme and The Tropenbos Foundation, Wageningen

Nwaga D (1988) Intérêt du polymorphisme protéique et enzymatique pour la caractérisation des Pythiacées (*Phytophthora* et *Pythium*) et de leur interaction avec le tabac et le haricot. PhD Thesis, University of Rennes I, France

Nwaga D (1997) Production pilote de biofertilisants rhizobien et mycorhizien au Cameroun: intérêt et contraintes. Cam J Biol Biochem Sci 7:16–23

Nwaga D (2001) Why should we develop a programme for biological agriculture to improve food production and soil fertility by a low input technology. Biosci Proc 8:13–33

Nwaga D, Ngo Nkot L (1998) Tolérance à l'acidité in vitro de rhizobia isolés du niébé (*Vigna unguiculata*) en comparaison avec *Bradyrhizobium japonicum*. Cahiers Agricultures 7:407–410

Nwaga D, Ngonkeu MEL, Oyong MM, Ngakou A, Abelong M-P, Foko J (2000) Soil beneficial micro-organisms and sustainable agricultural production in Cameroon: current research and perspectives. In: UNESCO-TSBF (ed) The biology and fertility of tropical soils, TSBF Report 1998, Nairobi, pp 62–65

Nwaga D, Bindzi M, Biye H, Atangana Eteme R, Ngonkeu MEL (2003) Diversité des champignons mycorhiziens arbusculaires et biomasse microbienne des sols des zones de forêt humide du sud-Cameroun: effet des systèmes d'utilisation des terres. Ann Fac Sci Univ Yaoundé I 35:14–23

Nwaga D, The C, Ambassa-Kiki R, Ngonkeu Mangapché EL, Tchiegang-Megueni C (2004) Selection of arbuscular mycorrhizal fungi for inoculating maize and sorghum, grown in oxisol/ultisol and vertisol in Cameroon. In: Bationo A (ed) Managing nutrient cycles to sustain soil fertility in sub-Saharan Africa. Academy Science Publishers and TSBF Institute of CIAT, Nairobi, pp 467–486

Nwaga D, Hamon S, Djiéto Lordon C, Engelmann F (2007a) Biotechnologies et maîtrise des intrants agricoles en Afrique centrale. Réseau 'BIOVEG', AUF/IRD/Université de Yaoundé I/IRAD, Yaoundé, Cameroon. http://www.bioveg.auf.org

Nwaga D, Fankem H, Essono Obougou G, Ngo Nkot L, Randrianangaly JS (2007b) Pseudomonads and symbiotic micro-organisms as biocontrol agents against fungal disease caused by *Pythium aphanidermatum*. Afr J Biotechnol 6:190–197

Nwaga D, Tenkouano A, Tomekpe K, Fogain R, Kinfack DM, Tsané G, Yombo O (2009) Multifunctional properties of mycorrhizal fungi for crop production: the case study of banana development and drought tolerance. In: Bationo A et al (eds) Innovations as key to Green Revolution in Africa: exploring the scientific facts. Springer, Berlin

Nyobe T, Nwaga D (2008) Country report for Cameroon. Soil health program: alliance for a Green Revolution in Africa (AGRA), Accra, Ghana

Obaton M (1992) Facteurs pédoclimatiques limitant la fixation biologique de l'azote chez les légumineuses. In: Mulongoy K, Gueye M, Spencer DSC (eds) Biological nitrogen fixation and sustainability of tropical agriculture. Wiley, Chichester, pp 57–66

Oberson A, Bünemann EK, Friesen DK, Rao IM, Smithson PC, Turner BL, Frossard E (2006) Improving phosphorus fertility in tropical soils through biological interventions. In: Uphoff N et al (eds) Biological approaches to sustainable soil systems. CRC, Boca Raton, FL, pp 531–546

Oehl F, Sieverding E, Ineichen K, Ris E, Boller T, Wiemken A (2004) Community structure of arbuscular mycorrhizal fungal communities at different soil depths in extensively and intensively managed agroecosystems. New Phytol 106:273–283

Oloke JK, Odeyemi O (1988) Effect of some Nigerian *Bradyrhizobium* inoculants on the performance of three cowpea cultivars (*Vigna unguiculata* L. Walp.) in two field plots. Biol Fertil Soils 6:178–182

Opik M, Moora M, Liira J, Zobel M (2006) Composition of root-colonizing arbuscular mycorrhizal fungi communities in different ecosystems around the globe. J Ecol 94:778–790

Pimentel D, Wilson C, Mc Cullum C, Huang R, Dwen P, Flack J, Tran Q, Saltman T, Cliff B (1997) Economic and environmental benefits of biodiversity. Bioscience 47:747–757

Plenchette C, Clermont-Dauphin C, Meynard JM, Fortin JA (2005) Managing arbuscular mycorrhizal fungi in cropping systems. Can J Plant Sci 85:31–40

Reddy BV, Chinnappa AS, Gopal V, Shankar S (2005) Economic valuation of production impact of legume nodulating bacteria on groundnut production. National workshop on conservation and sustainable management of belowground biodiversity. KFRI, Peechi

Sanchez PA (2002) Soil fertility and hunger in Africa. Science 295:219–220
Sanchez PA et al (1997) Soil fertility replenishment in Africa: an investment in natural resource capital. In: Buresh RJ, Sanchez PA, Calhoun F (eds) Replenishing soil fertility in Africa. Soil Science Society of America, Madison, WI, p 1 Special publication No. 51
Sanginga N, Okogun JA, Vanlauwe B, Diels J, Carsky RJ, Dashield K (2001) Contribution of promiscuous soybeans in maize-based cropping systems. In: Tian G, Ishida F, Keatinge JDH (eds) Soil fertility maintenance in West Africa. Soil Science Society of America, Madison, WI, pp 157–177 Special publication No. 58
Sow HA, Diop TA, Ndiaye F, Manga AGB (2008) Influence de la mycorhization arbusculaire sur la culture intensive de l'oignon (*Allium cepa* L) au Sénégal. J Sci 8:1–6
Sprent JI (2001) Nodulation in legumes. Royal Botanic Gardens, Kew
Swift MJ (1998) Towards the second paradigm: integrated biological management of soil. In: Siqueira JO et al (eds) Soil fertility, soil biology and plant nutrition interrelationships. SBCS/UFLA/DCS, Lavras, pp 11–24
Swift MJ, Shepherd KD (2007) Saving Africa's soils: Science and technology for improved soil management in Africa. Word Agroforestry Centre, Nairobi
Swift MJ, Bignell DE, Moreira FMS, Huising EJ (2008) The inventory of soil biological diversity: concepts and general guidelines. In: Moreira FMS, Huising J, Bignell DE (eds) A handbook of tropical soil biology: Sampling and characterization of below-ground biodiversity. Earthscan, London, pp 1–16
Tchabi A, Coyne D, Hountondji F, Lawouin L, Wiemken A, Oehl F (2008) Arbuscular mycorrhizal fungal communities in sub-Saharan savannas of Benin, West Africa, as affected by agricultural land use intensity and ecological zone. Mycorrhiza 18:181–195
Thies JE, Singleton PW, Bohlool BB (1991) Influence of the size of indigenous rhizobial populations on establishment and symbiotic performance of introduced rhizobia on field grown legumes. Appl Environ Microbiol 57:19–28
TSBF (1999) Integrating soils, systems and society strategic directions for TSBF. TSBF, Nairobi
UNEP (1995) Global biodiversity assessment. United Nation Environment Programme, Nairobi
Van Straaten P (2007) The geological basis of farming in sub-Saharan Africa. In: Bationo A et al (eds) Innovations as key to Green Revolution in Africa: exploring the scientific facts. AfNet-TSBF CIAT/SOFECSA, Arusha
Woomer PL, Karanja NK (1996) The *Rhizobium* ecology network of East and South Africa (RENESA). TSBF, UNESCO, Nairobi

Chapter 6
Microbial Populations of Arid Lands and their Potential for Restoration of Deserts

Yoav Bashan and Luz E. de-Bashan

6.1 Introduction

The rapid expansion of deserts in recent decades as a result of human actions combined with climatic disasters has highlighted the necessity to understand biological processes in arid environments. Whereas physical processes and the biology of flora and fauna have been relatively well studied in marginally used arid areas, knowledge of desert soil micro-organisms remains fragmentary.

This chapter describes several biological phenomena in hot deserts related to microbial populations and the potential use of micro-organisms for restoring hot desert environments. A few relevant examples from colder deserts are also provided.

6.2 Resource Islands (Fertility Islands)

The most common phenomenon of natural vegetation pattern in deserts is the resource island phenomenon, a complex feature of microhabitat buildup. Many plants attenuate the adverse environmental growth conditions to which they are submitted by modifying their habitat, causing changes in microclimate and soil properties that may benefit future vegetation. Dryland ecosystems have a highly

Y. Bashan (✉)
Environmental Microbiology Group, Northwestern Center for Biological Research (CIBNOR), La Paz, B.C.S., Mexico; Department of Soil, Water and Environmental Science, The University of Arizona, Tucson, AZ, USA; Bashan Foundation, Corvallis, OR, USA
e-mail: bashan@cals.arizona.edu

L.E. de-Bashan
Environmental Microbiology Group, Northwestern Center for Biological Research (CIBNOR), La Paz, B.C.S., Mexico; Bashan Foundation, Corvallis, OR, USA

P. Dion (ed.), *Soil Biology and Agriculture in the Tropics*, Soil Biology 21,
DOI 10.1007/978-3-642-05076-3_6, © Springer-Verlag Berlin Heidelberg 2010

heterogeneous distribution of resources, with greater nutrient concentrations and microbial densities occurring in vegetated than in bare soils. This process of habitat "engineering" facilitates the establishment of other plant species, with the result that species richness is increased under the canopy of some shrubs and trees. This nursing of various plant species by a pioneer, or "engineer," species is presumably aided by a large community of soil micro-organisms. A well-known nurse plant–seedling association occurs between leguminous small trees and shrubs and desert succulents. Specifically, these legume plants create, within the zone of their roots and overhanging branches, slight mounds that form from the deposition and stabilization of windborne soil particles under the canopy. Shade, nutrients, and organic matter also contribute to this process of stabilization. These fine-textured soils become microhabitats for entire communities of organisms at all scales. These slight mounds are called resource islands or fertility islands (Burquez and Quintana 1994; Garner and Steinberger 1989; Nabhan and Suzan 1994; Suzan et al. 1994, 1996a, b, Tewksbury and Petrovich 1994; West et al. 2000) (Figs. 6.1 and 6.2). Over time, soils under shrubs and trees accumulate resources at the expense of open space soils and also stimulate microbial activity. This leads to improved vegetation in these limited areas. Associations of establishing seedlings with nurse plants are thought to be examples of commensalism in which seedlings benefit from the microhabitat created by nurse plants, with no adverse effect for the latter (Flores and Jurado 2003).

Examples of this phenomenon in hot and cold deserts are many. Sagebrush (*Artemisia tridentata*) in the western USA is a major dominant species that produces resource islands, with higher C and N soil content than in the open areas. Whereas sagebrush is commonly removed, resource islands are persistent and were still present 6 years after removal (Bechtold and Inouye 2007).

Perennial shrubs strongly influence Mojave Desert (USA) soil characteristics and result in higher nutrient levels. Small mammal burrows further enhance the mineral nutrient content of soils (Titus et al. 2002). The distribution of soil N, P, K, and S in desert ecosystems of the southwestern USA is strongly associated with the presence of shrubs. Shrubs concentrate the biogeochemical cycles of these elements in resource islands located beneath their canopies, while adjacent barren, intershrub spaces are comparatively devoid of biotic activity. Both physical and biological processes are involved in the formation of resource islands. The loss of semi-arid grassland in favor of invading desert shrubs initiates changes in the distribution of soil nutrients, which may promote further invasion and persistence of shrubs. Hence, the invasion of semi-arid grasslands by desert shrubs is associated with development of resource islands under those shrubs.

Monitoring changes in the distribution of soil properties can serve as a useful index of desertification in arid and semi-arid grasslands (Schlesinger et al. 1996; Schlesinger and Pilmanis 1998). In the central Monte Desert of Argentina, the mesquite *Prosopis flexuosa* is the most common tree. Shrubs were more common under their canopies, and concentrations of soil organic matter (OM) and of N, K, and P were significantly higher. Mesquite trees modify the spatial arrangement

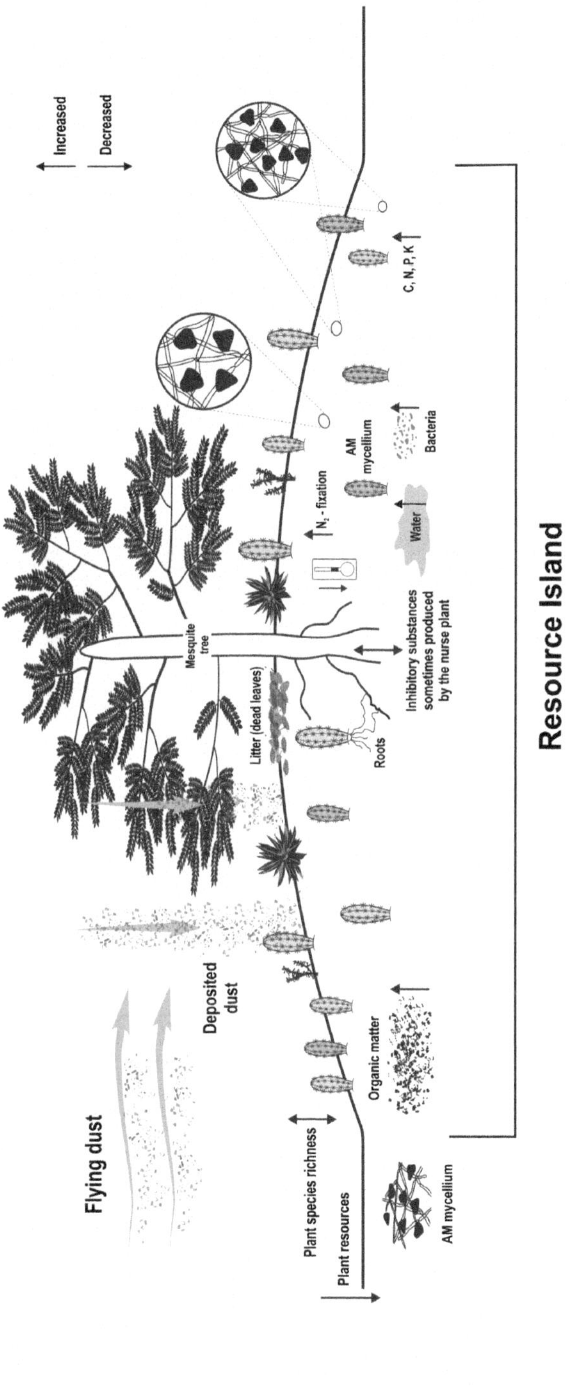

Fig. 6.1 Schematic representation of the role of resource islands in re-vegetation of disturbed arid environment, such as the southern Sonoran Desert, Baja California Peninsula, Mexico

Fig. 6.2 (**a**) Typical resource island under mesquite (*Prosopis chilensis*) tree in a desert in northwest Argentina. The nurslings (*arrows*) seen are of the cactus *Opuntia sulphurea*, as identified by Roberto Kiesling, Mendoza, Argentina. (**b**) Typical resource island in the southern Sonoran Desert under mesquite amargo tree in Baja California, Mexico. The nurslings seen are the cacti *Stenocereus thurberi*, *Lophocereus schottii*, and pitaya agria (*Stenocereus gummosus*)

of shrubs and herbaceous plant species and the chemical conditions of the soil, generating local vegetative heterogeneity (Rossi and Villagra 2003). Similarly, in resource islands in the southern Sonoran Desert in the Baja California Peninsula of Mexico, pachycereid cacti grow under mesquite amargo (*Prosopis articulata*) and *Agave datilyo* grows under palo fiero (*Olneya tesota*) canopies. Of the nine species of trees and arborescent shrubs in this area, the mature (>20 year) nurse legumes *P. articulata* and *O. tesota* supported the largest number of understory plants.

Younger legume trees had only occasional understory associates (Carrillo-Garcia et al. 1999).

The content of resource islands may be affected by landscape parameters, such as geographic location of the tree and patterns of local precipitation. In the Sonoran Desert in Arizona (USA), velvet mesquite trees (*Prosopis velutina*) clearly created resource islands under their canopy that contained higher microbial biomass relative to areas outside the canopy. While these effects were generally consistent across the landscape and showed little temporal variability, the magnitude and direction of effect of mesquite on soil moisture changed with location, ranging from positive to negative (Schade and Hobbie 2005).

The availability of extra resources under the canopy is not always supportive of seedling proliferation. Soil nutrients, as well as density and biomass of annual plants, underneath and outside the canopy of *Porlieria chilensis* shrubs were measured at the end of the growing season in an arid coastal site in Chile. Levels of soil nitrogen, phosphorus, and organic matter were significantly higher underneath than outside the canopies of shrubs, as is common in resource islands. In contrast with this, almost four times as many plants occurred outside than underneath shrub canopies, although no significant difference in total aboveground biomass was found. Several species had higher densities and/or biomass outside rather than underneath shrub canopies, whereas others showed the opposite trend. Species richness was lower underneath *P. chilensis* canopies. The micro distribution of ephemeral species may be explained by differential water and nutrient requirements (Gutiérrez et al. 1993). Another example of a negative influence of resource islands was described in a mixed desert shrub community in the Mojave Desert. The influence of shrub canopies of three species of shrubs on the creation of resource islands and the survival of transplanted *Ambrosia dumosa* seedlings was studied. On the positive side, in the wet season, soils under shrubs had higher pH, water content, organic matter, total nitrogen, and mineralizable nitrogen than soil in adjacent open areas, confirming the widely established pattern of resource islands in arid lands. Removing the shrubs did not affect most of the improved soil parameters in the long dry season occurring after shrub removal, except for water content. These positive effects notwithstanding, over a 1-year period, transplanted *A. dumosa* seedlings had highest survival where shrub removal had occurred in open areas and died out under the canopy of all three shrub species. This indicates that these shrubs have a strong negative effect on seedling survival, despite the creation of resource islands. The negative effects might be due to shading or root competition (Walker et al. 2001).

6.2.1 Arbuscular Mycorrhizal Fungi within Resource Islands

The resource island phenomenon significantly affects the microbial biota of the desert soil; arbuscular mycorrhizal (AM) fungi and bacteria are pivotal for the well-being of the desert vegetation and may facilitate re-vegetation and reforestation of eroded desert soil. The abundance of micro-organisms in desert soil environments is closely related to the accumulation of resources in the resource island soil. AM fungi stabilize the soil and enhance plant growth by alleviating nutrient and drought stress. Their contributions to agriculture are well-known, but their role in desert ecosystems has received far less attention. Usually, an evaluation of the mycorrhizal status of threatened desert ecosystems is recommended as a first step in rehabilitation and restoration.

The AM status of perennial plants in disturbed and undisturbed desert plots was investigated in the southern Sonoran Desert in Baja California Sur, Mexico, to determine if AM fungi contribute to resource-island stability and establishment of seedlings. Roots of all 46 species of perennial plants in the study plots contained AM, but incidence of root colonization varied widely (from $<10\%$ to $>70\%$). Roots of pioneer colonizers, which become established in originally plant-free, disturbed areas, had high biomass of AM. Plants with very little ($<10\%$) AM root colonization, such as Pachycereae cacti (*Pachycereus pringlei, Machaerocereus gummosus,* and *Lemaireocereus thurberi*) and the succulent *Agave datilyo* occurred preferentially under the canopy of nurse trees. AM propagule densities in plant-free areas were lower than under plant canopies. Resource-island soils were enmeshed with AM fungal hyphae, especially in the upper layer. Seedlings of the giant cardon cactus, *P. pringlei*, growing under a shade roof for 6 months in resource-island soil collected under the mesquite *Prosopis articulata*, had a biomass ten times greater than plants growing in bare-area soil. These findings are consistent with the proposition that AM fungi contribute to the plant-soil system of this desert area by: (1) stabilizing windborne soil that settles under dense plant canopies, (2) enhancing establishment of colonizer plants in bare soils of disturbed areas, and (3) influencing plant associations through differences in the mycotrophic status of the associates (Carrillo-Garcia et al. 1999). Some experiments were conducted in our laboratory to evaluate the importance of AM fungal inoculum for the establishment of six species of cactus under native mesquite (*P. articulata*) trees. The results suggested that AM fungal inoculum potential in these hot desert soils, although relatively low, is probably maintained in the upper soil layers by means of hyphal fragments rather than spores. Since the soil AM fungal inoculum density was not different under and away from the mesquite tree canopy, it was concluded that AM fungal inoculum density is not the primary factor for the establishment of cactus seedlings and that favorable edaphic factors probably play a more important role (Bashan et al. 2000).

In Israel, the spatial dynamics of AM fungi were investigated under the canopy of the halophyte *Zygophyllum dumosum*. Soil moisture was positively correlated with colonization of AM fungi. Vesicular and arbuscular root colonization was positively correlated with soil organic matter and total soluble N. Spore density was also positively correlated with total soluble N, making spore density and the extent of vesicular and arbuscular root colonization useful indicators for evaluating changes in desert soil ecosystems (He et al. 2002a). Measuring the mycorrhizal potential in resource islands was done in degraded semi-arid areas in southeast Spain where the general grassy vegetation includes patches of shrub species.

Diversity of AM fungi species in the study area was rather low, clearly indicating the high degree of degradation of the ecosystem. There were more AM fungal propagules in the rhizosphere of the shrub species compared with nearby grassy soils, indicating that AM propagules can be considered as a functional component of these resource islands. The number of spores of the most representative AM fungal species, *Glomus constrictum*, and the total length of extra-radical AM mycelium were correlated with the mycorrhizal potential, measured in terms of the number of "infective" AM propagules (Azcón-Aguilar et al. 2003).

Interaction of different plant species can negatively affect the rate of AM colonization in the resource-island soil. In the desert of the Junggar Basin of China, the AM fungal status of the four most common ephemeral plant species, *Chorispora tenella*, *Ceratocephalus testiculatus*, *Eremopyrum orientale*, and *Veronica campylopoda*, growing in an area dominated by tamarisk shrubs (*Tamarix* spp.), was measured. Available N and P, total P, organic matter content, total salt content, and electrical conductivity were all higher in the rhizosphere of the four ephemerals when these were located under the shrub canopies than when they were situated away from the canopies. However, the number of AM fungal species, the mycorrhizal colonization rates of the ephemerals, and spore densities in their corresponding rhizosphere soils were significantly lower under the shrub canopies than beyond. This may indicate that the canopies of *Tamarix* spp. exerted some negative influence on AM status of the ephemerals and on the AM fungal communities (Shi et al. 2006b).

6.2.2 Bacteria in the Resource Islands

The resource-island hypothesis predicts that soil resources such as nitrogen, phosphorus and water will be distributed evenly in grasslands, but have a patchy distribution focused around plants in shrublands. This hypothesis predicts that the population of micro-organisms will follow resources and be (1) evenly distributed in grasslands, (2) concentrated around individual perennial plants in shrublands, and (3) higher where resources are higher when comparing the same vegetation type. This fundamental hypothesis was tested in the Chihuahuan Desert (USA), by measuring densities of heterotrophic bacteria in grassland vs shrublands. Populations of heterotrophs followed the distribution pattern predicted by the resource-island hypothesis, where the population size was uniform in grasslands, and higher under shrubs than between shrubs at the three shrub sites tested (Herman et al. 1995). In a Chilean coastal desert, bacteria and fungi, organic matter, and nitrogen content were significantly higher underneath the shrub canopy of *Adesmia bedwellii*. Microbial abundances were positively correlated with nitrogen levels and soil moisture (Aguilera et al. 1999). These studies suggest that abundance of nutrient resources in resource islands predicts the distribution of heterotrophic bacteria.

Resource islands formed under the mesquite, *P. laevigata*, in Zapotitlán Salinas, Mexico, are the most important nutrient depository in this environment, enhancing microbial activity in degraded soils. Resources are more evenly distributed in preserved soil than in degraded soil. Total organic carbon accumulation, water content, and nitrogen fixation have significantly higher values in the preserved area, as does the number of culturable bacteria, including *Rhizobium* sp. and *Streptomyces* sp. This bacterial abundance in resource islands suggests that plant–microbe interactions delay formation of badlands in the area (González-Ruiz et al. 2008). The same mesquite species and palo verde (*Parkinsonia praecox)* are the most abundant perennial shrubs in the Tehuacán Desert in Mexico, forming numerous resource islands that dominate alluvial terraces. Both species contribute significant amounts of leaf litter and thus affect microbial activity under their canopies. Total bacteria, fungi, and heterotrophic diazotrophs were found to be significantly higher under shrubs than in the intervening soil. Nitrogen-fixing bacteria were isolated from soil located beneath the shrubs, but not from interplant soil. Heterotrophic diazotrophs were not more numerous, but fixed significantly more nitrogen under palo verde shrubs than under mesquite shrubs. Greater nitrogenase activity under palo verde shrubs may influence nitrogen distribution in this arid environment (Rodríguez-Zaragoza et al. 2008).

6.2.3 Micro-Organisms in Rocky Resource Islands

Desert resource islands of a different type are formed by pioneering plants capable of growing on rocks in the absence of soil. In the arid region of the southern part of the Baja California Peninsula, unweathered and weathered volcanic rocks contained rock-colonizing plants, mostly cacti that grow on volcanic rocks without a cover or even a miniscule amount of soil. Many are at the seedling stage. At least four cacti species (*Pachycereus pringlei, Stenocereus thurberi, Mamillaria fraileana, Opuntia cholla*), and two trees (wild fig, *Ficus palmeri* and elephant tree, *Pachycormus discolor*) were capable of cracking, wedging, growing in and colonizing cliffs and rocks formed from ancient lava flows. In turn, they produce weathered mineral matter or soil for succession by other plant species (Bashan et al. 2002, 2006; Lopez et al. 2009) (Fig. 6.3). An abundance of micro-organisms was found to reside in the roots of these plants. Dense layers of bacteria and fungi in the rhizoplane of these species were also revealed. The dominant bacterial groups colonizing the rhizoplane were fluorescent pseudomonads and bacilli. Unidentified fungal and actinomycete species were also present in large quantities. Some of the root-colonizing micro-organisms fixed N_2 in vitro and produced volatile and non-volatile organic acids that subsequently reduced the pH of the rock medium in which the bacteria grew. These bacterial exudates significantly dissolved insoluble phosphates, extrusive igneous rock, marble, and limestone. The bacteria released significant amounts of minerals, such as P, K, Mg, Mn, Fe, Cu, and Zn from the rocks and were tolerant to high temperature, salt, and drought. The microbial community survived on the rhizoplane of cacti during the annual 10-month dry season (Puente et al. 2004a).

Fig. 6.3 Rock colonization by plant, a mini resource island in rocky arid environments. (**a**) cardon (*Trichocereus pagacana*) in northwest Argentina. (**b**) *Mammillaria fraileana* and *Echinocereus* sp. (*arrows*) growing on a rock in Baja California, Mexico. (**c**) *Agave* sp. growing on overhang rock in "Organ Pipe National Monument" Arizona, USA. (**d**) Elephant tree (*Pachycormus discolor*) growing in an ancient lava flow in Baja California, Mexico

Similarly, plant–bacteria associations between the cardon cactus and endophytic bacteria promoted establishment of seedlings and growth on igneous rocks without soil. In the same way as the rhizoplane bacteria found on the surface of the roots of this plant species, endophytic bacteria dissolved several rock types and minerals, released significant amounts of useful minerals for plants, fixed N_2 in vitro, produced volatile and non-volatile organic acids, and reduced rock particle size to form mineral soil. Large populations of culturable endophytic bacteria were found inside the seeds obtained from wild plants, from seeds extracted from the guano of bats feeding on cactus fruit, in seedlings growing from these seeds, in the pulp of the fruit, and in small, mature wild plants. The endophytic populations of cardon cactus were comparable in size to endophytic populations in some agricultural crops.

The dominant culturable endophytes were isolates of the genera *Bacillus* spp., *Klebsiella* spp., *Staphylococcus* spp., and *Pseudomonas* spp. Seedlings grown from seeds containing particular species of endophytes harbored those same species of endophytes in their shoots, possibly derived from the seeds. Cacti capable of acquiring diverse populations of endophytes may benefit from an evolutionary advantage on a highly uncompromising terrain (Puente et al. 2009a). The search for culturable endophytic bacteria within the rock-dwelling, small cactus *Mammillaria fraileana,* responsible for massive rock degradation, revealed that endophytes are virtually absent in stems and fruits, found in seeds, and abundant in roots. Five months after the inoculation of sterile seedlings, the endophytes were found in the dermal system of roots, mainly in peridermal cells, and in the root–stem transition zone. Some endophytic isolates showed capacity to degrade particles of rock by decreasing the pH of liquid culture, and a diazotrophic strain apparently supplied nitrogen for the growth of other bacteria (Lopez et al. unpublished).

Taken together, these studies show that plant colonization of volcanic rocks may assist soil formation, which eventually leads to accumulation of soil, water, and nutrients in a desert ecosystem that otherwise lacks these essential plant-growth elements. The formation of soil is promoted by rhizoplane and endophytic bacteria residing on and in cacti roots.

6.3 Micro-Organisms in Naturally Vegetated and Impaired Arid Land Soils

Several microbial phenomena can be distinguished in desert environments: formation of biological soil crusts, colonization of plant roots by AM fungi, and the action of plant growth-promoting bacteria (PGPB).

6.3.1 Biological Soil Crust

Deserts are frequently characterized by low productivity and substantial unvegetated space, which is often covered by micro-organisms creating biological soil crust communities (hereafter called crusts, but also known as cryptogamic, microbiotic, cryptobiotic, and microphytic crusts); these are mainly formed in open desert spaces (Belnap 2003; Belnap and Lange 2001; Belnap et al. 2001, 2004; Rosentreter et al. 2007; www.soilcrust.org). Crusts are gaining recognition as an important ecological component of desert ecosystems that significantly contributes to plant well-being by stabilizing sand and finer particulates, promoting moisture retention, and supplying soil nitrogen via nitrogen fixation (Yeager et al. 2003).

Biological crusts are a highly specialized, complex community of cyanobacteria, green and brown algae, mosses, and lichens and their metabolic by-products (Lange et al. 1992). Liverworts, fungi, and bacteria can also be important components in some crusts. Together, microbial and plant communities form a structured, gelatinous matrix that binds the uppermost layers of the soil. The complex creates a relatively thin, hard soil cover, usually 1–4 mm thick, but sometimes reaching 10 cm in thickness. Because crusts are made of living organisms, their appearance varies, yet they can be easily distinguished from one another by color and surface topography (Rosentreter et al. 2007). Crusts generally cover all soil spaces not occupied by vascular plants, and can reach up to 70% or more of the living cover of a desert (Yeager et al. 2003). Members of the community may protect each other from environmental aggressions. For example, sunscreen pigments produced by *Nostoc* sp. and *Scytonema* sp. at the surface of crusts protect other, less-pigmented taxa from the damaging effects of intense solar radiation. Most pigments associated with UV radiation protection or repair activity are at their highest concentrations near the exterior of the crusts and at their lowest concentrations in the part contacting the soil (Bowker et al. 2002).

Availability of water is the major environmental limit for crusts. The crust is photosynthetically and metabolically active mainly in the presence of liquid water, but can also function by hydration in equilibrium with high humidity (Lange et al. 1994). Rainfall in southern Arizona and in many other deserts occurs as small, localized events. Following these brief rainfalls, crusts contribute 80% of the soil-level CO_2 fluxes to the atmosphere. However, following a large and rarer rainfall, roots and soil microbes contribute nearly 100% of the soil-level CO_2 flux. Therefore, after typically small, pulsed precipitation, crusts may significantly contribute to desert ecosystem production capacity (Cable and Huxman 2004). The crusts, both in cold and hot deserts, undergo continuous growth and surface increase. This happens because, in the rainy season, filamentous cyanobacteria and green algae swell when they are wet and migrate out of their sheaths. After each migration, a new sheath material is formed, a process that extends the length of the sheath. Repeated swelling leaves a complex network of empty sheath material that maintains soil structure after the micro-organisms have dehydrated and decreased in size (www.soilcrust.org).

Although most crusts are dominated by various species of cyanobacteria, in more acidic soils, the crusts are dominated by green algae. Crusts dominated by lichens and/or mosses are particularly stable. The nature of the organism that dominates the crust is partly determined by microclimate and may also vary along successional stages of crust development.

Prokaryotic communities in crusts of the Sonoran Desert in Arizona were much less diverse than those of typical bulk arid soils or agricultural soils. No difference in microbial diversity or composition was detected between crusts under plant canopies and those in plant interspaces, suggesting a crust independence from higher plant resources. However, samples within a site were more similar than samples between sites. The most common bacteria included cyanobacteria, proteobacteria, actinobacteria and acidobacteria (Nagy et al. 2005).

Because of their strategic location at the soil–air interface of the desert, crusts significantly affect major environmental processes, such as soil stability and erosion, atmospheric nitrogen fixation, nutrient contributions to plants, soil–plant–water relations, water infiltration, seedling germination, and plant growth.

6.3.1.1 Soil Stability and Water Infiltration

Filamentous crust-forming cyanobacteria bind soil particles and increase soil aggregation by exuding sticky polysaccharide sheaths around their cells that cement particles. Free-living fungi increase soil stability by binding soil particles with their hyphae, whereas lichens and mosses support soil stability by binding soil particles with their rhizines and rhizoids. In addition to stabilizing the soil, the various crust community members also modify local topography, and their combined effect is to significantly increase resistance to wind and water erosion. The rough crust surface increases water infiltration by slowing the flow of flood water (Abu-Awwad 1997).

6.3.1.2 Contribution to Nitrogen Input

Crusts contribute fixed nitrogen to the desert environment in various ways. Mature soil crusts harbor a greater abundance of lichens and mosses than do earlier succession crusts dominated by free-living cyanobacteria. Mature crusts, which have greater biomass, fixation rates, and stabilizing activity, contribute more to soil fertility than pioneer crusts. For example, in a study in Utah (USA), a comparison between "young, light" (cyanobacteria-dominated) and more "developed mature, dark" crusts, that also contained a variety of visible lichens and mosses, showed that nitrogen fixation rates were far higher in the "developed" crust than in the "young" crusts. Molecular fingerprinting profiles showed that *nifH* sequence types were very similar between the two crust types, with *Nostoc* spp. as the dominant microorganism in all crusts. However, quantitative PCR showed that "developed" crusts contained approximately 30-fold more *nifH* gene copies than the "young" crusts. This suggests that the difference in nitrogen fixation rates between the two crust types is most likely a consequence of the number of nitrogen-fixing organisms, rather than the result of a difference in diazotroph species composition (Yeager et al. 2003). Similarly, in the Negev Desert of Israel, higher rates of nitrogen fixation occurred in crusts made of mixed community of cyanobacteria, lichens, and mosses than in cyanobacteria-dominated crusts. These results suggest that nitrogen fixers in mixed communities are important to the nitrogen budget in the Negev (Zaady et al. 1998).

6.3.1.3 Effects on Plant Germination and Growth and on Insects and Animals

There is no strong evidence that crusts enhance germination of plants and any actual effect may well be plant-species related. On the one hand, increased soil surface and topography created by the crust provides seeds with safer sites and protected locations against herbivores, while darker surfaces of the crust increase soil temperatures for germination earlier in the season, which coincides with spring water availability in some colder deserts. On the other hand, large-seeded plants that often require burial for germination, use self-drilling mechanisms, or are at risk of being consumed by rodents, have a severe disadvantage. As soil crusts reduce soil movement, this may limit passive burial and germination of those seeds. The variable effect of soil crusts on seed germination is illustrated by the variable response of three plant species from the Negev desert. Germination of *Plantago coronopus* was promoted by disturbing and removing the crust biological and structural components. *Reboudia pinnata* and *Carrichtera annua* germination was less affected by crust disturbance, but was generally inhibited after removing the live components of the cyanobacterial crust (Zaady et al. 1997).

The positive effects of soil crust on plant growth are more obvious. Increases in survival and nutritional advantage in crust-covered environments, as opposed to bare soil, are repeatedly demonstrated; N, P, K, Fe, Ca, Mg, and Mn increased in tissues of plants grown on crusts. For example, common crusts dominated by cyanobacteria and cyanolichens (such as *Collema* sp.) altered and influenced the mineral uptake by six plant species in desert areas of Utah. The presence of crust always increased N, Cu, K, Mg, and Zn content of the associated plants. This increase was greatest for short-lived herbs that are rooted primarily within the surface soil, the horizon most influenced by crust organisms. The mineral content of a deeply rooted shrub (*Coleogyne ramosissima*) was less influenced by the presence of crust (Harper and Belnap 2001).

Crusts in the plant-free areas and resource islands under the tree canopy may be metabolically interconnected by fungal networks. Data gathered in New Mexico suggested that in this particular environment the dominant fungi of roots, rhizosphere soil, and crusts are dark septate ascomycetes, mostly classified as *Pleosporales*, and that AM fungi are rare (but see also below for more on AM fungi). Phylogenetic analyses indicated substantial overlap in fungal community composition between plant roots and crusts, which may facilitate nutrient transfers. Thus, resource islands and crusts may be functionally integrated by exchanges of C and N through a symbiotic fungal network (Green et al. 2008). However, crusts may counteract the formation of resource islands, at least to some extent. By stabilizing the soil surface, crusts reduce or prevent redistribution of soil, organic matter, and seeds from the open areas to sites covered with plants.

The effect of crusts on secondary consumers, such as arthropods and reptiles, is hardly understood. Lichen-dominated soil crusts in the Namib Desert (Namibia) are important supporters of secondary producers, such as arthropods (Lalley et al. 2006). Rodents and tortoises in some arid western USA localities suffer mineral imbalances when crusts are damaged (Gillis 1994).

6.3.1.4 Responses to Disturbance

Soil compaction and disruption of crusts can result in decreased water availability to vascular plants through decreased water infiltration. Surface disturbance may also cause accelerated soil loss through wind and water erosion and decreased diversity and abundance of soil biota. Furthermore, it may also alter nutrient cycles by diminishing nitrogen and carbon inputs and retarding decomposition of soil organic matter (Belnap 1995).

As a direct outcome of their thin, fibrous structure, crusts are extremely fragile. A small mechanical disturbance, such as single footprint or tire track, is sufficient to disrupt the soil crust and damage its organisms for long periods of time. When heavily disturbed, all crusts lose their lichen–moss component and their microtopography and resemble flat, poorly developed crusts found in extremely arid deserts. Crusts recover very slowly from surface disturbances, resulting in increased vulnerability of the disturbed areas to desertification for prolonged periods. Recovery depends on climate and local physical conditions, including soil structure, topography, and solar radiation. While some species within the soil crust ecosystem may re-grow within a few years of a disturbance and form a new, emerging crust, the damage to slow-growing species may require more than a century before the delicate soil returns to its former productivity. Full recovery from compaction and decreased soil stability is estimated to take several hundred years. To recover their nitrogen fixation capability, crusts may require at least 50 years. Re-establishment can be extremely difficult or impossible in some areas (Belnap 1995).

Mechanical Disturbances

Although crusts are extremely well adapted to the harsh growing conditions in the driest deserts, they have no adaptability to compressional disturbances (Belnap and Gillette 1998). Grazing domestic livestock, military activities and recreational activities (hiking, mountain biking, and off-road driving) greatly increase the vulnerability of crusts. Compression disturbances of crusts break their sheaths and filaments. Subsequent sandblasting by wind can quickly remove parts of the crust, thereby reducing N and C inputs to the soil. This results in reduction in productivity, as well as in exposure of unprotected subsurface sediments to further wind and water erosion. Such mechanical disturbances of the crust cause severe loss of moss and lichen cover, reduction of cyanobacterial presence in the crust, increases in runoff by 50% or more, with the rate of soil loss increasing six times without apparent damage to the nearby vegetation. Thus, disturbance of crusts in arid regions leads to large soil losses (Belnap and Gillette 1998).

Crust disturbance represents a serious ecological issue in the dryland agricultural landscape of north-west Victoria, Australia. There, isolated remnants of crusts are exposed to ongoing disturbance from sheep grazing and agriculture. A statistical model identified relationships between crust abundance and available P, soil C and perennial grass presence, and showed that disturbances from grazing and camping are the main causes of loss (Read et al. 2008).

Indirect Disturbances

1. Loss of Nitrogen. The contribution of crusts to nitrogen input in semi-arid and arid landscapes is of special interest because these ecosystems usually contain low amounts of nitrogen. Several native rangeland shrubs in the southwestern

United States, such as *Artemisia tridentata*, *Atriplex confertifolia*, and *Ceratoides lanata* may have allelopathic effects on the nitrogen-fixing capabilities of cyanobacteria in the crusts, potentially lowering nitrogen fixation. Human activities that indirectly increase shrub population, such as overgrazing, can have an unexpected impact on crust nitrogen fixation.

2. Crust Burial. When the integrity of the crust is broken through trampling, the soil underneath is far more susceptible to wind and water erosion. This soil can be moved over long distances by wind and water, burying intact crusts. Crusts can tolerate shallow burial by extending sheaths to the surface of the covering sand so as to reinitiate photosynthesis. However, deep burial by moving soil will eliminate crusts. A shallow burial effect by sand was tested on four types of crust (with different moss species) collected from a revegetated area of the Tengger Desert in Northern China. Burial significantly decreased the respiration rate, and elongation of moss shoots was significantly increased. Both of these responses may have acted as compensatory mechanisms that favored recovery of crust after burial. The recovery of the four crusts occurred in the successional order of these crusts in their original area, suggesting that sand burial is a factor driving the succession of crusts in desert ecosystems (Jia et al. 2008).
3. Global Warming. Because biological soil crust organisms are only metabolically active when they are wet, and because soil surfaces dry quickly in deserts, the amount and timing of precipitation are likely to have significant impact on the physiological functioning of these communities. Various experiments have aimed to predict the effect of global warming. Photosynthetic performance, nitrogenase activity, and ability of crusts to maintain concentrations of radiation-protective pigments were diminished after exposure to increased precipitation frequency for 6 months, and over this period most crusts died out. In another test involving field conditions of high air temperatures and frequent, small precipitation events, crusts were unable to produce protective pigments, as reduced activity time probably resulted in less carbon being available to produce or repair chlorophyll and protective pigments (Belnap et al. 2004). Therefore, it is likely that crusts will show accelerated indirect damage in the future as the result of climate change.
4. Soil Crusts as Indicators of Functioning in Desert Ecosystems. A model that involves the use of soil crusts as indicators of the quality of the desert ecosystem function was recently developed. It is based on the ability of crusts to trap and retain soil and water resources and to function as a major carbon and nitrogen fixer (Bowker et al. 2008).

6.3.2 Arbuscular-Mycorrhizal (AM) Fungi in Arid Areas Apart from Resource Islands

AM fungi are common in harsh and limiting environments because they mitigate plant stress. Their hyphae permeate large volumes of soil, interconnect the root systems of adjacent plants to facilitate exchange of nutrients between them, and contribute to soil structure. AM fungi are an essential component of plant–soil systems of deserts and have been detected worldwide.

Despite the importance of AM fungi associated with the root system of plants, the distribution and activity of AM fungi in desert ecosystems are poorly known. Perennial plants in 19 families were surveyed for colonization by AM fungi at four sites in the Anza–Borrego Desert State Park, California. Collectively, the 38 plant species studied were colonized by six species of AM fungi. The distribution of the AM mycoflora was related to factors pertaining to the host plants and edaphic and climatic conditions at each site (Bethlenfalvay et al. 1984). In the northern part of the Baja California Peninsula, the mycorrhizal association with the endemic boojum tree, *Fouquieria columnaris*, covered the entire extent of the plant distribution. The roots of the boojum tree contained all the structures of AM fungi. Morphologically, 23 species of AM fungi were identified in close vicinity to the boojum tree, a dramatic example of a promiscuous association of several AM fungi genera and families with a plant host (Bashan et al. 2007). Detailed surveys of the AM association with perennial plants in disturbed and undisturbed areas in the southern Sonoran Desert in the State of Baja California Sur showed that 46 species of perennial plants had AM associations, but that the extent of root colonization varied widely. Roots of plants that were present in greater numbers in otherwise plant-free zones (pioneer plants of disturbed areas) had more AM fungi. Plants with trace amounts of AM fungi were rather established in association with nurse trees, and AM propagule densities in plant-free areas were lower than under plant canopies (Bethlenfalvay et al. 2007; Carrillo-Garcia et al. 1999).

The AM status of 73 springtime, ephemeral plant species from the desert ecosystem of the Junggar Basin in China was examined. AM fungal spores were isolated from rhizosphere samples of all 73 plant species and colonization rates ranged from 7% to 73%. The AM fungi belonged to six genera, *Acaulospora*, *Archaeopora*, *Entrophospora*, *Glomus*, *Paraglomus*, and *Scutellospora*, with *Glomus* as the dominant genus. It appears that spring ephemerals may be highly dependent on AM associations for survival in this infertile and arid ecosystem (Shi et al. 2006a). In the Negev Desert, higher spore density correlated with higher AM colonization of *Artemisia herba-alba* and *Atriplex halimus*. Spore density was positively correlated with vesicular colonization and negatively correlated with arbuscular colonization (He et al. 2002b). AM fungi were surveyed among the dominant annual and perennial plants of the Chihuahuan desert. Annuals were determined to have significantly less AM colonization than perennials, indicating that the dominant plant species fall into two categories: (1) short-lived annuals with thin roots and low AM dependency, and (2) longer-lived perennials with thick mycorrhizal roots (Collier et al. 2003). A survey of the Chilean fog-free, Pacific coastal desert, one of the driest desert regions of the world, examined endemic plant species and endangered and rare geophytes. More than 90% of 38 species (belonging to 19 different families) formed associations with AM fungi. Six species of mycorrhizal fungi, of which four were undescribed, were isolated from the root zones of sampled plants (Dhillion et al. 1995). A preliminary survey of AM fungi in four different habitats of the Cholistan Desert of Pakistan found that some plant species were mycorrhizal at one site but non-mycorrhizal at other sites (Chaudhry et al. 2005).

Mycorrhizal colonization apparently enhances water and nutrient uptake in dry environments for the succulent *Agave deserti* and the cacti *Ferocactus acanthodes* and *Opuntia ficus-indica*. Artificial inoculation of these plants with field-collected AM fungi increased the phosphorus content of roots and shoots compared with uninoculated plants. Lateral root hydraulic conductivity in *A. deserti* was significantly higher for inoculated plants (Cui and Nobel 1992).

6.3.3 Bacteria

Heterotrophic bacteria in arid lands have been studied for decades, but mostly in the context of agricultural crops. Bacteria in native desert habitats remain far less studied.

The desert varnish (or colorization of rocks) is a product of microbial activity. The varnish is composed of micro-organisms that concentrate manganese from their surroundings to produce manganese-rich films that eventually form brown-to-black coatings. These microbes are culturable and in the laboratory produce manganese-rich biofilms (Dorn and Oberlander 1981).

Few studies have examined the diversity of bacterial populations in deserts. The phylogenetic diversity of prokaryotic communities exposed to arid conditions in the hot Tataouine Desert in southern Tunisia was estimated. Bacteria in this hot desert are, as expected, tolerant to desiccation and a few strains are also tolerant to radiation (Chanal et al. 2006). The hyper-arid Atacama Desert of northern Chile is almost devoid of plant life and may harbor low or absent microbial populations, especially in the hyper-arid core where conditions for photosynthetic life and thus primary production reach their lower limits (Gómez-Silva et al. 2008; Maier et al. 2004). Samples from the Mars-like soils of this zone only have traces of organic matter and extremely low levels of culturable bacteria (Navarro-González et al. 2003). However, a later molecular analysis compared the structure of bacterial communities along a transect through sectors of this desert that varied from arid to extremely arid. The profiles from each of the samples revealed that microbial communities from the extreme hyper-arid core clustered separately from all of the remaining communities and were dominated by bacteria from the *Gemmatimonadetes* and *Planctomycetes* phyla (Drees et al. 2006). Furthermore, hypolithic cyanobacteria that colonized translucent stones were quantified along an aridity gradient in this desert. With increasing aridity, abundance of these cyanobacteria dropped significantly, molecular diversity declined threefold, and organic carbon residence time increased by three orders of magnitude. Together with heterotrophic associates, cyanobacteria belonging to the *Chroococcidiopsis* morphospecies formed specific communities, each stone supporting a number of unique genotypes. Warren-Rhodes et al. (2006) proposed that in the hyper-arid core, hypolithic cyanobacteria are rare and exist in small, spatially isolated patches amidst a microbially bare soil. Microhabitats are widely dispersed, difficult to detect, and are millimeters away from virtually lifeless surroundings.

Nitrogen fixation is a common event in most desert ecosystems, especially in association with leguminous plants or in bacterial populations residing on the surface and in the interior of roots of plants growing in rocks without soil (Puente et al. 2004a; Puente et al. 2009a; Lopez et al. unpublished). Diazotrophs also occur in epiphytic plants. The desert moss ball epiphyte *Tillandsia recurvata* colonizes many desert plants and even electrical wires. It grows in the foggy Pacific plains of the Baja California Peninsula and harbors the nitrogen-fixing bacterium *Pseudomonas stutzeri* (Puente and Bashan 1994).

A large variety of Gram-negative and Gram-positive bacteria were isolated and identified from the roots of *Lasiurus sindicus,* a perennial grass endemic to the Thar Desert of Rajasthan, India. The occurrence of *nifH* sequences in some isolates such as *Azospirillum* sp., *Rhizobium* sp., and *Pseudomonas pseudoalcaligenes* indicates that root-associated diazotrophs may supply biologically-fixed nitrogen to the host plant (Chowdhury et al. 2007).

Whereas various diazotrophs contribute nitrogen to desert soils, other species of soil bacteria may be responsible for nitrogen loss from the ecosystem after infrequent major rainfalls. Denitrification was directly measured in the Sonoran Desert dominated by the mesquite *Prosopis glandulosa* after artificial wetting, equivalent to a major rain, under the canopy of the tree and in the area devoid of vegetation. Denitrification rates were 50-fold higher under the mesquite, probably as a result of high available organic carbon under the tree and other chemical and physical changes produced by the mesquite in its resource island (Virginia et al. 1982).

The effect of soil moisture on the contribution of bacteria and fungi to the soil microbial biomass under scattered rainfall events in the Negev Desert was measured. Wetting of the dry desert soil triggered a rapid outburst of microbial activity, which resulted in a significant increase in soil microbial biomass. In the top 0–10 cm soil layer, fungal biomass was significantly higher than bacterial biomass. In the 10–20 cm soil layer, bacterial biomass significantly exceeded fungal biomass. These phenomena may suggest that the detritus food web composition differs along the desert soil profile (Vishnevetsky and Steinberger 1997).

Alkaline desert soils are frequently high in insoluble calcium phosphates, but deficient in soluble orthophosphates, a compound that is essential for plant growth. Many phosphate-solubilizing bacteria were identified on the surface and interiors of the roots of plants growing on rocks without soil in the Baja California Peninsula (Puente et al. 2004a; Puente et al. 2009a; Lopez et al. unpublished). A unique phosphate-solubilizing bacterial population was isolated from the roots of *Helianthus annuus* ssp. *jaegeri* growing at the edge of an alkaline dry lake in the Mojave Desert. A bacterium from this population produced high levels of gluconic acid, a compound involved in phosphate solubilization, but only in the presence of undefined root substrates washed from plant roots (Goldstein et al. 1999).

6.4 Restoration of Desert Environments

During desertification, physical, chemical, and biological soil damage is known to occur simultaneously with disturbance of plant communities. Such soil degradation limits reestablishment of the natural plant cover. Specifically, desertification causes disturbance of plant–microbe interactions that are a fundamental ecological factor in restoration of degraded ecosystems. Natural revegetation in deforested deserts is extremely slow. Reforestation programs are one of the proposed solutions to combat encroaching deserts. Some projects, such as the tree-belt of northeastern China, are notable for their very large scale (Moore and Russell 1990). Many times, restoration to "original" conditions is considered impossible and the only alternatives are rehabilitation or construction of a new ecosystem without considering the previous ecosystem (Aronson et al. 1993).

6.4.1 Restoration of Arid Lands without Intentional Use of Micro-Organisms

Revegetation of arid lands has been done routinely on a small scale for years without attention to the micro-organisms involved, assuming that these will appear once the trees and shrubs are established. Two modes of operation are chosen: (1) natural recovery of vegetation and plant succession (Bolling and Walker 2000; Burke 2001; Castro et al. 2002), and (2) planting seedlings or applying seeds to empty lands.

One of the most important ways of combatting desertification is to stabilize sandy land and facilitate natural revegetation. It is commonly believed that continually drifting sand from strong winds is the limiting factor for plant survival. Supposedly, sand barriers would effectively control drifts and native plant species could gradually colonize shifting dunes, and hence revegetation would be facilitated. A study was conducted on shifting sand dunes in a cold desert on the Tibetan Plateau to explore the relationships between mechanical sand control, vegetation restoration, and seed plant diversity dynamics. In this case, mechanical sand stabilization facilitated revegetation by local plants, resulting in near total vegetation cover with *Leymus secalinus*, a dominant local species (Yang et al. 2006). In large desert areas in the southwestern USA that have been severely impacted by military activities or are abandoned farmlands, revegetation efforts commonly seek to optimize growing conditions by increasing irrigation and mulching to enhance moisture retention (Roundy et al. 2001; McDonald 2002). In a desert section of the Nile Valley, 32 multi-purpose tree species used to generate wood biomass and nutritious leaves were screened for planting under irrigated conditions (Stewart et al. 1993). Reseeding of degraded areas, although effective in moister climates, is ineffective in desert areas. For example, a remediation program aiming to remove surface contamination at an abandoned uranium mill in the Great Basin Desert of

Arizona failed to produce reasonable revegetation 10 years after reseeding (Lash et al. 1999; Glenn et al. 2001).

As explained earlier, early survival and growth of some plants in arid environments depend on the formation of a resource island by a nurse plant and associated organisms. Reducing high soil temperature through shading and accumulation of mineral nutrients near nurse plants are facilitation mechanisms. In a study of the effects of shading and soil type on survival and growth of the giant cardon cactus (*Pachycereus pringlei*) of the Baja California Peninsula, cardon were grown in a resource island soil or in soil from bare areas that surround the resource islands. Survival and growth in the resource island soils were significantly higher than in soils from bare areas, and were further enhanced by shading. In this study, root/shoot ratios of plants grown in the bare area soil indicated increased resource allocation to roots under limiting conditions (Carrillo-Garcia et al. 2000a). Because resource island soils are in short supply in the arid areas of the southern Baja California Peninsula, a substitute was sought. Since the benefits of this soil come mainly from its higher organic matter and nutrient content, common organic compost was tested as a supplement. Soil that otherwise could not support perennial plant growth was amended with various quantities of common agricultural compost. Cardon seedlings were grown for 18 months under nursery conditions. During this prolonged period, the decisive factor for seedling development in barren soil was the addition of small amounts of common compost. Compost added to barren soil increased the dry weight parameters of the plant to levels almost similar to those obtained in resource island soil; furthermore, as compared to resource island soil, the compost amendment supported a more voluminous and greener plant with higher pigment levels (Bacilio et al. 2006).

6.4.2 Restoration of Soil Microbial Communities in Disturbed Arid Lands

6.4.2.1 Restoration of Biological Soil Crusts

Crusts are rarely addressed in the restoration literature. Rehabilitation of crusts is attainable and may be required for the recovery of some desert ecosystems. Full recovery of crusts after disturbance is always a very slow process, especially for the moss and lichen components. A plausible shortcut might be to first achieve recovery of the cyanobacteria and green algae. This will give the initial appearance of a healthy and functional crust, even though these crusts would be less effective as restorers of desert fertility. This initial recovery might take only a few years. However, a crust may require up to 50 years to regain its original thickness and up to 250 years for mosses and lichens to reappear and flourish. Another factor that facilitates recovery is to limit the size of the disturbed area, provided that there is a nearby source of inoculum. Because crusts are ecosystem "engineers" in abiotic, high stress systems, their loss may be synonymous with crossing degradation thresholds. Conversely, the assisted recovery of crusts may provide a transition from a degraded steady state to a more desired alternative steady state (Bowker 2007).

Some attempts at restoration of crusts have been done in China. As a preliminary step for restoration, observations of formation of soil crusts in the Gurbantunggut desert showed that crusts serve as one of the biological factors contributing to stability of sand. One year after removing the local crusts, the exposed sandy surface could be fixed by heterotrophic bacteria adhering to the sand particles with exopolysaccharides. After 4 years, the crusts on the exposed sandy surface were mainly composed of communities of filamentous cyanobacteria dominated by *Microcoleus* sp., which occurs as a cluster of filaments surrounded by a gelatinous sheath. Microscopic examination at this time revealed an intricate network of filamentous cyanobacteria and extracellular polymer secretions, which bound and entrapped mineral particles and finer particles on the filament surface, creating a stable crust (Zhang 2005). The next step in creating crusts is to restore the plant component because the establishment of moss-dominated crusts is crucial for ecological restoration. A micropropagation technology has been developed using the desert moss *Tortula desertorum* which is the main component of crusts in the Gurbantunggut Desert. The explants cultivated with in situ soil produced more protonema and shoots than those cultivated on synthetic growth media. Based on the life cycle of *T. desertorum* under cultivation, a procedure for artificial reconstruction of moss-dominated soil crusts was developed. A large number of protonema were induced to grow by breeding detached green leaves in a specific medium. These protonema were harvested and moss-dominated crusts formed through transplanting protonema into sand that was supplemented with the same growth medium (Xu et al. 2008). Finally, a successful example of restoration of cryptogam species diversity in the Tengger Desert highlights the importance of long-term action. Crust formation occurred after 44 years of revegetation of sand dunes stabilized with straw in a checkerboard pattern. Apparently, revegetation accelerated the improvement of environmental conditions leading to establishment and propagation of cryptogams on the sand dunes (Li et al. 2003).

6.4.2.2 Restoration Using PGPB and AM Fungi

PGPB are well known in agriculture, but only in recent years has their value been demonstrated for restoration of vegetation in abandoned, dryland habitats (Bashan et al. 2005; 2008). AM fungi have been recognized for a long time as contributors to desert plant growth and primary succession (Allen 1989; Peter 2003), but the information available is meager. Artificial inoculation with PGPB was done mainly with cacti and to a lesser extent with several legume trees.

Cardon cacti have a widespread, finely branched, subsurface root system which stabilizes desert soil. They were inoculated with strains of the PGPB *Azospirillum brasilense* of agricultural origin. Differential germination responses were observed, as *A. brasilense* Cd significantly reduced seed germination, whereas *A. brasilense* Sp245 significantly increased germination of seedlings at the same inoculum concentration. Both strains improved several plant-growth parameters of cardon related to seedling survivability such as height, diameter, volume, the volume/surface ratio, and persistence of seedling spines. Both strains survived in the cactus rhizosphere for about 300 days after seed inoculation (Puente and Bashan 1993). The effect of *A. brasilense* increased linearly as soil nutrients declined. In the best resource island soil, *A. brasilense* had no effect on cardon growth, but in the soil from bare areas, shoot dry mass increased by almost 60% and root length by more than 100% as a result of inoculation (Carrillo-Garcia et al. 2000b). A mechanism was proposed to explain how this plant–bacterium association increased growth of seedlings in poor desert soils, largely based on acidification of the rhizosphere of cactus seedlings after inoculation, and a consequent increase in phosphorus availability (Carrillo et al. 2002). Similarly, when used in conjunction with compost, *A. brasilense* Cd had little effect on cardon growth (Bacilio et al. 2006).

Studies using PGPB obtained from roots of several species of cacti growing in rocks where soil is absent open another avenue for desert restoration. When cardon seeds were inoculated with several rhizoplane or endophytic PGPB strains capable of weathering rocks, the seedlings grew in pulverized rock for at least a year without fertilization and without showing distress. Cacti that were not inoculated grew less vigorously and several died. Both types of bacteria–plant associations released significant amounts of nutrients from the pulverized rock substrate. When endophytic bacteria were eliminated from the seeds using antibiotics, the development of seedlings stopped. Upon reinoculation with the endophytes, plant growth was restored. Similar results occurred with endophytic bacteria and the small cactus *Mammillaria fraileana* (Puente et al. 2004b; Puente et al. 2009a,b; Lopez et al. unpublished).

The effects of PGPB inoculation was demonstrated on a number of desert plants under greenhouse conditions. The mesquite amargo, *Prosopis articulata*, one of the main nurse trees of the Sonoran Desert in northwestern Mexico, produces root exudates containing gluconic acid, a favorite carbon source for the PGPB *A. brasilense*. Two enzymes participating in the phosphogluconate pathway, gluconokinase and 6-phosphogluconate dehydrogenase, are active in this bacterium and also in young, uninoculated mesquite seedlings growing under hydroponic conditions. Upon inoculation of the root system with *A. brasilense* Cd, the roots showed much higher activity of gluconokinase, but not 6-phosphogluconate dehydrogenase. Mesquite roots exhibited high root colonization by the bacteria, and the seedlings grew taller, were greener, had longer leaves and were heavier than uninoculated plants (Leyva and Bashan 2008).

Three slow-growing legume trees used for desert reforestation and urban gardening in the Sonoran Desert in northwestern Mexico and the southwestern USA were evaluated for their response to treatment with PGPB, AM fungi, and common compost under regular screen-house cultivation. Mesquite amargo and the yellow palo verde *Parkinsonia microphylla* showed positive responses to several of the treatments, while the blue palo verde *P. florida* did not respond. When these trees were cultivated without water restrictions, inoculation with several growth-promoting micro-organisms induced significant effects on the gas exchange in the leaves, measured as transpiration and diffusive resistance (Bashan et al. 2009a).

So far, there have been few attempts to move beyond greenhouse studies and test PGPB in the field. Survival and development of cactus transplants in urban, disturbed areas in the State of Baja California Sur was monitored. Young plants of three species of pachycereid cacti (*Pachycereus pringlei, Stenocereus thurberi,* and *Lophocereus schottii*) inoculated with the PGPB *A. brasilense* had a high rate of survival and developed more rapidly, compared to uninoculated control plants over a 3.5-year period. Soil erosion in the inoculated experimental area diminished. Small, but significant amounts of soil accumulated in association with the growth of cactus roots in the form of wind-deposited dust. One demonstrated mechanism for stabilizing the dust was by upward growth of small roots during the rainy season. *A. brasilense* survived well in the rhizosphere of these cacti for 2 years, but not in root-free soil (Bashan et al. 1999). In two long-term experiments in a desertified Mediterranean ecosystem, inoculation with indigenous AM fungi and rhizobia not only enhanced establishment of key plant species, but also increased soil fertility and quality. The dual symbioses increased soil nitrogen and organic matter content, stabilized soil aggregates, and enhanced nitrogen transfer from nitrogen-fixing to nonfixing species associated within the natural succession (Herrera et al. 1993; Requena et al. 2001).

Reforestation of highly eroded desert land in the southern Sonoran Desert, after it had lost its natural capacity for self-revegetation, was attempted with three native legume trees (mesquite amargo, yellow palo verde, and blue palo verde) in seven field trials. Reforestation was aided by inoculation with two species of PGPB, *A. brasilense* and *Paenibacillus chitinolyticus,* native mycorrhizal fungi, small amounts of compost, and limited irrigation. Survival of the trees was marginally affected by the various treatments after 30 months and was in the range of 60%–90%, depending on the plant species. With respect to growth, mesquite amargo and yellow palo verde responded positively to inoculation with PGPB, AM fungi, and supplements of compost, whereas blue palo verde did not respond to most treatments. Some combinations of tree/inoculant/amendment resulted in small negative effects or no response when evaluated over extended periods of time (Bashan et al. 2009b).

To evaluate the feasibility of long-term desert reforestation technology using mixed vegetation, cardon cactus seedlings from indoor and outdoor nurseries were planted in the field adjacent to one seedling of various potential legume nurse trees: mesquite amargo, yellow palo verde, and blue palo verde. Additionally, the combinations of legume tree and cactus were inoculated with either a consortium of desert AM fungi, PGPB (the diazotroph *A. brasilense* Cd, and the phosphate solubilizer *P. chitinolyticus*), or a mixture of all. Some of the planting holes were also supplemented with common dairy cow compost. The field trials were observed periodically over 30 months for survival and growth. Cardon cactus that had been reared in an outdoor screen-house survived better once in the field than cardon that was reared in a controlled growth chamber and later hardened outdoors. Association with any legume nurse tree increased survival and enhanced growth of untreated cardon. For cardon growing alone, application of compost, AM fungi, or all the treatments combined increased survival. Assessment after 30 months of cultivation showed that all microbial treatments positively affected cardon growth when growing alone or in combination with mesquite (Bashan et al. 2009b).

These studies demonstrate that reforestation of severely eroded desert lands is possible with native legume trees aided by microbial agents and compost to increase soil fertility. Introduction of indigenous plants associated with a managed microbial community should prove a successful biotechnological approach to aid recovery of desertified ecosystems and degraded deserts.

6.5 Concluding Remarks

As a result of expanding human population, encroachments into deserts, and military activities, degradation of arid areas seems inevitable. Water deficiency delays natural rehabilitation of arid lands over a timescale that is exceedingly long compared to a human life. This enhances desert vulnerability and makes man-made rehabilitation and restoration necessary to allow for recovery over a reasonable time span (Bainbridge 2007).

Two natural floral–microbial phenomena dominate deserts: (a) formation of resource islands, where a single tree or a few shrubs modify the land and accumulate enough plant resources under their canopy to allow other plant species to establish, and (b) formation of biological soil crusts in the open areas between the patches of vegetation. Both phenomena have an integral microbial component upon which the ecosystem depends for its function and even its creation. Knowledge of these microbial components of the ecosystems, although expanding, is small compared to what is known about agricultural lands. Nonetheless, as rehabilitation and restoration of arid lands is paramount, it is imperative that the microbial components of these arid ecosystems be studied and used for the benefit of people living in deserts or otherwise affected by the negative effects of desertification.

References

Abu-Awwad AM (1997) Water infiltration and redistribution within soils affected by a surface crust. J Arid Environ 37:231–242

Aguilera LE, Gutiérrez JR, Meserve PL (1999) Variation in soil microorganisms and nutrients underneath and outside the canopy of *Adesmia bedwellii* (Papilionaceae) shrubs in arid coastal Chile following drought and above average rainfall. J Arid Environ 42:61–70

Allen EB (1989) The restoration of disturbed arid landscapes with special reference to mycorrhizal fungi. J Arid Environ 17:279–286

Aronson J, Floret C, LeFloch E, Ovalle C, Pontanier R (1993) Restoration and rehabilitation of degraded ecosystems in arid and semi-arid lands. I. A view from the south. Restor Ecol 1:8–17
Azcón-Aguilar C, Palenzuela J, Roldán A, Bautista S, Vallejo R, Barea JM (2003) Analysis of the mycorrhizal potential in the rhizosphere of representative plant species from desertification-threatened Mediterranean shrublands. Appl Soil Ecol 22:29–37
Bacilio M, Hernandez J-P, Bashan Y (2006) Restoration of giant cardon cacti in barren desert soil amended with common compost and inoculated with *Azospirillum brasilense*. Biol Fertil Soils 43:112–119
Bainbridge DA (2007) A guide for desert and dryland restoration: new hope for arid lands. Island Press, Washington DC, p 391
Bashan Y, Rojas A, Puente ME (1999) Improved establishment and development of three cacti species inoculated with *Azospirillum brasilense* transplanted into disturbed urban desert soil. Can J Microbiol 45:441–451
Bashan Y, Davis EA, Carrillo-Garcia A, Linderman RG (2000) Assessment of VA mycorrhizal inoculum potential in relation to the establishment of cactus seedlings under mesquite nurse-trees in the Sonoran desert. Appl Soil Ecol 14:165–175
Bashan Y, Li CY, Lebsky VK, Moreno M, de-Bashan LE (2002) Primary colonization of volcanic rocks by plants in arid Baja California, Mexico. Plant Biol 4:392–402
Bashan Y, Puente ME, Salazar B, de-Bashan LE, Bacilio M, Hernandez J-P, Leyva LA, Romero B, Villalpando R, Bethlenfalvay GJ (2005) Reforestation of eroded lands in the desert: the role of plant growth-promoting bacteria and organic matter. Suelos Ecuatoriales 35:70–77 In Spanish
Bashan Y, Vierheilig H, Salazar BG, de-Bashan LE (2006) Primary colonization and breakdown of igneous rocks by endemic, succulent elephant trees (*Pachycormus discolor*) of the deserts in Baja California, Mexico. Naturwissenschaften 93:344–347
Bashan Y et al (2007) Mycorrhizal characterization of the boojum tree, *Fouquieria columnaris*, an endemic ancient tree from the Baja California Peninsula, Mexico. Trees Struct Funct 21:329–335
Bashan Y, Puente ME, de-Bashan LE, Hernandez J-P (2008) Environmental uses of plant growth-promoting bacteria. In: Ait Barka E, Clément C (eds) Plant-microbe interactions. Research Signpost, Trivandrum, pp 69–93
Bashan Y, Salazar B, Puente ME (2009a) Responses of native, legume desert trees used for reforestation in the Sonoran Desert to plant growth-promoting microorganisms in screen house. Biol Fertil Soils 45:655–662
Bashan Y, Salazar B, Puente ME, Bacilio M, Linderman RG (2009b) Enhanced establishment and growth of giant cardon cactus in an eroded field in the Sonoran Desert using native legume trees as nurse plants aided by plant growth-promoting microorganisms and compost. Biol Fertil Soils 45:585–594
Bechtold HA, Inouye RS (2007) Distribution of carbon and nitrogen in sagebrush steppe after six years of nitrogen addition and shrub removal. J Arid Environ 71:122–132
Belnap J (1995) Surface disturbances: their role in accelerating desertification. Environ Monit Assess 37:39–57
Belnap J (2003) The world at your feet: desert biological soil crusts. Front Ecol Environ 1:181–189
Belnap J, Gillette DA (1998) Vulnerability of desert biological soil crusts to wind erosion: the influences of crust development, soil texture, and disturbance. J Arid Environ 39:133–142
Belnap J, Lange OL (eds) (2001) Biological soil crusts: structure, function, and management. Springer, Berlin
Belnap J, Kaltenecker JH, Rosentreter R, Williams J, Leonard S, Eldridge D (2001) Biological soil crusts: ecology and management. Technical Reference 1730-2, United States Department of the Interior, Bureau of Land Management, National Science and Technology Center, Denver, CO
Belnap J, Phillips SL, Miller ME (2004) Response of desert biological soil crusts to alterations in precipitation frequency. Oecologia 141:306–316
Bethlenfalvay GJ, Dakessian S, Pacovsky RS (1984) Mycorrhizae in a southern California desert: ecological implications. Can J Bot 62:519–524

Bethlenfalvay GJ, Bashan Y, Carrillo-Garcia A, Stutz J (2007) Mycorrhizae as biological components of resource islands in the Sonoran desert. In: Montaño NM, Camargo-Ricalde SL, García-Sánchez R, Monroy-Ata A (eds) Micorrizas arbusculares en ecosistemas áridos y semiáridos. Mundi Prensa the National Institute of Ecology-SEMARNAT and the Autonomous Metropolitan University, Iztapalapa, Mexico, pp 147–169

Bolling JD, Walker LR (2000) Plant and soil recovery along a series of abandoned desert roads. J Arid Environ 46:1–24

Bowker MA (2007) Biological soil crust rehabilitation in theory and practice: an underexploited opportunity. Restor Ecol 15:13–23

Bowker MA, Reed SC, Belnap J, Phillips SL (2002) Temporal variation in community composition, pigmentation, and F_v/F_m of desert cyanobacterial soil crusts. Microb Ecol 43:13–25

Bowker MA, Miller ME, Belnap J, Sisk TD, Johnson NC (2008) Prioritizing conservation effort through the use of biological soil crusts as ecosystem function indicators in an arid region. Conserv Biol 22:1533–1543

Burke A (2001) Determining landscape function and ecosystem dynamics: contribution to ecological restoration in the southern Namib Desert. AMBIO 30:29–36

Burquez A, Quintana MA (1994) Islands of diversity: ironwood ecology and the richness of perennials in a Sonoran Desert biological preserve. In: Nabhan GP, Carr JL (eds) Ironwood: An Ecological and Cultural Keystone on the Sonoran Desert. Conservation International, Washington, DC, pp 9–27

Cable JM, Huxman TE (2004) Precipitation pulse size effects on Sonoran Desert soil microbial crusts. Oecologia 141:317–324

Carrillo AE, Li CY, Bashan Y (2002) Increased acidification in the rhizosphere of cactus seedlings induced by *Azospirillum brasilense*. Naturwissenschaften 89:428–432

Carrillo-Garcia A, Leon de la Luz J-L, Bashan Y, Bethlenfalvay GJ (1999) Nurse plants, mycorrhizae, and plant establishment in a disturbed area of the Sonoran Desert. Restor Ecol 7:321–335

Carrillo-Garcia A, Bashan Y, Bethlenfalvay GJ (2000a) Resource-island soils and the survival of the giant cactus, cardon, of Baja California Sur. Plant Soil 218:207–214

Carrillo-Garcia A, Bashan Y, Diaz-Rivera E, Bethlenfalvay GJ (2000b) Effects of resource-island soils, competition, and inoculation with *Azospirillum* on survival and growth of *Pachycereus pringlei*, the giant cactus of the Sonoran Desert. Restor Ecol 8:65–73

Castro J, Zamora R, Hódar JA, Gómez JM (2002) Use of shrubs as nurse plants: a new technique for reforestation in Mediterranean mountains. Restor Ecol 10:297–305

Chanal A et al (2006) The desert of Tataouine: an extreme environment that hosts a wide diversity of microorganisms and radiotolerant bacteria. Environ Microbiol 8:14–525

Chaudhry MS, Batool Z, Khan AG (2005) Preliminary assessment of plant community structure and arbuscular mycorrhizas in rangeland habitats of Cholistan Desert, Pakistan. Mycorrhiza 15:606–611

Chowdhury SP, Schmid M, Hartmann A, Tripathi AK (2007) Identification of diazotrophs in the culturable bacterial community associated with roots of *Lasiurus sindicus*, a perennial grass of Thar Desert, India. Microb Ecol 54:82–90

Collier SC, Yarnes CT, Herman RP (2003) Mycorrhizal dependency of Chihuahuan Desert plants is influenced by life history strategy and root morphology. J Arid Environ 55:223–229

Cui M, Nobel PS (1992) Nutrient status, water uptake and gas exchange for three desert succulents infected with mycorrhizal fungi. New Phytol 122:643–649

Dhillion SS, Vidiella PE, Aquilera LE, Friese CF, De Leon E, Armesto JJ, Zak JC (1995) Mycorrhizal plants and fungi in the fog-free Pacific coastal desert of Chile. Mycorrhiza 5:381–386

Dorn RI, Oberlander TM (1981) Microbial origin of desert varnish. Science 213:1245–1247

Drees KP et al (2006) Bacterial community structure in the hyperarid core of the Atacama Desert, Chile. Appl Environ Microbiol 72:7902–7908

Flores J, Jurado E (2003) Are nurse-protege interactions more common among plants from arid environments? J Veg Sci 14:911–916

Garner W, Steinberger Y (1989) A proposed mechanism for the formation of 'Fertile Islands' in the desert ecosystem. J Arid Environ 16:257–262

Gillis AM (1994) Crustful of nutrients. Bioscience 44:735–736

Glenn EP, Waugh WJ, Moore D, McKeon C, Nelson SG (2001) Revegetation of an abandoned uranium millsite on the Colorado plateau, Arizona. J Environ Qual 30:1154–1162

Goldstein AH, Braverman K, Osorio N (1999) Evidence for mutualism between a plant growing in a phosphate-limited desert environment and a mineral phosphate solubilizing (MPS) rhizobacterium. FEMS Microbiol Ecol 30:295–300

Gómez-Silva B, Rainey FA, Warren-Rhodes KA, McKay CP, Navarro-González R (2008) Atacama Desert soil microbiology. In: Dion P, Nautiyal CS (eds) Microbiology of extreme soils. Springer, Berlin, pp 117–132

González-Ruiz T, Rodríguez-Zaragoza S, Ferrera Cerrato R (2008) Fertility islands around *Prosopis laevigata* and *Pachycereus hollianus* in the drylands of Zapotitlán Salinas, México. J Arid Environ 72:1202–1212

Green LE, Porras-Alfaro A, Sinsabaugh RL (2008) Translocation of nitrogen and carbon integrates biotic crust and grass production in desert grassland. J Ecol 96:1076–1085

Gutiérrez JR, Meserve PL, Contreas LC, Vásquez H, Jaksic FM (1993) Spatial distribution of soil nutrients and ephemeral plants underneath and outside the canopy of *Porlieria chilensis* shrubs (Zygophyllaceae) in arid coastal Chile. Oecologia 95:347–352

Harper KT, Belnap J (2001) The influence of biological soil crusts on mineral uptake by associated vascular plants. J Arid Environ 47:347–357

He X, Mouratov S, Steinberger Y (2002a) Temporal and spatial dynamics of vesicular–arbuscular mycorrhizal fungi under the canopy of *Zygophyllum dumosum* Boiss. in the Negev Desert. J Arid Environ 52:379–387

He X, Mouratov S, Steinberger Y (2002b) Spatial distribution and colonization of arbuscular mycorrhizal fungi under the canopies of desert halophytes. Arid Land Res Manag 16:149–160

Herman RP, Provencio KR, Herrera-Matos J, Torrez RJ (1995) Resource islands predict the distribution of heterotrophic bacteria in Chihuahuan Desert soils. Appl Environ Microbiol 61:1816–1821

Herrera MA, Salamanca CP, Barea JM (1993) Inoculation of woody legumes with selected arbuscular mycorrhizal fungi and rhizobia to recover desertified Mediterranean ecosystems. Appl Environ Microbiol 59:129–133

Jia RL, Li XR, Liu LC, Gao YH, Li XJ (2008) Responses of biological soil crusts to sand burial in a revegetated area of the Tengger desert, Northern China. Soil Biol Biochem 40:2827–2834

Lalley JS, Viles HA, Henschel JR, Lalley V (2006) Lichen-dominated soil crusts as arthropod habitat in warm deserts. J Arid Environ 67:579–593

Lange OL, Kidron GJ, Büdel B, Meyer A, Kilian E, Abeliovich A (1992) Taxonomic composition and photosynthetic characteristics of the biological soil crusts covering sand dunes in the western Negev Desert. Funct Ecol 6:519–527

Lange OL, Meyer A, Buedel B (1994) Net photosynthesis activation of a desiccated cyanobacterium without liquid water in high air humidity alone: experiments with *Microcoleus sociatus* isolated from a desert soil crust. Funct Ecol 8:52–57

Lash DW, Glenn EP, Waugh WJ, Baumgartner DJ (1999) Effects of grazing exclusion and reseeding on a former uranium mill site in the Great Basin Desert, Arizona. Arid Soil Res Rehabil 13:253–264

Leyva LA, Bashan Y (2008) Activity of two catabolic enzymes of the phosphogluconate pathway in mesquite roots inoculated with *Azospirillum brasilense* Cd. Plant Physiol Biochem 46:898–904

Li X-R, Zhou H-Y, Wang X-P, Zhu Y-G, O'Conner PJ (2003) The effects of sand stabilization and revegetation on cryptogam species diversity and soil fertility in the Tengger Desert, Northern China. Plant Soil 251:237–245

Lopez BR, Bashan Y, Bacilio M, De la Cruz-Agüero G (2009) Rock-colonizing plants: abundance of the endemic cactus *Mammillaria fraileana* related to rock type in the southern Sonoran Desert. Plant Ecol 200:575–588

Maier RM et al (2004) Microbial life in the Atacama desert. Science 306:1289–1290
McDonald EV (2002) Numerical simulations of soil water balance in support of revegetation of damaged military lands in arid regions. Arid Land Res Manag 16:277–290
Moore R, Russell R (1990) The 'Three Norths' forest protection system – China. Agroforest Syst 10:71–88
Nabhan GP, Suzan H (1994) Boundary effects on endangered cacti and their nurse plants in and near a Sonoran desert biosphere reserve. In: Nabhan GP, Carr JL (eds) Ironwood: An ecological and cultural keystone on the Sonoran Desert. Conservation International, Washington, DC, pp 55–68
Nagy ML, Pérez A, Garcia-Pichel F (2005) The prokaryotic diversity of biological soil crusts in the Sonoran Desert (Organ Pipe Cactus National Monument, AZ). FEMS Microbiol Ecol 54:233–245
Navarro-González R et al (2003) Mars-like soils in the Atacama Desert, Chile, and the dry limit of microbial life. Science 302:1018–1021
Peter M (2003) Volcanic deserts and primary succession: when and how do mycorrhizal fungi participate? New Phytol 159:534–536
Puente M-E, Bashan Y (1993) Effect of inoculation with *Azospirillum brasilense* strains on the germination and seedlings growth of the giant columnar Cardon cactus (*Pachycereus pringlei*). Symbiosis 15:49–60
Puente M-E, Bashan Y (1994) The desert epiphyte *Tillandsia recurvata* harbors the nitrogen-fixing bacterium *Pseudomonas stutzeri*. Can J Bot 72:406–408
Puente ME, Bashan Y, Li CY, Lebsky VK (2004a) Microbial populations and activities in the rhizoplane of rock-weathering desert plants. I. Root colonization and weathering of igneous rocks. Plant Biol 6:629–642
Puente ME, Li CY, Bashan Y (2004b) Microbial populations and activities in the rhizoplane of rock-weathering desert plants. II. Growth promotion of cactus seedlings. Plant Biol 6:643–650
Puente ME, Li CY, Bashan Y (2009a) Rock-degrading endophytic bacteria in cacti. Environ Expt Bot 66:389–401
Puente ME, Li CY, Bashan Y (2009b) Endophytic bacteria in cacti seeds can improve the development of cactus seedlings. Environ Expt Bot 66:402–408
Read CF, Duncan DH, Vesk PA, Elith J (2008) Biological soil crust distribution is related to patterns of fragmentation and land use in a dryland agricultural landscape of southern Australia. Landscape Ecol 23:1093–1105
Requena N, Perez-Solis E, Azcón-Aguilar C, Jeffries P, Barea J-M (2001) Management of indigenous plant–microbe symbioses aids restoration of desertified ecosystems. Appl Environ Microbiol 67:495–498
Rodríguez-Zaragoza S, González-Ruíz T, González-Lozano E, Lozada-Rojas A, Mayzlish-Gati E, Steinberger Y (2008) Vertical distribution of microbial communities under the canopy of two legume bushes in the Tehuacán Desert, Mexico. Eur J Soil Biol 44:373–380
Rosentreter R, Bowker M, Belnap J (2007) A field guide to biological soil crusts of Western U.S drylands. U.S. Government Printing Office, Denver, CO
Rossi BE, Villagra PE (2003) Effects of *Prosopis flexuosa* on soil properties and the spatial pattern of understorey species in arid Argentina. J Veg Sci 14:543–550
Roundy BA, Heydari H, Watson C, Smith SE, Munda B, Pater M (2001) Summer establishment of Sonoran desert species for revegetation of abandoned farmland using line source sprinkler irrigation. Arid Land Res Manag 15:23–39
Schade JD, Hobbie SE (2005) Spatial and temporal variation in islands of fertility in the Sonoran Desert. Biogeochemistry 73:541–553
Schlesinger WH, Pilmanis AM (1998) Plant–soil interactions in deserts. Biogeochemistry 42:169–187
Schlesinger WH, Raikes JA, Hartley AE, Cross AF (1996) On the spatial pattern of soil nutrients in desert ecosystems. Ecology 77:364–374
Shi ZY, Feng G, Christie P, Li XL (2006a) Arbuscular mycorrhizal status of spring ephemerals in the desert ecosystem of Junggar Basin, China. Mycorrhiza 16:269–275

Shi Z, Zhang L, Feng G, Christie P, Tian C, Li X (2006b) Diversity of arbuscular mycorrhizal fungi associated with desert ephemerals growing under and beyond the canopies of tamarisk shrubs. Chinese Sci Bull 51(Suppl 1):132–139

Stewart JL, Dunsdon AJ, Kishk MA, Aly MK (1993) Screening of non-industrial tree species on reclaimed desert land in Middle Egypt. J Arid Environ 25:375–385

Suzan H, Nabhan GP, Patten DT (1994) Nurse plants and floral biology of a rare night-blooming Cereus, *Peniocereus striatus* (Brandagee) F. Buxbaum. Conserv Biol 8:461–470

Suzan H, Nabhan GP, Patten DT (1996a) The importance of *Olneya tesota* as a nurse plant in the Sonoran desert. J Veg Sci 7:635–644

Suzan H, Malda G, Patten DT, Nabhan GP (1996b) Effects of exploitation and park boundaries on legume trees in the Sonoran desert. Conserv Biol 13:1497–1501

Tewksbury J, Petrovich CA (1994) The influence of ironwood as a habitat-modifier species: a case study on the Sonoran desert coast of the Sea of Cortes. In: Nabhan GP, Carr JL (eds) Ironwood: An ecological and cultural keystone on the Sonoran Desert. Conservation International, Washington, DC, pp 29–54

Titus JH, Nowak RS, Smith SD (2002) Soil resource heterogeneity in the Mojave Desert. J Arid Environ 52:269–292

Virginia RA, Jarrell WM, Franco-Vizcaino E (1982) Direct measurement of denitrification in a *Prosopis* (Mesquite) dominated Sonoran Desert ecosystem. Oecologia 53:120–122

Vishnevetsky S, Steinberger Y (1997) Bacterial and fungal dynamics and their contribution to microbial biomass in desert soil. J Arid Environ 37:83–90

Walker LR, Thompson DB, Landau FH (2001) Experimental manipulations of fertile islands and nurse plant effects in the Mojave Desert, USA. West N Am Nat 61:25–35

Warren-Rhodes KA et al (2006) Hypolithic cyanobacteria, dry limit of photosynthesis, and microbial ecology in the hyperarid Atacama Desert. Microb Ecol 52:389–398

West P, Nabhan GP, Suzan H, Monti L (2000) Ironwood diversity study. In: Nabhan GP, Behan M (eds) Desert ironwood primer. Arizona–Sonora Desert Museum, Tucson, AZ, pp 46–61

Xu S, Yin C, He M, Wang Y (2008) A technology for rapid reconstruction of moss-dominated soil crusts. Environ Eng Sci 25:1129–1138

Yang H, Lu Q, Wu B, Yang H, Zhang J, Lin Y (2006) Vegetation diversity and its application in sandy desert revegetation on Tibetan Plateau. J Arid Environ 65:619–631

Yeager CM, Kornosky JL, Grote EE, Housman DC, Belnap J, Kuske CR (2003) Characterization of the diazotrophic community associated with desert soil crusts from Utah, USA. In: Abstracts of the general meeting of the American Society for Microbiology, p N-018

Zaady E, Gutterman Y, Boeken B (1997) The germination of mucilaginous seeds of *Plantago coronopus, Reboudia pinnata,* and *Carrichtera annua* on cyanobacterial soil crust from the Negev desert. Plant Soil 190:247–252

Zaady E, Groffman P, Shachak M (1998) Nitrogen fixation in macro- and microphytic patches in the Negev Desert. Soil Biol Biochem 30:449–454

Zhang Y (2005) The microstructure and formation of biological soil crusts in their early developmental stage. Chin Sci Bull 50:117–121

Chapter 7
Exploring the Ecological Significance of Microbial Diversity and Networking in the Rice Ecosystem

Radha Prasanna, Lata Nain, Alok Kumar Pandey, and Saswati Nayak

7.1 Introduction

Rice is the world's most important food crop in terms of ensuring global food security and the livelihood of millions of producers. Broadly speaking, rice is grown in more than 100 countries, with a total harvested area of about 153 million ha, and production of more than 600 million tons annually. Yields range from less than 1 t ha^{-1} under very poor rainfed conditions to 10 t ha^{-1} in intensive temperate irrigated conditions (Zeigler and Barclay 2008). The consumption of rice exceeds 100 kg per capita annually in many Asian countries, providing 20% of the direct human calorie intake. Asia accounts for over 90% of the world's production of rice, where the poorest of the poor spend up to 50% of their total income on rice alone.

Rice is grown in a wide range of ecological environments (from dryland to flooded land, even under water layers several meters deep, and at varying altitudes (from sea level to 3,000 m above sea level), diverse climates (from tropical to temperate) and different soil types (from saline soils along sea coast to upland soil with toxic aluminium levels). In general, rice cultivation systems are classified as irrigated, rainfed lowland, upland and others. About 50% of the area under rice comprises intensive irrigated systems, which account for 75% of global rice production and represent the home of the Green Revolution in rice.

The rice ecosystems constitute artificial biotopes of a dynamic character. They represent one of the most extensive freshwater ecosystems on Earth, while as "wetlands" they are characterized by unique biogeochemical cycles and variable microbial communities. The biocenoses in them are formed by the association of two micro-organizations – that of water and that of soil, which partly overlap.

R. Prasanna (✉), L. Nain, and A.K. Pandey
Division of Microbiology, Indian Agricultural Research Institute, New Delhi 110012, India
e-mail: radhapr@gmail.com

S. Nayak
National Institute of Science, Technology and Developmental Studies, New Delhi 110012, India

P. Dion (ed.), *Soil Biology and Agriculture in the Tropics*. Soil Biology 21,
DOI 10.1007/978-3-642-05076-3_7,

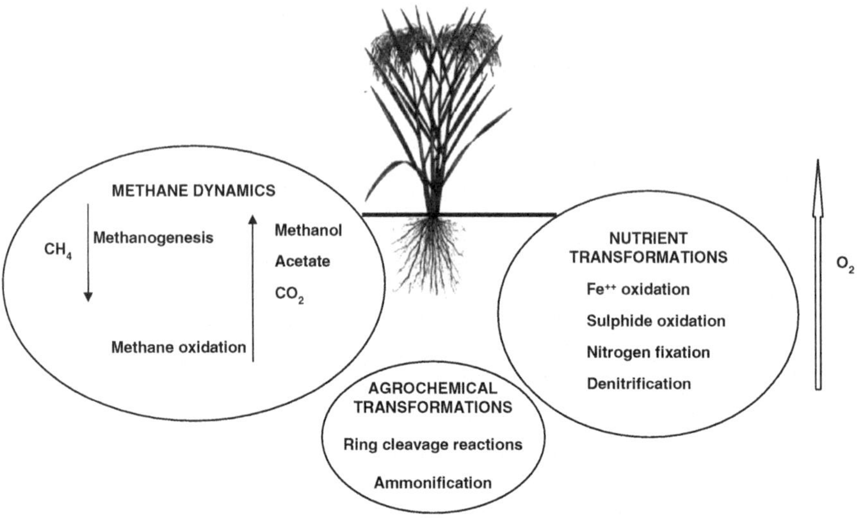

Fig. 7.1 Overview of major microbiological transformations in the rice rhizosphere. *Arrows* are drawn with their *pointed end* oriented towards area of increased activity or concentration

The sustainability of the rice production system over long periods of time is intricately related to the chemistry, physics and biology of rice soils, the basic properties of which differ considerably between wetland and dryland soils. Some of the significant attributes of wetland soils which have been fundamental to the long-term sustainability of rice farming systems in Asia include adequate replenishment of nutrients from flood water and irrigation, biological nitrogen fixation and minimization of soil acidification, erosion and nutrient leaching. The dynamic and complicated nature of this wetland ecosystem makes it an intriguing subject of research, especially in relation to its microbiology. The present compilation focuses on some defining characteristics of the rice rhizosphere, especially nutrient transformations and taxonomic/functional diversity, so as to assess their ecological and agricultural importance in the sustenance of rice ecosystems in a global context (Fig. 7.1).

7.2 Microbial Diversity and Population Dynamics in the Rice Rhizosphere

A variety of agroclimatic conditions and agronomic practices such as puddling, transplanting, application of chemical fertilizers, biofertilizers and pesticides, drying and flooding during crop growth, influence the functional and taxonomic diversity of micro-organisms. Considerable differences are also observed in the composition of microflora in wetland and upland rice soil. An upland rice soil is essentially aerobic except for rare localized anaerobic pockets (confined to pores of soil particles or soil aggregates), with a microflora essentially composed of aerobic bacteria and fungi. Facultative anaerobes show a sudden population upsurge after flooding. On the other hand, some genera commonly observed in submerged soil include *Bacillus, Pseudomonas, Mycobacterium* and *Arthrobacter*, in addition to nonsulfur and sulfur photosynthetic bacteria. Tremendous differences, both qualitative and quantitative, are also observed among microbial populations of tropical and temperate soils (Sethunathan et al. 1982).

The rhizosphere, which represents the soil area under the direct influence of the roots, serves as a dynamic circumjacent site with intense microbial activity, and represents a favorable aerobic–anaerobic interface suitable for diverse groups of micro-organisms. It functions much like a central commodities exchange – where organic C flux from roots fuels microbial and faunal decomposers, which in turn, make nutrients available to these roots. Rhizospheric microbial populations comprise nutrient mobilizers such as diazotrophs, phosphate solubilizers, sulfate oxidizers, as well as bacterial and fungal pathogens and a hydrolytic/catabolic flora. In the case of aerobic and facultative micro-organisms, the rhizosphere effect (corresponding to differences in microbial populations of rhizospheric and non-rhizospheric soils) has been observed to be weaker under submerged conditions than in upland conditions (Kimura et al. 1979). The key organisms in the rhizosphere, including microbes (bacteria and fungi) and their predators (e.g., protozoa, nematodes) strongly influence soil structure and respond to conditions and resources at micrometer to millimeter scales that are difficult to characterize. Now that the importance of a microbially active rhizosphere to the nutrition and growth of the rice plant is well established, current research focuses on specific aspects of rhizosphere ecology such as quorum sensing and cross-talk between root and microbes, nitrogen and carbon cycling, and detoxification of pesticides and other agro-chemicals (Fig. 7.2).

The aerobic–anaerobic interface of the rhizosphere provides a favorable environment for diverse groups of facultative and obligate aerobes/anaerobes, the most prominent of which are diazotrophs, methanogens, methanotrophs, plant growth promoters, heterotrophs and biological control agents. The use of microscopic examination and culture methods is insufficient to characterize microbial communities, as non-cultivable micro-organisms are not evaluated in this manner, nor can we always distinguish between active and inactive micro-organisms. Therefore, current research efforts employ polyphasic approaches. Using polymerase chain reaction amplification coupled to denaturing gradient gel electrophoresis (PCR-DGGE) and fluorescent in situ hybridization (FISH), Doi et al. (2007) observed a greater diversity of bacteria associated to the rice rhizosphere in upland soil than in lowland soil. *Klebsiella planticola* and *Bacillus fusiformis* dominated upland field rhizospheres while *Clostridium bifermentans* and an unknown bacterium were the dominant bacteria in lowland soil. FISH studies also showed a predominance of gram-positive low GC bacteria in both types of rhizospheres. Restriction fragment length polymorphisms (RFLP) analyses of 16S rDNA revealed succession of the microbial community from higher to lower nodal roots.

Burkholderia vietnamiensis was discovered in association with roots of rice plants grown in Vietnamese soil (Gillis et al. 1995) and has attracted attention

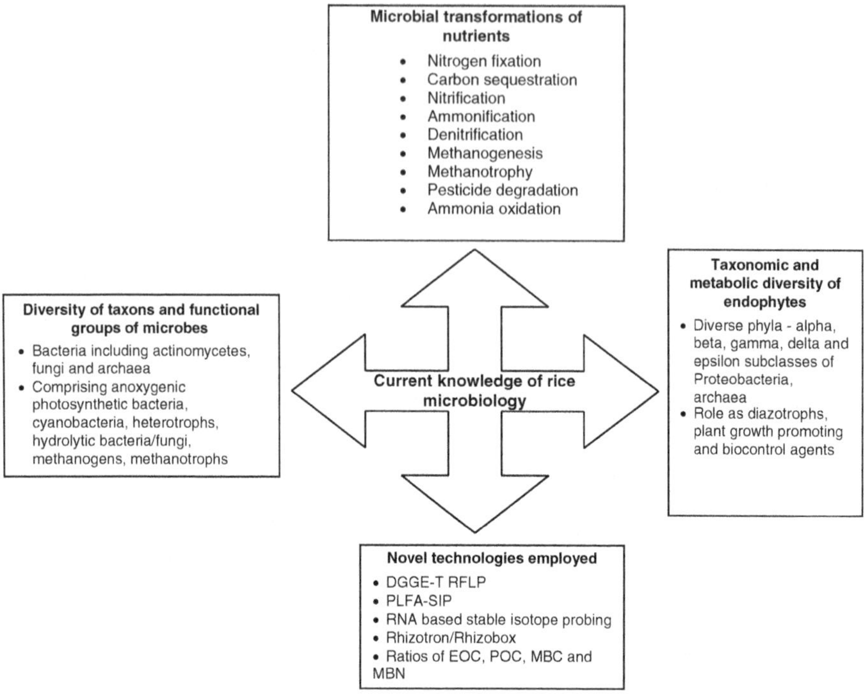

Fig. 7.2 Current scenario on research aspects of microbiology of the rice rhizosphere. Abbreviations: *DGGE*, Denaturing gradient gel electrophoresis; *T-RFLP*, Terminal restriction fragment length polymorphism; *PLFA-SIP*, Phospholipid fatty acid-stable isotope probing; *EOC*, Extracellular organic carbon (easily oxidizable organic *carbon); POC*, Particulate organic carbon (soluble C consumed by persulfate oxidation); *MBC*, Microbial biomass carbon; *MBN*, Microbial biomass nitrogen.

because of its ability to promote plant growth and yields (Van Tran et al. 2000). *Leeia oryzae* gen. nov., sp. nov., a strictly aerobic non-spore-forming gram-negative bacterium classified into a novel genus and species of the *Betaproteobacteria*, was isolated from a rice field in Korea (Lim et al. 2007). A number of *Pseudomonas* species isolated from *Oryza sativa* cv. Pathum Thani 1 exhibited plant growth-promoting abilities, in addition to fluorescent pigment production and biocontrol potential against bacterial and fungal root pathogens (Lawongsa et al. 2008). Rangarajan et al. (2002) assessed the diversity of *Pseudomonas* in rice cultivated along a salinity gradient. Fluorescent pseudomonads were the dominant species found in nonsaline sites, while in the saline sites they were replaced by salt-tolerant species, in particular *Pseudomonas alcaligenes* and *Pseudomonas pseudoalcaligenes*.

Rice plants not only promote microbial activity and growth through photosynthate rhizodeposition, but also bring about a change in microbial community structure. PLFA-SIP (phospholipid fatty acid-based stable isotope probing) was utilized to detect the spatial variation of active micro-organisms associated with rhizosphere carbon flow (Lu et al. 2006). The results suggested that in the rhizosphere, gram-negative and eukaryotic micro-organisms were most active in assembling root-derived C, whereas in the bulk soil gram-positive micro-organisms were relatively more important. Systematic changes were also observed in terms of activity of soil microbiota, depending on their distance from roots.

Numerous nondestructive studies have been conducted using belowground rhizosphere images to examine root dynamics, such as the mini rhizotron method (Hendrick and Pregitzer 1992) or rhizobox (Watt et al. 2006). Achieving a major advance in the use of rhizotrons, Nakaji et al. (2008) developed an automatic image analysis technique of VIS-NIR (visible-near infra red) spectral images captured at four spectral bands, which can be used to classify the growth stage in both the live and dead stages of roots.

Microbial biomass carbon (MBC) is also recognized as a sensitive indicator of changes in both soil organic carbon (SOC) and biological determinants of soil quality and soil health (see Chap. 2). It is therefore a key to understanding the nutrient dynamics of rice fields. Reichardt et al. (1997) observed that a reduction of microbial biomass in the second half of the cropping season may contribute to the declining productivity in irrigated, continuous rice cropping systems.

Protozoan grazing exerts a significant effect on soil bacteria (Schwarz and Frenzel 2005), with flagellates playing a greater role than ciliates in rice fields. Murase et al. (2006) utilized DGGE and terminal-RFLP (T-RFLP) analyses for microcosm-based studies to evaluate the impact of protists on the activity and structure of the bacterial community in a rice-field soil. They observed that grazing effects were more obvious in the partially oxic surface layer as compared to the lower anoxic layers, pointing out to the significant role of O_2 availability.

7.2.1 Diversity in Diazotrophs and Other Bacteria of the Nitrogen Cycle

Among the reduction reactions taking place in rice rhizosphere, biological nitrogen fixation is of tremendous significance. This is evidenced by several reports from all over the world on consistent and stable rice yields, even without addition of nitrogenous fertilizers, and is further supported by reliable gas chromatographic and isotope tracer analyses. The relative contributions of free-living cyanobacteria and heterotrophs, and of symbiotic associations involving cyanobacteria and endophytes, have been investigated in depth.

A number of methodologies have been developed for evaluating nitrogen fixation in the surface, sub-surface, rhizosphere and water column zones of flooded rice soils, through the use of soil augers, sampling devices, portable chambers and automated samplers (Lee and Watanabe 1977). In general, microbial contribution to nitrogen economy has been determined by the indirect acetylene reduction technique or the direct ^{15}N technique, as well as nitrogen balance studies (using Kjeldahl assays of total nitrogen). Diverse active sites are associated with nitrogen fixation in the rice ecosystem, including the soil, floodwater, rhizosphere, roots, leaf sheath, and basal parts of the shoot. Nitrogen fixation is also carried out by the epiphytic and phyllospheric microflora. The free-living and associative nitrogen fixers are implicated in the rhizosphere component, which contributes 0.5–63 kg N ha^{-1} (Charyulu et al. 1981). An improved technique for measurement of nitrogen fixation by cyanobacteria has been developed for field level use. This involves the use of soil auger, and is suitable for studies at surface level or sub-surface level in rice paddies (Prasanna et al. 2002, 2003; Nayak et al. 2004). A microtitre plate-based method has also been developed for enumeration of cyano-bacteria, especially nitrogen fixers, from large sets of soil samples, using agar-based medium as a support system (Prasanna et al. 2006).

With regard to heterotrophic N_2 fixers, most of the work has concentrated on *Azotobacter* and *Azospirillum* and to a lesser extent on *Pseudomonas, Rhizobium* and *Planctomycetes*. The incidence of nitrogen-fixing aerobic and heterotrophic bacteria in rhizosphere soil, root and basal shoots was much higher in wetland than in dryland rice (Barraquio et al. 1997). Using novel 16S rDNA-targeted PCR assays, followed by T-RFLP fingerprint analyses of bulk/anoxic soil samples and rice roots, several novel *Planctomycetes* sub-lineages have been detected, which suggests that extensive microbial diversity might exist in the anoxic bulk soil of rice crop (Derakshani et al. 2001). Xie et al. (2003) investigated the dominant cultivable aerobic and heterotrophic N_2-fixing bacteria in paddy fields along the Yangtze rice plains in central China. The number of culturable N_2-fixing bacteria ranged between 1.41×10^6 and 1.24×10^8 colony-forming units (cfu) g^{-1} of soil. The most prominent strains belonged to the genera *Bacillus, Burkholderia, Pseudomonas, Derxia, Alcaligenes*, and *Citrobacter*. Several such diazotrophs have also shown potential for plant growth promotion and biocontrol (Thakuria et al. 2004; Verma et al. 2004).

Molecular methods involving the use of primer sets based on *nifH* sequences have shown promise for analyzing the physiologically and phylogenetically diverse groups of free-living diazotrophs in soil. The genetic diversity of diazotrophs associated with rice using PCR-RFLP analysis of various *nifH* clones has revealed that two RFLP patterns in the washed roots and rhizospheric soil were 100% similar; while three patterns were exclusive to the rice roots (Bin et al. 2007). Ueda et al. (1995) were among the earliest researchers to recognize the remarkable diversity of nitrogen fixers existing in rice roots, through molecular evolutionary analysis of *nifH* sequences. These nitrogen-fixing populations were found to be dominated mainly by proteobacteria. Using an mRNA-based profiling of *nifH* genes to study the impact of lowland rice genotypes on the functional diversity of root-associated diazotrophs, Knauth et al. (2005) observed remarkable varietal differences in root-associated *nifH*-expressing communities. Other cultivation-independent studies have also suggested that roots of rice are colonized by a diverse community of N_2-fixing bacteria, which are known to be affected by *Oryza* species affiliation and environmental conditions (Tan et al. 2003). Yan et al. (2008) reported a nitrogen fixation island and rhizosphere competence traits in the genome of root-associated *Pseudomonas stutzeri* A1501, and identified several genes required for optimum nitrogenase activity.

Flooded soils are generally not considered a favorable environment for rhizobia. However, rhizobial inoculation has been shown to improve nutrient uptake, seedling vigor and yield of lowland rice (Biswas et al. 2000). Inoculation of rice with nitrogen-fixing *A. caulinodans* contributed 14% of plant N at the vegetative growth stage under low N conditions, and stimulated overall nitrogenase activity (Van Nieuwenhove et al. 2001). Tan et al. (2001) developed a specific detection method using IGS sequences for *Bradyrhizobium* and *Rhizobium* species colonizing rice roots.

Cyanobacteria are known to inhabit a wide range of habitats, including plant surfaces (Whitton et al. 1988; Mano and Morisaki 2008), but information on cyanobacteria in relation to rhizosphere biology is lacking. Studies on cyanobacterial diversity of rice fields belonging to diverse agroecologies of India have shown the predominance of certain genera, particularly *Nostoc, Anabaena* and *Phormidium*, irrespective of chemical/biofertilizer treatments and stage of crop growth (Singh and Bisoyi 1989; Nayak et al. 2001, 2004; Nayak and Prasanna 2007; Prasanna and Nayak 2007). An investigation was undertaken to characterize the abundance, genus-level diversity and metabolic capabilities of cyanobacteria isolated from the rice rhizosphere of different rice varieties belonging to diverse soil types, which revealed that the genera *Nostoc* and *Anabaena* comprised 80% of the rhizosphere isolates (Nayak et al. 2009; Prasanna et al. 2009).

Diversity and activities of ammonia-oxidizing bacteria (AOB) in the rice root environment showed marked intervarietal differences, which could be partly explained by differences in the oxygen concentrations around roots (Briones et al. 2002). The two prominent ammonia-oxidizing bacterial strains in flooded rice soil were *Nitrosospira multiformis* and *Nitrosomonas europaea*. Ammonia-oxidizing archaea (AOA) are reported to be more abundant in the rhizosphere than in bulk soil (Chen et al. 2008). Rice cultivation led to a greater abundance of AOA than AOB, and generated differences in AOA and AOB community composition. Wang et al. (2009) evaluated the effect of nitrogen fertilizer on the composition of AOB and AOA communities in rice soil from microcosm experiments in which three levels of N fertilizer, i.e., 50, 100, and 150 mg N kg^{-1} soil, had been added. Cloning and sequencing of the gene for ammonia monooxygenase (*amoA*) showed that the AOB community in this microcosm soil consisted of three major groups, the *Nitrosomonas communis* cluster and the *Nitrosospira* clusters 3a and 3b. N fertilizer application and soil depth had a significant effect on AOB community composition, whereas the AOA community composition remained unchanged. These studies showed that AOA are dominant in the rhizosphere paddy soil and are more responsive than AOB to exudation from rice root.

7.2.2 Diversity of Methanogens and Methanotrophs

Flooded rice paddy fields can be considered as a system with three compartments (oxic surface soil, anoxic bulk soil and rhizosphere) characterized by different physicochemical conditions. Anaerobic micro-organisms such as fermentative bacteria and methanogenic archaea predominate within the microbial community, with methane as the final product of anaerobic degradation of organic matter. Both methane- and ammonia-oxidizing bacteria can act as a sink for methane in rice soils, due to the homology of the key enzymes methane monooxygenase and ammonia monooxygenase.

In situ stable isotope probing of methanogenic archaea in rice rhizosphere by pulse labeling of rice plants with $^{13}CO_2$ resulted in incorporation of ^{13}C into ribosomal RNA of rice cluster I (RC-I) archaea in the soil. This abundant and ubiquitous group of methanogens plays a key role in methane production from plant-derived carbon.

Sakai et al. (2007) isolated a methanogen (strain SANAE) belonging to RC-I from Japanese rice by co-culture with the propionate-oxidizing and hydrogen-producing syntroph, *Syntrophobacter fumaroxidans*. Most interestingly, root colonization with RC-I methanogens was denser than with *Methanomicrobiales*, so that the plant colonized with RC-I produced and emitted relatively more methane (Conrad et al. 2008). Such observations may be important for developing effective mitigation strategies for reducing methane emission from rice fields.

Methanotrophs are a sub-set of the physiological group formed by the methylotrophs, which utilize a variety of one-carbon compounds (Hanson and Hanson 1996). Two types of methanotrophs have been distinguished on the basis of various criteria including phylogeny, internal membrane ultrastructure and carbon assimilation pathways. The diversity of methanotrophic bacteria associated with roots of submerged rice plants was assessed by Horz et al. (2001), using cultivation-independent techniques based on the T-RFLP analysis of *pmo*A, which encodes the α-subunit of the particulate methane monooxygenase. In this analysis, eight T-RFLP profiles with high relative abundance (>1%) were retrieved. Individual profiles were tentatively assigned to different methanotrophic populations (e.g., *Methylococcus/Methylocaldum/Methylomicrobium, Methylobacter, Methylocystis/Methylosinus)*. Dubey et al. (2003) observed a patchy distribution of populations of methanotrophs in the rice rhizosphere. They also tracked methanotrophs and their diversity in paddy soil using amplified ribosomal DNA restriction analysis (ARDRA), which revealed the presence of the type-I methanotrophs *Methylomonas* and *Methylocaldum* in the rhizosphere soil and co-existence of both type-I and type-II methanotrophs *Methylobacterium* and/or *Methylocella* in non-rhizosphere soil. Shrestha et al. (2008) measured the activity and composition of methanotrophic bacterial communities in planted rice soil by flux measurements of CH_4, T-RFLP analyses of *pmoA* gene and stable isotope probing of phospholipid fatty acid (PLFA-SIP). They showed that type-I and type-II methanotrophic populations changed over time with respect to activity and population size in the rhizospheric soil and the rice roots. However, type-I methanotrophs were more active than type-II methanotrophs, indicating that they are of particular importance in the rhizosphere.

7.3 Nutrient Transformations in the Rice Rhizosphere and their Effects on Soil Fertility and Plant Growth

7.3.1 Nitrogen Cycling

The quantity of nitrogen added through biological nitrogen fixation (BNF) to the soil and to the rice crop has been a subject of much study, and cyanobacteria are widely assumed to make an important contribution to the sustainability of rice-based farming systems. It would be reasonable to assume that most of the world's rice paddies contain free-living cyanobacteria. On the other hand, the cyanobacteria-*Azolla* symbiosis is present in only about 2% (3 million ha) of paddies (Galal 1997; Pabby et al. 2004). Therefore, the extent of cyanobacterial N fixation in rice paddies may be conservatively estimated to be of the order of 30 kg N ha^{-1} $year^{-1}$ and amount to a total of 5 Tg N $year^{-1}$. After the legume–rhizobia system, the rice–cyanobacteria and the rice–*Azolla*–cyanobacteria systems exhibit the highest efficiency in terms of the proportion of N derived from the atmosphere (Ndfa) (Table 7.1).

Since rice plants require large amounts of mineral nitrogen for their growth and for grain production, fertilizer N applications are required to meet the rice crop N demands. Rice crops remove around 16–17 kg N for the production of each ton of rough rice, including straw (De Datta 1981; Ponnamperuma and Detruck 1993; Sahrawat 2000). The low N-use efficiency, coupled with long-term use of urea, depletes the soil organic matter, which may be replenished through the combined application of biofertilizers. BNF by some diazotrophic bacteria such as *Azotobacter, Clostridium, Azospirillum, Herbaspirillum* and *Burkholderia* can substitute for urea-N, while *Rhizobium* can promote the growth physiology or improve the root morphology of the rice plant (Choudhury and Kennedy 2004). Free-living *Azotobacter* can bring about 7–20% of yield increase and 11–15 kg ha^{-1} increase in N accumulated by the rice plants (Yanni and El-Fattah 1999). Combined application of *Azotobacter armeniacus* and *Azotobacter nigricans* showed growth-promoting effect on rice in China (Piao et al. 2005).

Table 7.1 Nitrogen fixed by various biological sources in rice-based cropping systems[a]

Biological sources of N	Quantities of N fixed (kg ha^{-1} $year^{-1}$) in cropping systems:		
	Irrigated paddies	Rainfed/naturally flooded paddies	Upland rice
Associative nitrogen fixation in rice rhizosphere	5–15	0–10	0–4
Free-living cyanobacteria	10–140	2–30	1–10
Azolla–Anabaena symbiosis	20–140	5–50	0
Legume green manure	50–150	10–80	20–80

[a]Data compiled from Roger 1986; Tripathi et al. 1997; Kannaiyan 1998; Liesack et al. 2000; Choudhury and Kennedy 2004

In tropical rice cultivation practices, variations in N use efficiency, N uptake patterns and N absorption/recovery efficiency of different cultivars influence the microbiological activity of the rhizosphere (Charyulu et al. 1981; Bronson et al. 2000). Kanungo et al. (1997) evaluated the nitrogenase activity and the distribution of nitrogen-fixing bacteria associated with the rhizosphere of rice cultivars with varying N absorption efficiencies at three N fertilizer regimes. Microbiological analyses indicated that at 60 kg N ha^{-1}, anaerobic nitrogen-fixing bacteria were stimulated to a greater extent than nitrogen-fixing *Azospirillum* sp. and *Azotobacter* sp. The activity of N-fixing bacteria is generally higher in cultivated rice strains than in wild varieties (Sano et al. 1981), which suggests that the association between rice plants and N-fixing bacteria is controlled by the plant genotype. Rice plants can also utilize more biologically fixed N in soils with a relatively low fertility than in soils with high fertility status. Shrestha and Ladha (1996) evaluated the variation among 70 rice genotypes of diverse origin and growth duration for N_2 fixation using the ^{15}N isotope dilution technique. They found that % N dfa varied from 1.5 to 21.0 depending on the rice variety and the plant organ. Enrichment was lowest in roots, intermediate in the grain, and highest in shoots.

Nitrogen fixation rate is more pronounced in lowland rice that is not fertilized with nitrogen, which is indicative of a potential contribution of photosynthetic and rhizosphere bacteria. Habte and Alexander (1980) reported that photosynthetic bacteria contributed significantly to the overall nitrogen fixation in the rice rhizosphere. They also observed that the nitrogenase activity of heterotrophic rhizospheric bacteria, but not that of cyanobacteria, was enhanced in flooded soil by the presence of rice plant. A long-term multifactorial pot experiment was carried out by Trolldenier (1987) under greenhouse conditions, to assess the N economy of the rice plant and N fixation by heterotrophic soil micro-organisms under different water regimes and varying N and K nutrition, for 10 consecutive years. The experiment comprised two N and two K levels with wet (WF) and dry fallow (DF) between the cropping seasons. The highest N gain was found in DF treatment with low N and high K application, and under these conditions nearly one-quarter of the N taken up by the aboveground parts of the plants could be ascribed to associative N_2 fixation in the rice rhizosphere.

Watanabe et al. (1988) summarized the nitrogen cycling process in wetland rice soil, and suggested that microbial biomass and aquatic photosynthetic communities are the main N source for flooded rice, when mineral N is deficient. Roger (1986) showed that the photosynthetic aquatic biomass is usually a few 100 kg dry weight per hectare, and rarely exceeds 1 t dry weight per hectare. At 2.5% N content, an average aquatic biomass of 200 kg dry matter per hectare would correspond to only 5 kg biomass N ha^{-1}. Biomass N rarely exceeds 10–20 kg ha^{-1}. The *Azolla–Anabaena* symbiotic system can partially fulfill the N requirements of the plant, replacing 30%–50% of the required urea-N (Galal 1997; Kannaiyan 1998). The level of activity of cyanobacteria is dependent upon various survival factors, including soil type, water quality, temperature, and agronomic production practices. Successful establishment of promising strains of cyanobacteria has been reported to enhance the crop yield up to 30% (Venkataraman 1972; Kaushik and

Prasanna 1989), and to provide 15–25 kg ha^{-1} biologically fixed nitrogen/season, in addition to many other beneficial effects on soil quality (Venkataraman 1972; Nayak and Prasanna 2007; Prasanna and Nayak 2007).

Apart from indigenous cyanobacteria, planktonic, filamentous, and macrophytic algae, as well as vascular macrophytes, develop during different phases of rice growth, and compete with rice for nutrients and light (Quesada et al. 1997). Estimates of the N balance in the presence and absence of light indicate that, on average, photo-dependent nitrogen fixation contributes two-thirds of the balance (Roger 1986). The photo-dependent nitrogen fixation ranged from 0.23 to 75.5 kg of N ha^{-1} $year^{-1}$ in the absence of cyanobacterial blooms; however, this activity was not correlated with the number of N-fixing cyanobacteria in water and soil. To explain this absence of association, the distribution of nitrogenase activity in the rice soil system and the possible N contribution of epiphytic cyanobacteria to rice plants and to the macrophyte *Chara vulgaris* was studied by Ariosa et al. (2004) in Spain. The largest proportion of photo-dependent N fixation was associated with epiphytic cyanobacteria on *Chara*, contributing more than 45% of the nitrogenase activity measured in rice field, equivalent to 27 kg of N ha^{-1}. These results indicate that the *Chara* macrophyte, which is usually considered a weed in the context of rice cultivation, may help to maintain soil N fertility in rice ecosystem.

Since rice paddies are typically maintained under flooded conditions, ammonium constitutes the major form of mineral N in bulk soil. However, considerable amounts of nitrate formed through nitrification do accumulate in the oxygen-rich surface layer of irrigated paddy soil, as well as in well-drained upland rice fields and in rainfed environments during the dry season. As compared to ammonium, which is usually bound to the soil matrix, nitrate is mobile within soil, making it susceptible to loss due to leaching. Nitrate can also be lost by denitrification (Tripathi et al. 1997). Nitrification is therefore critical to the supply of N in rice fields as well as to the balance of ammonium and nitrate. The initial rate-limiting step of nitrification is carried out by the chemolitho–autotrophic ammonia-oxidizing bacteria and archaea (see Sect. 7.2.1). Comparison of the activities and diversities of ammonia-oxidizing bacteria in the root environment of different cultivars of rice indicated marked differences despite identical environmental conditions during growth. Gross nitrification rates obtained by the ^{15}N dilution techniques were significantly higher in the presence of a modern rice variety IR 63087-1-17 than in the presence of two traditional varieties.

7.3.2 Other Nutrients

The micronutrients required by plants are iron (Fe), manganese (Mn), zinc (Zn), copper (Cu), boron (B), chlorine (Cl) and possibly vanadium (V) (Tiller 1981). Most of the micronutrients are present in sufficient amounts in soil to meet the crop demand. Zinc is the micronutrient that has most commonly been reported to be deficient. The soils at risk are commonly neutral to alkaline, although the pH is less critical under permanently flooded conditions. Soil with a high organic matter content may also have low available Zn (Yoshida et al. 1973). Iron also plays an important role in the lowland rice rhizosphere (Begg et al. 1994). In most flooded soils, oxygen released from rice roots reacts with mobile ferrous ion, producing insoluble ferric hydroxide and acidity. The pH changes caused by these processes and accumulation of insoluble ferric hydroxide together strongly influence processes in rice rhizosphere. It has been shown that K application on K-deficient soils lowered the content of active reducing substances and ferrous ions in the paddy soil, raised the soil redox potential in the rhizosphere, increased the Eh values of rice roots and lowered the content of iron in the rice plants (Chen et al. 1997).

7.3.3 Carbon Sequestration and Plant Growth Promotion

In wetland soil, rhizodeposition occurs during paddy growth in the form of water-soluble exudates, secretion lysates of sloughed-off cells and decaying roots which serve as carbon source, indirectly or directly, for the production of methane by micro-organisms. Emission of methane from flooded fields is governed by soil, plant, atmosphere, and agro-climatic variables (Conrad et al. 2000; Satpathy et al. 1997). It is the net result of contrasting microbial processes, methane production (methanogenesis, a strictly anaerobic process) and methane oxidation (methanotrophy, essentially an aerobic process). A significant negative correlation was found between methane emission under different water regimes and rhizosphere redox potential. Extractable Fe^{2+}, readily mineralizable carbon and root biomass presented a significant positive correlation with cumulative methane emission (Mishra et al. 1997). These studies suggest the possibility that methane emissions may be reduced through appropriate water management in a rainfed rice ecosystem.

Hou et al. (2000) studied methane and nitrous oxide emission from a rice field in relation to soil redox and microbiological processes, and observed a strong correlation with changes in soil redox potential. Methane emission was significantly related to the logarithm number of zymogenic bacteria as well as to soil redox potential; hence, both parameters appear to be predictors of methane emission potential.

A substantial part of the nitrification potential in the rhizosphere of rice was attributed to the activity of methanotrophs, as demonstrated using the inhibitors methyl fluoride and acetylene (Bodelier et al. 2000), while the contribution of nitrifiers to CH_4 oxidation was insignificant.

Urea fertilization selectively stimulated type I methanotrophs in rice fields, as revealed by RNA-based stable isotope probing (Noll et al. 2008). Potassium application also promoted an increase in redox potential, as it reduced the contents of active reducing substances and ferrous content in the rhizosphere soil (see also above), leading to inhibition of methanogenic bacteria and stimulation of methanotrophs (Babu et al. 2006).

Cyanobacteria are known to liberate a wide array of extracellular substances, such as plant growth regulators, vitamins, amino acids and sugars, which have direct or indirect impact on plant growth (Misra and Kaushik 1989; Prasanna et al. 2008a). A set of rhizosphere cyanobacterial isolates from diverse agroecologies of India (Prasanna et al. 2009) were found to efficiently enhance the germination of wheat and rice seeds. Extensive diversity was observed with respect to the taxonomic genera isolated (Fig. 7.3) and nitrogenase activity levels (measured as acetylene-reducing activity, ARA). Comparatively, the *Anabaena* and *Nostoc* isolates exhibited the highest nitrogenase activity. The penetration and presence of cyanobacterial cells and short filaments in the root cortical region has been revealed by electron microscopic investigations (Jaiswal et al. 2008; Karthikeyan et al. 2009), which is indicative of the potential of these isolates for developing effective associations with roots. A set of cyanobacterial rhizosphere isolates exhibited biocidal activity and high values of chitosanase and xylanase activity, when tested against selected phytopathogenic fungi (Prasanna et al. 2008b). Metabolites of *Calothrix elenkenii*, a rice rhizosphere isolate, exhibited potential for biological control against *Pythium aphanidermatum* in pot experiments (Manjunath et al. 2009). The general but inaccurate belief that all cyanobacteria are obligate photoautotrophs has been, perhaps, the major reason for the paucity of information available on the diverse roles of these micro-organisms in the rhizosphere.

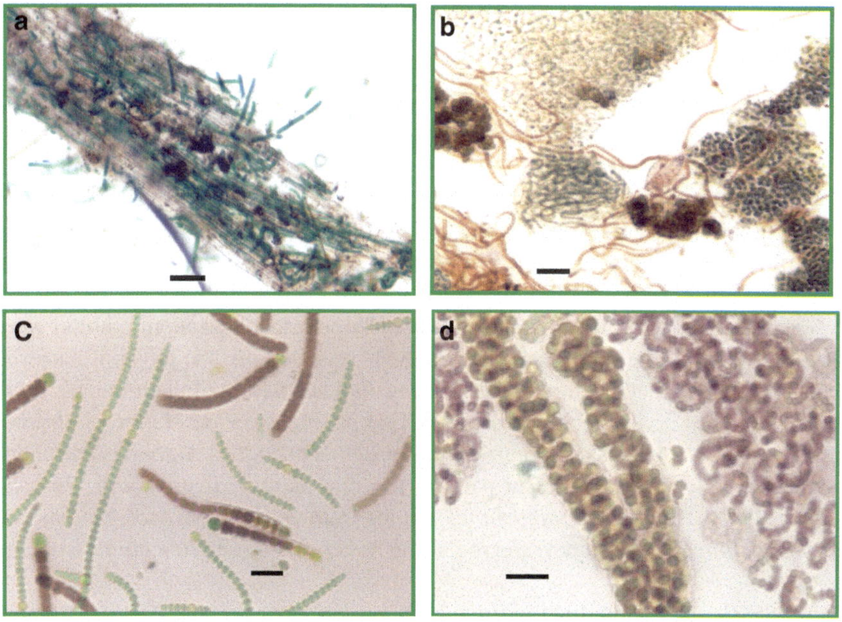

Fig. 7.3 Photomicrographs of cyanobacteria isolated from rhizosphere of rice. **a** Root fragments with cyanobacterial growth. **b** Cyanobacterial diversity in enrichment of rhizosphere soil samples. **c** *Anabaena* sp. and *Calothrix* sp. **d** *Nostoc* sp. (*Scale bar* 20 μm)

7.3.4 *Effect of Pesticides on Rice Rhizosphere*

Agrochemicals applied in the rice field affect the rhizospheric micro-organisms (Sethunathan et al. 1982). Several pesticides undergo fairly rapid decomposition in anaerobic ecosystems such as flooded soil. Ramakrishna and Sethunathan (1982) reported stimulation of autotrophic ammonium oxidation in rice rhizosphere soil by application of the insecticide carbofuran.

The number and growth of nitrogen-fixing bacteria was also modified by agrochemicals such as carbofuran, thiobencarb and butachlor (Jena et al. 1990). Long-term application of agrochemicals modified the nitrogen-fixing flora of rice rhizosphere (Hirano et al. 2001) and growth and distribution of micro-organisms in rhizosphere soil of wetland rice (Das et al. 2003).

7.4 Endophytes Associated with Rice: The Inside Story

Rice endophytic bacteria originate from the outside environments, and invade the host through stomata, lenticels, wounds, areas of emergence of lateral roots and germinating radicles (Mano et al. 2007).

Yanni et al. (1997) documented a beneficial natural endophytic association of rhizobia and rice grown in rotation with clover, using a legume trap host to specifically enumerate and isolate endophytic rhizobia from rice roots (see Sect. 8.6 for a complementary discussion of endophytic rhizobia). Whereas several other reports also describe the isolation and phylogenetic classification of nitrogen-fixing bacteria from the rice rhizosphere (Ladha 1986; Lakshmi Kumari et al. 1976; Barraquio et al. 1997), few present successful field experimental data.

In their review on endophytic bacteria in the rice plant, Mano and Morisaki (2008) have provided a very comprehensive survey of the diversity of cultivable and uncultivable bacteria from various organs of the rice host plant. They observed more population divergence in rice roots than in the seeds and leaves (Table 7.2). Using a set of bacterial PCR primers for selective amplification of bacterial 16S rDNA sequences from root tissues, 52 OTUs (operational taxonomic units) were identified (Sun et al. 2007). Sequence analyses revealed diverse phyla of bacteria, comprising alpha, beta, gamma, delta, and epsilon subclasses of the proteobacteria, the *Cytophaga/Flexibacter/Bacteroides* (CFB) phylum, low (G+C) gram-positive bacteria, the *Deinococcus/Thermus* group, *Acidobacteria*, and *Archaea*. The endophytic bacterial flora in different parts of the rice plant is diverse, with *Pantoea* being isolated from seeds, *Methylobacterium* from the shoots, *Azospirillum* and *Herbaspirillum* from the stems and roots, and *Burkholderia* and *Rhizobium* from the roots.

Among the large and diverse range of N_2-fixing heterotrophic bacteria isolated from surface-sterilized rice plants, only a few – notably *Alcaligenes faecalis* (later identified as *Pseudomonas stutzeri* A16), *Azoarcus* spp., *Herbaspirillum* spp., *Rhizobium* spp., and *Serratia marcescens* IR BG 500 – have been confirmed to be

Table 7.2 Culturable and unculturable bacterial endophytic diversity of *Oryza sativa* plant parts[a]

Rice organ	Bacterial taxa isolated and cultured	Bacterial taxa as revealed by culture-independent methods
Seed	*Acidovorax* sp., *Bacillus cereus*, *B. pumilus*, *B. subtilis*, *Curtobacterium* sp., *Klebsiella oxytoca*, *Ochrobactrum anthropi*, *Methylobacterium aquaticum*, *Micrococcus luteus*, *Paenibacillus amylolyticus*, *Pantoea agglomerans*, *P. ananatis*, *Pseudomonas boreopolis*, *Sphingomonas echinoide*, *S. parapaucimobilis*, *S. melonis*, *S. yabuuchiae*, *Xanthomonas translucens*	ND[b]
Leaf	*Aurantimonas altamirensis*, *Bacillus gibsonii*, *B. pumilus*, *Curtobacterium* sp., *Diaphorobacter nitroreducens*, *Methylobacterium aquaticum*, *Methylobacterium* sp., *Pantoea ananatis*, *Sphingomonas echinoides*, *S. melonis*, *S. yabuuchiae*, *Stenotrophomonas maltophilia*, *Streptomyces* sp.	ND
Stem	*Azosprillum lipoferum*, *Agrobacterium vitis*, *Azorhizobium caulinodans*, *Azospirillum* sp., *Bacillus megaterium*, *B. subtilis*, *Cytotophagales* str. *MBIC4147*, *Ideonella dechloratans*, *Pseudomonas cepacia*, *Methylobacterium* sp.	ND
Leaves, stem, root	*Herbaspirillum seropedicae*	ND
Root	*Azoarcus indigens*, *Azoarcus* sp., *Azorhizobium caulinodans*, *Azospirillum brasilense*, *A. irakense*, *A. lipoferum*, *Bacillus luciferensis*, *B. megaterium*, *Bradyrhizobium elkanii*, *Bradyrhizobium japonicum*, *Brevibacillus agri*, *Burkholderia cepacia*, *B. kururiensis*, *Burkholderia* sp., *Caulobacter crescentus*, *Chryseobacterium taichungense*, *Enterobacter cloacae*, *E. ludwigii*, *Herbaspirillum seropedicae*, *Herbaspirillum* sp., *Hyphomicrobium sulfonivorans*, *Klebsiella pneumoniae*, *Methylocapsa acidiphila*, *Micrococcus luteus*, *Mycobacterium*	*Achromobacter xylosoxidans*, *Acidaminobacter hydrogenoformas*, *Acidovorax faecalis*, *Acinetobacter baumannii*, *Bdellovibrio bacteriovorus*, *Brevundimonas diminuta*, *Burkholderia fungorum*, *Burkholderia* sp., *Caulobacter* sp., *Clostridium* sp., *Comamonas testosterone*, *Curvibacter gracilis*, *Delftia acidovorans*, *Deinococcus indicus*, *Delftia tsuruhatensis*, *Duganella violaceinigra*, *Enterobacter* sp., *Flavobacterium frigoris*, *F. psychrophilum*, *Gallionella ferruginea*, *Geobacter* sp., *Herbaspirilum frisingense*, *Holophaga foetida*, *Hydrogenophaga taeniospiralis*,

(*continued*)

Table 7.2 (continued)

Rice organ	Bacterial taxa isolated and cultured	Bacterial taxa as revealed by culture-independent methods
	petroleophilum, *Ochrobactrum* sp., *Paenibacillus alvei*, *Rhizobium leguminosarum*, *R. loti*, *Roseateles depolymerans*, *Sphingomonas paucimobilis*	*Kaistina koreensis*, *Lachnospiraceae* sp. 19gly4, *Methylophaga marina*, *Methyloversatilis universalis*, *Pantoea* sp., *Planomicrobium mcmeekinii*, *P. okeanokoites*, *Plesiomonas shigelloides*, *Pseudomonas stutzeri*, *Sphingobacterium* sp., *Sterolibacterium denitrificans*, *Sulfurospirillum multivorans*, *Methylobacterium* sp., *Novosphingobium tardaugens*, *Sinorhizobium terangae*, *Variovorax* sp.

[a]Compiled from Barraquio et al. 1997; Elbeltagy et al. 2000; Sun et al. 2007; Mano et al. 2007; Mano and Morisaki 2008
[b]*ND* No data available

genuine endophytes of rice (Reinhold-Hurek and Hurek 1998; James 2000), as they can only be isolated from host tissues and do not survive well in soil. *Herbaspirillum seropedicae* exhibits significantly high nitrogen-fixing potential, as shown by ARA and ^{15}N isotope dilution data: this bacterium contributes as much as 54% N to rice seedlings (Baldani et al. 2000; Mirza et al. 2000). *Herbaspirillum* sp. isolated from wild rice species exhibited endophytic colonization of intercellular spaces of root and shoot tissues and in planta nitrogen fixation (Elbeltagy et al. 2001). James et al. (2002) observed the aggressive entry of *H. seropedicae* 267 through cracks at points of lateral root emergence and its colonization of leaves, stem epidermal cells, and stomatal cells. Barraquio et al. (1997) reported that the diazotrophic endophytes in the root of a rice plant grown in an unfertilized field were most numerous at or near heading stage. Stoltzfus et al. (1997) reported successful internal colonization of rice tissues by putative endophytic bacteria, which had been isolated from diverse rice varieties grown on different soil types. Several strains of *Serratia marcescens* have shown promise for endophytic colonization of rice (Gyaneshwar et al. 2001), based on detailed studies on light and transmission electron microscopy and diazotrophic potential. Further work on the molecular mechanism of entry is needed to establish whether these strains possess hydrolytic enzymes (such as chitinase), as observed in *Azoarcus* sp. (Engelhard et al. 2000), or ligninase as in other *Serratia marcescens* strains. A degree of host specificity, especially in relation to invasiveness and endophytic persistence, was also observed (Stoltzfus et al. 1997; Gyaneshwar et al. 2001).

Azoarcus sp. from Kallar grass is known to abundantly colonize both its original host and rice roots, and show *nif* gene expression (Elbeltagy et al. 2000). Nitrogen fixation rate by endophytic *Azoarcus* was highest in stems, as *Azoarcus* possesses the ability to spread systemically and reach aerial tissues. Hurek et al. (2002) provided evidence for involvement of endoglucanase in active intracellular colonization of root epidermis cells by *Azoarcus*.

Despite the widespread occurrence of natural endophytic *Rhizobium*–cereal association (see Chap. 8), the mode of rhizobial colonization of rice tissues is poorly understood. Chi et al. (2005) analyzed the influence of the growth physiology of rice on the colonization process of *gfp*-tagged rhizobia. A dynamic infection process was observed, beginning with surface colonization of rhizoplane, followed by endophytic colonization of roots and ascending migration to stem base, leaf sheath and leaves, wherein high populations were observed. The inoculated rice plants exhibited higher photosynthetic rates and efficiency, enhanced shoot and root biomass, and phytohormone production, as compared to uninoculated plants. The expression of *nifH* genes was used as a marker to investigate the expression of nitrogenase in leaf and stem of wild rice *Oryza officinalis*. Light irradiation was observed to enhance *nifH* expression, providing evidence for a daily rhythm determined by the light variation and photosynthate synthesis cycles.

Endophytic bacteria are also increasingly being investigated for their role in plant growth promotion and plant disease suppression, which they achieve by mechanisms similar to those of the plant growth-promoting rhizobacteria. *Pantoea agglomerans*, an endophyte of rice, produces four categories of phytohormones (Feng et al. 2006) and influences host photosynthesis. *P. agglomerans* forms multicellular structures called symplasmata, which improve the stability of endophytic bacteria in the host plant by providing a buffering environment. Mukhopadhyay et al. (1996) have reported that several endophytic bacteria isolated from rice seed exhibit strong antagonistic activity against plant pathogenic fungi, such as *Rhizoctonia solani* and *Pythium* spp., and also against pests such as *Meloidogyne graminicola*.

7.5 Conclusions and Future Perspectives

A definite need exists for enhancing our understanding of the microbiologically active rice rhizosphere – an environment which serves as an effective medium for nitrogen fixation, detoxification of agrochemicals, and immobilization/mobilization of several inorganic and organic ions relevant to the nutrition and growth of rice plant and soil fertility. Plant–microbe interactions and agronomic practices are known to play central roles in carbon and nitrogen sequestration and nutrient cycling processes in rice fields, and they can bring about spatial and temporal changes in microbial community structure, leading to improved soil health. The endophytes of rice are gaining importance as promising candidates in integrated nutrient and pest management strategies, as they achieve a close interaction in the protected environment of root and other tissues, where they are shielded from carbon competition with other micro-organisms.

Future research should be directed towards:

- The formulation of microbial consortia designed for action in problematic areas; some of these sets of micro-organisms would be effective in biocontrol and carbon sequestration in organic-carbon-depleted or pathogen-infested soils, whereas others would comprise diazotrophs and plant growth-promoting microbes for resource-poor soils.
- The optimization of cultural practices and management strategies to decrease the populations of methanogens in the rice rhizosphere as a means of regulating methane emissions.
- The development of effective inoculants, by means of in-depth studies on culturable microbial diversity and plant-microbe interactions.

Acknowledgments We wish to acknowledge with gratitude the authorities of the Indian Agricultural Research Institute, New Delhi and Indian Council of Agricultural Research (ICAR), New Delhi for providing through various projects [AP Cess Fund, Network Project on Micro-organisms (AMAAS) and All India Coordinated Research Project on Biofertilizers (AINPB)] the funding and facilities necessary for undertaking the research activities described as part of this compilation. We also wish to thank our students/project staffs who have provided their research inputs.

References

Ariosa Y, Quesada A, Aburto J, Carrasco D, Carreres R, Leganis F, Valiente EF (2004) Epiphytic cyanobacteria on *Chara vulgaris* are the main contributors to N_2 fixation in rice fields. Appl Environ Microbiol 70:5391–5397

Babu JY, Li C, Frolking S, Nayak DR, Adhya TK (2006) Field validation of DNDC model for methane and nitrous oxide emissions from rice-based production systems of India. Nutr Cycl Agroecosyst 74:157–174

Baldani VLD, Baldani JI, Dobereiner J (2000) Inoculation of rice plants with the endophytic diazotrophs *Herbaspirillum seropedicae* and *Burkholderia spp.* Biol Fertil Soils 30:485–489

Barraquio WL, Revilla L, Ladha JK (1997) Isolation of endophytic diazotrophic bacteria from wetland rice. Plant Soil 194:15–24

Begg CBM, Kirk GJUD, Mackenjia AF, Neue HU (1994) Root induced iron oxidation and pH changes in the lowland rice rhizosphere. New Phytol 128:469–477

Bin C, Si-Ping Z, Li-Juan Z, Zhi-Min L, Ya-Na S, Wei-Wen Z (2007) Genetic diversity analysis of diazotrophs in the rice rhizosphere. Chin J Agric Biotechnol 4:253–258

Biswas JC, Ladha JK, Dazzo FB, Yanni YG, Rolfe BG (2000) Rhizobial inoculation influences seedling vigor and yield of rice. Agron J 92:880–886

Bodelier PLE, Hahn AP, Arth IR, Frenzel P (2000) Effects of ammonium-based fertilization on microbial processes involved in methane emission from soils planted with rice. Biogeochemistry 51:225–257

Briones AM, Okabe S, Umemiya Y, Ramsing N-B, Reichardt W, Okuyama H (2002) Influence of different cultivars on populations of ammonia-oxidizing bacteria in the root environment of rice. Appl Environ Microbiol 68:3067–3075

Bronson KF, Hussain F, Pasuquin E, Ladha JK (2000) Use of ^{15}N-labeled soil in measuring nitrogen fertilizer recovery efficiency in transplanted rice. Soil Sci Soc Am J 64:235–239

Charyulu PBBN, Nayak DN, Rao VR (1981) $^{15}N_2$ incorporation by rhizosphere soil: influence of rice variety, organic matter and combined nitrogen. Plant Soil 59:399–405

Chen J, Xuan J, Du C, Xie J (1997) Effect of potassium nutrition of rice on rhizosphere redox status. Plant Soil 188:131–137

Chen X-P, Zhu Y-G, Xia Y, Shen J-P, He J-Z (2008) Ammonia-oxidizing archaea: important players in paddy rhizosphere soil. Environ Microbiol 10:1978–1987

Chi F, Shen S-H, Cheng H-P, Jing Y-X, Yanni YG, Dazzo FB (2005) Ascending migration of endophytic rhizobia, from roots to leaves, inside rice plants and assessment of benefits to rice growth physiology. Appl Environ Microbiol 71:7271–7278

Choudhury ATMA, Kennedy IR (2004) Prospects and potentials for systems of biological nitrogen fixation in sustainable rice production. Biol Fertil Soils 39:219–227

Conrad R, Klose M, Claus P (2000) Phosphate inhibits acetotrophic methanogenesis on rice roots. Appl Environ Microbiol 66:828–831

Conrad R, Klose M, Noll M, Kemnitz D, Bodelier PLE (2008) Soil type links microbial colonization of rice roots to methane emission. Global Change Biol 14:657–669

Das AC, Chakravarty A, Sukul P, Mukherjee D (2003) Effect of HCH and Fenvalerate on growth and distribution of micro-organisms in relation to persistence of the insecticides in the rhizosphere soils of wetland rice. Bull Environ Contam Toxicol 70:1059–1064

De Datta SK (1981) Principles and practices of rice production. Wiley, New York

Derakshani M, Lukow T, Liesack W (2001) Novel bacterial lineages at the (Sub) division level as detected by signature nucleotide-targeted recovery of 16S rRNA genes from bulk soil and rice roots of flooded rice microcosms. Appl Environ Microbiol 67:623–631

Doi T, Hagiwara Y, Abe J, Morita S (2007) Analysis of rhizosphere bacteria of rice cultivated in andosol lowland and upland fields using molecular biological methods. Plant Root 1:66–74

Dubey SK, Padmanabhan P, Purohit HJ, Upadhyay SN (2003) Tracking of methanotrophs and their diversity in paddy soil: a molecular approach. Curr Sci 85:92–95

Elbeltagy A et al (2000) Isolation and characterization of endophytic bacteria from wild and traditionally cultivated rice varieties. Soil Sci Plant Nutr 46:617–629

Elbeltagy A et al (2001) Endophytic colonization and in planta nitrogen fixation by a *Herbaspirillum* sp. isolated from wild rice. Appl Environ Microbiol 67:5287–5293

Engelhard M, Hurek T, Reinhold-Hurek B (2000) Preferential occurrence of diazotrophic endophytes, *Azoarcus* sp. in wild rice species and land races of *Oryza sativa* in comparison with modern races. Environ Microbiol 2:131–141

Feng Y, Shen D, Song W (2006) Rice endophyte *Pantoea agglomerans* YS19 promotes host plant growth and affects allocations of host photosynthates. J Appl Microbiol 100:938–945

Galal YGM (1997) Estimation of nitrogen fixation in an *Azolla*–rice association using the ^{15}N isotope dilution technique. Biol Fertil Soils 24:76–80

Gillis M et al (1995) Polyphasic taxonomy in the genus *Burkholderia* leading to an amended description of the genus and proposition of *Burkholderia vietnamiensis* sp. nov. for N_2-fixing isolates from rice in Vietnam. Int J Syst Bacteriol 45:274–289

Gyaneshwar P, James EK, Mathan N, Reddy PM, Reinhold-Hurek B, Ladha JK (2001) Endophytic colonization of rice by a diazotrophic strain of *Serratia marcescens*. J Bacteriol 183: 2634–2645

Habte M, Alexander M (1980) Nitrogen fixation by photosynthetic bacteria in lowland rice cultivar. Appl Environ Microbiol 39:342–347

Hanson RS, Hanson TE (1996) Methanotrophic bacteria. Microbiol Rev 60:439–471

Hendrick RL, Pregitzer KS (1992) The demography of fine roots in a northern hardwood forest. Ecology 73:1094–1104

Hirano K, Hayatsu M, Nioh I, Nakai H (2001) Comparison of nitrogen-fixing bacterial flora of rice rhizosphere in the fields treated long-term with agrochemicals and non-agrochemicals. Microbes Environ 16:155–160

Horz H-P, Yimga MT, Liesack W (2001) Detection of methanotroph diversity on roots of submerged rice plants by molecular retrieval of *pmo*A, *mmo*X, *mxa*F, and 16S rRNA and ribosomal DNA, including *pmo*A-based terminal restriction fragment length polymorphism profiling. Appl Environ Microbiol 67:4177–4185

Hou AX, Chen GX, Wang ZP, Cleemput OV, Patrick WH Jr (2000) Methane and nitrous oxide emissions from a rice field in relation to soil redox and microbiological processes. Soil Sci Soc Am J 64:2180–2186

Hurek T, Handley L, Reinhold-Hurek B, Piche Y (2002) *Azoarcus* grass endophytes contribute fixed nitrogen to the plant in an unculturable state. Mol Plant Microbe Interact 15:233–242

Jaiswal P, Prasanna R, Nayak S, Sood A, Suseela MR (2008) Characterization of rhizo-cyanobacteria and their associations with wheat seedlings. Egypt J Biol 10:20–27

James EK (2000) Nitrogen fixation in endophytic and associative symbiosis. Field Crops Res 65:197–209

James EK, Gyaneshwar P, Mathan N, Barraquio WL, Reddy PM, Iannetta PPM, Olivares FL, Ladha JK (2002) Infection and colonization of rice seedlings by the plant growth promoting bacterium *Herbaspirillum seropedicae* Z67. Mol Plant Microbe Interact 15:894–906

Jena PK, Adhya TK, Rao VR (1990) Nitrogen-fixing bacterial populations as influenced by butachlor and thiobencarb in rice soils. Microbiol Res 145:457–460

Kannaiyan S (1998) Nitrogen contribution by *Azolla* to rice crop. Proc Indian Natl Sci Acad Part B Biol Sci 59:309–314

Kanungo PK, Panda D, Adhya TK, Ramakrishnan B, Rao VR (1997) Nitrogenase activity and nitrogen fixing bacteria associated with rhizosphere of rice cultivars with varying N absorption efficiency. J Sci Food Agric 73:485–488

Karthikeyan N, Prasanna R, Sood A, Jaiswal P, Nayak S, Kaushik BD (2009) Physiological characterization and electron microscopic investigations of cyanobacteria associated with wheat rhizosphere. Folia Microbiol 54:43–51

Kaushik BD, Prasanna R (1989) Status of biological nitrogen fixation by cyanobacteria and *Azolla* in India. In: Dadarwal KR, Yadav KS (eds) Biological nitrogen fixation research status in India: 1889–1989. Society of Plant Physiologists and Biochemists, New Delhi, pp 141–208

Kimura M, Wada H, Takai Y (1979) The studies on the rhizosphere of paddy rice IV. The effects of anaerobiosis on microbes. Soil Sci Plant Nutr 25:145–153

Knauth S, Hurch T, Brar D, Reinhold-Hurek B (2005) Influence of different *Oryza* cultivars on expression of *nif*H gene pools in roots of rice. Environ Microbiol 7:1725–1733

Ladha JK (1986) Studies on nitrogen fixation by free-living and rice-plant associated bacteria in wetland rice field. Bionature 6:47–58

Lakshmi Kumari M, Kavimandan SK, Subba Rao NS (1976) Occurrence of nitrogen-fixing *Spirillum* in roots of rice, sorghum, maize and other plants. Indian J Exp Biol 14:638–639

Lawongsa P, Boonkerd N, Wongkaew S, O'Gara F, Teaumroong N (2008) Molecular and phenotypic characterization of potential plant growth promoting *Pseudomonas* from rice and maize rhizospheres. World J Microbiol Biotechnol 24:1877–1884

Lee KK, Watanabe I (1977) Problems of the acetylene reduction technique applied to water-saturated paddy soils. Appl Environ Microbiol 34:654–660

Liesack W, Schnell S, Revsbech NP (2000) Microbiology of flooded rice paddies. FEMS Microbiol Rev 24:625–645

Lim J-M, Jeon CO, Lee GS, Park D-J, Kang U-G, Park C-Y, Kim C-J (2007) *Leeia oryzae* gen. nov., sp. nov., isolated from a rice field in Korea. Int J Syst Evol Microbiol 57:1204–1208

Lu Y, Rosencrantz D, Liesack W, Conrad R (2006) Structure and activity of bacterial community inhabiting rice roots and the rhizosphere. Environ Microbiol 8:1351–1360

Manjunath M, et al. Biocontrol potential of cyanobacterial metabolites against damping off disease caused by *Pythium aphanidermatum* in solanaceous vegetables. Arch Phytopathol Plant Protect (in press).

Mano H, Morisaki H (2008) Endophytic bacteria in the rice plant. Microbes Environ 22:175–185

Mano H, Tanaka F, Nakamura C, Kaga H, Morisaki H (2007) Culturable endophytic bacterial flora of the maturing leaves and roots of rice plants (*Oryza sativa*) cultivated in a paddy field. Microbes Environ 22:175–185

Mirza MS, Rasul G, Mehnaz S, Ladha JK, So RB, Ali S, Malik KA (2000) Beneficial effects of inoculated nitrogen-fixing bacteria on rice. In: Ladha JK, Reddy PM (eds) The quest for nitrogen fixation in rice. International Rice Research Institute, Makati City, Philippines, pp 191–204

Mishra S, Rath AK, Adhya TK, Rao VR, Sethunathan N (1997) Effect of continuous and alternate water regimes on methane efflux from rice under greenhouse conditions. Biol Fertil Soils 24:399–405

Misra S, Kaushik BD (1989) Growth promoting substances of cyanobacteria. I. Vitamins and their influence on rice plant. Proc Indian Sci Acad B55:295–300

Mukhopadhyay K, Garrison NK, Hinton DM, Bacon CW, Khush GS, Peck HD, Datta N (1996) Identification and characterization of bacterial endophytes of rice. Mycopathologia 134:151–159

Murase J, Noll M, Frenzel P (2006) Impact of protists on the activity and structure of the bacterial community in a rice field soil. Appl Environ Microbiol 72:5436–5444

Nakaji T, Noguchi K, Oguma H (2008) Classification of rhizosphere components using visible–near infrared spectral images. Plant Soil 310:245–261

Nayak S, Prasanna R (2007) Soil pH and its role in cyanobacterial abundance and diversity in rice field soils. Appl Ecol Environ Res 5:103–113

Nayak S, Prasanna R, Dominic TK, Singh PK (2001) Floristic abundance and relative distribution of different cyanobacterial genera in rice field soil at different crop growth stages. Phykos 40:14–21

Nayak S, Prasanna R, Pabby A, Dominic TK, Singh PK (2004) Effect of BGA–*Azolla* biofertilizers on nitrogen fixation and chlorophyll accumulation at different depths in soil cores. Biol Fertil Soils 40:67–72

Nayak S, Prasanna R, Prasanna BM, Sahoo DB (2008) Genotypic and phenotypic diversity of *Anabaena* isolates from diverse rice agroecologies of India. J Basic Microbiol 49:165–177

Noll M, Frenzel P, Conrad R (2008) Selective stimulation of type I methanotrophs in rice paddy soil by urea fertilization revealed by RNA-based stable isotope probing. FEMS Microbiol Ecol 65:125–132

Pabby A, Prasanna R, Singh PK (2004) Biological significance of *Azolla* and its utilization in agriculture. Proc India Natl Sci Acad Sect B 70:301–335

Piao Z et al (2005) Changes in acetylene reduction activities and effects of inoculated rhizosphere nitrogen fixing bacteria on rice. Biol Fertil Soils 41:371–378

Ponnamperuma FN, Detruck P (1993) A review of fertilization in rice production. Int Rice Commun News 42:1–12

Prasanna R, Nayak S (2007) Influence of diverse rice soil ecologies on cyanobacterial diversity and abundance. Wetlands Ecol Manag 15:127–134

Prasanna R et al (2002) Methane production in rice soils is inhibited by cyanobacteria. Microbiol Res 157:1–6

Prasanna R, Tripathi U, Dominic TK, Singh AK, Yadav AK, Singh PK (2003) An improvised technique for measurement of nitrogen fixation by blue green algae and *Azolla* using intact soil cores. Expl Agric 39:145–150

Prasanna R, Saxena AK, Jaiswal P, Nayak S (2006) Development of alternative support system for viable count of cyanobacteria by MPN method. Folia Microbiol 51:455–458

Prasanna R, Jaiswal P and Kaushik BD (2008a) Cyanobacteria as potential options for environmental sustainability – promises and challenges. Indian J Microbiol 48:89–94

Prasanna R, Lata, Tripathi RM, Gupta V, Middha S, Joshi M, Ancha R and Kaushik BD (2008b) Evaluation of fungicidal activity of extracellular filtrates of cyanobacteria - possible role of hydrolytic enzymes. J Basic Microbiol 48:186–194

Prasanna R, Jaiswal P, Nayak S, Sood A, Kaushik B D (2009) Cyanobacterial diversity in the rhizosphere of rice and its ecological significance. Indian J Microbiol 49:89–97

Quesada A, Leganis F, Valiente EF (1997) Environmental factors controlling N_2 fixation in Mediterranean rice fields. Microb Ecol 34:39–48

Ramakrishna C, Sethunathan N (1982) Stimulation of autotrophic ammonium oxidation in rice rhizosphere soil by the insecticide carbofuran. Appl Environ Microbiol 44:1–4

Rangarajan S, Saleena LM, Nair S (2002) Diversity of *Pseudomonas spp.* isolated from rice rhizosphere populations grown along a salinity gradient. Microb Ecol 43:280–289
Reichardt W, Mascarina G, Padre B, Doll J (1997) Microbial communities of continuously cropped, irrigated rice fields. Appl Environ Microbiol 63:233–238
Reinhold-Hurek B, Hurek T (1998) Life in grasses: diazotrophic endophytes. Trends Microbiol 6:139–144
Roger PA (1986) Effect of algae and aquatic macrophytes on nitrogen dynamics in wetland rice fields. Conference of International Soil Science Society, Hamburg, Germany, pp 13–21
Sahrawat KL (2000) Macro and micronutrients removed by upland and lowland rice cultivars in West Africa. Commun Soil Sci Plant Anal 31:717–723
Sano Y, Fujii T, Iyama S, Hirota Y, Komagata K (1981) Nitrogen fixation in the rhizosphere of cultivated and wild rice strains. Crop Sci 21:758–761
Sakai S, Imachi H, Sekiguchi Y, Ohashi A, Harada H, Kamagata Y (2007) Isolation of key methanogens for global methane emission from rice paddy fields: a novel isolate affiliated with the clone cluster rice cluster I. Appl Environ Microbiol 73:4326–4331
Satpathy SN, Rath AK, Ramakrishnan B, Rao VR, Adhya TK, Sethunathan N (1997) Diurnal variation in methane efflux at different growth stages of tropical rice. Plant Soil 195:267–271
Schwarz MVJ, Frenzel P (2005) Methanogenic symbionts of anaerobic ciliates and their contribution to methanogenesis in an anoxic rice field soil. FEMS Microbiol Ecol 52:93–99
Sethunathan N, Rao VR, Adhya TK, Raghu K (1982) Microbiology of rice soils. CRC Crit Rev Microbiol 10:125–172
Shrestha RK, Ladha JK (1996) Genotypic variation in promotion of rice dinitrogen fixation as determined by ^{15}N dilution. Soil Sci Soc Am J 60:1815–1821
Singh PK, Bisoyi RN (1989) Blue green algae in rice fields. Phykos 28:181–195
Stoltzfus JR, So R, Malarvizhi PP, Ladha JK, De Bruijn FJ (1997) Isolation of endophytic bacteria from rice and assessment of their potential for supplying rice with biologically fixed nitrogen. Plant Soil 194:25–36
Sun L, Qiu F, Zhang X, Dai X, Dong X, Song W (2007) Endophytic bacterial diversity in rice (*Oryza sativa* L.) roots estimated by 16S rDNA sequence analysis. Microb Ecol 55:415–424
Tan Z, Hurek T, Vinuesa P, Müller P, Ladha JK, Reinhold-Hurek B (2001) Specific detection of *Bradyrhizobium* and *Rhizobium* strains colonizing rice (*Oryza sativa*) roots by 16S–23S ribosomal DNA intergenic spacer-targeted PCR. Appl Environ Microbiol 67:3655–3664
Tan Z, Hurek T, Reinhold-Hurek B (2003) Effect of N-fertilization, plant genotype and environmental conditions on *nifH* gene pools in roots of rice. Environ Microbiol 5:1009–1015
Thakuria D, Talukdar NC, Goswami C, Hazarika S, Boro RC, Khan MR (2004) Characterization and screening of bacteria from rhizosphere of rice grown in acidic soils of Assam. Curr Sci 86:978–985
Tiller KG (1981) Micronutrients. Soils: an Australian viewpoint. CSIRO, Melbourne
Tripathi BP, Ladha JK, Timsina J, Pascua SR (1997) Nitrogen dynamics and balance in intensified rainfed lowland rice-based cropping systems. Soil Sci Soc Am J 61:812–821
Trolldenier G (1987) Estimation of association nitrogen fixation in relation to water regime and plant nutrition in a long term pot experiment with rice (*Oryza sativa* L.). Biol Fertil Soils 5:133–140
Ueda T, Suga Y, Yahiro N, Matsuguchi T (1995) Remarkable N_2-fixing bacterial diversity detected in rice roots by molecular evolutionary analysis of *nifH* gene sequences. J Bacteriol 177:1414–1417
Van Nieuwenhove C, Merck R, Van Holm L, Ulassak K (2001) Dinitrogen fixation activity of *Azorhizobium caulinodans* in the rice (*Oryza sativa* L.) rhizosphere assessed by nitrogen balance and N^{15} dilution method. Biol Fertil Soils 33:25–32
Van Tran V, Berge O, Ngo Ke S, Balandreau J, Heulin T (2000) Repeated beneficial effects of rice inoculation with a strain of *Burkholderia vietnamiensis* on early and late yield component in low fertility sulphate acid soils of Vietnam. Plant Soil 218:273–284
Venkataraman GS (1972) Algal biofertilizers and rice cultivation. Today and Tomorrow Publishers, Delhi

Verma SC, Singh A, Chawdhury SP, Tripathi AK (2004) Endophytic colonization ability of two deep-water rice endophytes, *Pantoea* sp. and *Ochrobactrum* sp. using green fluorescent protein reporter. Biotechnol Lett 26:425–429

Wang Y, Ke X, Wu L, Lu Y (2009) Community composition of ammonia-oxidizing bacteria and archaea in rice field soil as affected by nitrogen fertilization. Syst Appl Microbiol 32:27–36

Watanabe I, De Datta SK, Roger PA (1988) Nitrogen cycling in wetland rice soil. In: Wilson JR (ed) Advances in nitrogen cycling in agricultural ecosystems. CAB International, Wallingford, pp 239–256

Watt M, Silk WK, Paassioura JB (2006) Rates of root and organism growth, soil conditions, and temporal and spatial development of the rhizosphere. Ann Bot 97:839–855

Whitton BA, Aziz A, Kawecka B, Rother JA (1988) Ecology of deep water rice fields in Bangladesh 3. Associated algae and macrophytes. Hydrobiologia 169:31–42

Xie HG, Cai MY, Tao GC, Steinbergan Y (2003) Cultivable heterotrophic N_2-fixing bacterial diversity in rice fields in the Yangtze river plain. Biol Fertil Soils 37:29–38

Yan Y et al (2008) Nitrogen fixation island and rhizosphere competence traits in the genome of root-associated *Pseudomonas stutzeri*. Proc Natl Acad Sci USA 105:7564–7569

Yanni YG, El-Fattah FKA (1999) Towards integrated biofertilization management with free living and associative dinitrogen fixers for enhancing rice performance in the Nile delta. Symbiosis 27:319–331

Yanni YG et al (1997) Natural endophytic association between *Rhizobium leguminosarum* bv. *trifolii* and rice roots and assessment of its potential to promote rice growth. Plant Soil 194:99–114

Yoshida S, Ahn JS, Furno DA (1973) Occurrence, diagnosis and correction of zinc deficiency of lowland rice. Soil Sci Plant Nutr 101:83–89

Zeigler R, Barclay A (2008) The relevance of rice. Rice 1:3–10

Chapter 8
Rhizobial Symbioses in Tropical Legumes and Non-Legumes

Aline López-López, Mónica Rosenblueth, Julio Martínez, and Esperanza Martínez-Romero

8.1 Introduction: Legume Origin, Evolution and Domestication

Over 18,000 species have been described in the Leguminosae family (Legumes of the World 2005; Lewis et al. 2005). They include many important food crops, as well as species that are used for timber, oils, fibers, and medicines. Species richness is greatest in the tropics and legume evolution appears to have followed a tropical to temperate direction (Schrire et al. 2005). One of the three legume subfamilies, the Caesalpinioideae, is very diverse in tropical regions of the New World, Africa, and Southeast Asia. The subfamily Mimosoideae is also mostly tropical and is particularly abundant in arid and semi-arid regions. Many tropical legumes have been domesticated for human consumption (Table 8.1) and in most centers of domestication at least one legume has been developed as a main crop. The need for legumes in the human diet is probably related to their high protein content (Table 8.2). In spite of these nutritional advantages, legumes are currently little consumed in many countries.

Beans were domesticated in Mesoamerica from tiny, small-seeded, climbing wild beans around 7,000 years ago (Gepts 1998; Chacon et al. 2005). Currently, a large diversity of bean cultivars is found worldwide and many have been adopted for human nutrition. Soybean was originally domesticated in China where it has been grown for more than 5,000 years, and more than 23,000 soybean cultivars are currently found in Asia (Man et al. 2008). In the last 100 years, soybeans were introduced to the USA and Brazil, which both became important soybean producers, as well as to various other countries worldwide.

A. López-López, M. Rosenblueth, J. Martínez, and E. Martínez-Romero (✉)
Ecological Genomics Program, Centro de Ciencias Genómicas, UNAM, Av. Universidad SN, Chamilpa, Cuernavaca Mor. 62210, Mexico
e-mail: emartine@ccg.unam.mx

P. Dion (ed.), *Soil Biology and Agriculture in the Tropics*, Soil Biology 21,
DOI 10.1007/978-3-642-05076-3_8,

Table 8.1 List of some domesticated legumes

Plant species	Origin	Reference
Arachis hypogaea (peanut)	Tropical lowlands of South America	Pickersgill (2007)
Cajanus cajan (pigeon pea)	India	Gepts and Papa (2003); Smartt (1985a)
Canavalia ensiformis (jack bean)	Mesoamerica	Pickersgill (2007)
Canavalia plagiosperma (broad bean)	Andean region	Pickersgill (2007)
Cicer arietinum (chickpea)	Near East	Smartt (1984)
Glycine max (soybean)	East Asia	Gepts and Papa (2003)
Lathyrus sativus (grass pea)	Southwest to Central Asia	Smartt (1984)
Lens culinaris (lentils)	Eastern Mediterranean to Near East	Smartt (1984)
Leucaena esculenta (guaje colorado)	Mesoamerica	Casas et al. (2007)
Leucaena leucocephala (guaje or white lead tree)	Mesoamerica	Hughes et al. (2007)
Lupinus mutabilis (Andean lupin)	Andean region	Hernández Bermejo and León (1994)
Medicago sativa (alfalfa)	Near East to Central Asia	Muller et al. (2003)
Pachyrhizus ahipa (ajipo or Andean yam bean)	Andean region	Pickersgill (2007)
Pachyrhizus erosus (jícama or yam bean)	Mesoamerica	Pickersgill (2007); Hernández Bermejo and León (1994)
Pachyrhizus tuberosus (Amazonian yam bean)	Tropical lowlands of South America	Pickersgill (2007)
Phaseolus acutifolius (tepary bean)	Mesoamerica	Pickersgill (2007); Hernández Bermejo and León (1994)
Phaseolus coccineus (scarlet runner bean)	Mesoamerica	Pickersgill, (2007); Hernández Bermejo and León (1994)
Phaseolus dumosus	Mesoamerica	Pickersgill (2007)
Phaseolus lunatus (lima bean)	Mesoamerica and Andean region	Sauer (1993); Pickersgill (2007)
Phaseolus vulgaris (kidney bean)	Mesoamerica and Andean region	Pickersgill (2007)
Pisum sativum (pea)	Eastern Mediterranean coast	Smartt (1984)
Vicia ervilia (bitter vetch)	Mediterranean	Hernández Bermejo and León (1994)
Vicia faba (faba bean)	Near East	Smartt (1984)
Vigna aconitifolia (moth bean)	India	Smartt (1985b)
Vigna angularis (adzuki bean)	East Asia	Kaga et al. (2008)
Vigna mungo (urd bean)	India	Fuller (2007)
Vigna radiata (mung bean)	India	Fuller (2007)
Vigna umbellata (rice bean)	Indo-China to South-East Asia	Smartt (1985b)
Vigna subterranea (Bambara groundnut)	West Africa	Smartt (1985b)
Vigna unguiculata (cowpea)	West Africa	Smartt (1985b)

Table 8.2 Protein content in raw mature seeds of some legumes[a]

Legume	Percent of protein[b]
Arachis hypogaea (peanut)	25.80
Cajanus cajan (pigeon pea)	21.70
Cicer arietimun (chickpea)	19.30
Dolichos purpureus (hyacinth bean)	23.90
Glycine max (soybean)	36.49
Lens culinaris (lentils)	25.80
Lupinus albus (white lupin)	36.17
Phaseolus lunatus (lima bean)	21.46
Phaseolus vulgaris (kidney bean)	23.58
Pisum sativum (pea)	24.55
Psophocarpus tetragonolobus (winged bean)	29.65
Vicia faba (faba bean)	26.12
Vigna aconitifolia (moth bean)	22.94
Vigna angularis (adzuki bean)	19.87
Vigna mungo (urd bean)	25.21
Vigna radiata (mung bean)	23.86
Vigna unguiculata subsp. *cylindrica* (cowpea)	23.85
Vigna uniguiculata subsp. *sesquipedalis* (yardlong bean)	24.33

[a]*Source*: Agricultural Research Service, United States Department of Agriculture (http://www.ars.usda.gov/main/main.htm)
[b]Calculated as N content on a per dry weight basis, × 6.25

8.2 Rhizobial Legume Symbioses

The outstanding characteristic of legumes is their ability to establish symbiosis with nitrogen-fixing bacteria in nodules, formed mainly on roots but also on stems (Sprent 2001; Giller 2001). Tropical legume trees hold records for the amounts of nitrogen fixed per ha per year, with, for example, over 200 kg $ha^{-1}year^{-1}$ fixed by *Leucaena* trees (Peoples et al. 1995). These amounts are comparable to those used to fertilize high N-demanding crops. Biological nitrogen fertilization has always been considered as a substitute for chemical N fertilization, which makes it of great practical interest, since N fertilizers are the most widely used agrochemical products. "Obtaining N from legumes is potentially more sustainable than from industrial sources" (Crews and Peoples 2004). N fertilizers have become increasingly expensive and, therefore, research on N fixation and its uses in agriculture are relevant issues to pursue. Two books, "Tropical Legumes: Resources for the Future" (Anonymous 1979) and "*Leucaena*, promising forage and tree crop for the tropics" (Anonymous 1977) encouraged the use of legumes in developing countries, but in spite of their positive attributes and their potential to improve farming and nutrition in poor countries, most legumes are still underexploited.

Nodule-forming bacteria with specificity for local hosts are encountered at the centers of origin of legumes, thus indicating a selective effect by legumes on bacteria, and possibly a history of common evolutionary processes. Co-evolution of legume hosts and rhizobia has been suggested in several cases (Aguilar et al. 2004;

Silva et al. 2005; Lie et al. 1987), but recruitment of new symbiotic bacteria in areas where legumes have been introduced is documented as well. For example, some soybean varieties introduced in Africa may nodulate with native *Bradyrhizobium* strains and it was proposed that cultivars with better affinities for local bacteria should be selected and used in African regions where the availability of inoculants is low (Abaidoo et al. 2000). Legume trees such as *Leucaena* are nodulated mainly by native sinorhizobia in Mexico (Wang et al. 1999a, b) and *Sinorhizobium* strains were found to be the most efficient for nitrogen fixation in *Leucaena* plants (Bala and Giller 2006). However, in different areas, introduced *Leucaena* is nodulated by other bacterial species, such as *Rhizobium tropici* in Brazil (Martínez-Romero et al. 1991; Menna et al. 2006), *Mesorhizobium loti*-like bacteria in Malawi, and *Rhizobium gallicum*-like bacteria in Nigeria (Bala et al. 2003). Moreover, a strain related to *R. tropici, Rhizobium* TAL1145 (=CIAT 1967), was isolated from *Leucaena diversifolia* growing in Australian alluvial loam soils by Moawad and Bohlool (1984).

Although there is now considerable knowledge about the diversity of rhizobial species, the symbionts of many tropical legumes remain to be studied. Even less is known about the interaction of *Rhizobium* with tropical legumes, and the cellular and molecular symbiotic mechanisms are poorly characterized. It is often difficult for scientists in developing countries to identify nodule bacteria from native legumes, because bacterial classification is not an easy task and requires molecular techniques; in addition, the constantly changing taxonomy becomes confusing for researchers outside the field. With few exceptions, this situation has left the task of classifying nodule bacteria from legumes of Africa, Central America, and many other regions to Europeans and Americans (Barrett and Parker 2006; Qian et al 2003; Moulin et al. 2001; Haukka et al. 1998; de Lajudie et al. 1994; Dreyfus et al. 1988; Nick et al. 1999). A project to study underground diversity was financed by GEF (Global Environment Facility) and implemented by the United Nations Environmental Program, with the aim of allowing megadiverse countries to study the diversity of various soil organisms. India, Côte d'Ivoire, Kenya, Uganda, Indonesia, Mexico, and Brazil are participating in this project (Dance 2008). As a part of this project, we identified about 20 novel lineages of *Bradyrhizobium* in the rainforest of Los Tuxtlas in Mexico (Ormeño-Orrillo et al. unpublished).

As occurs for other bacterial groups, taxonomy of rhizobia is not without problems, and some named species may in fact correspond to polyphyletic groups. There has also been considerable discussion about some rhizobial names. Many scientists have rejected the name *Sinorhizobium*; for example, strain NGR 234, the rhizobial strain with the largest known host range (Pueppke and Broughton 1999), is frequently referred to as *Rhizobium* in spite of it clearly being a member of the genus *Sinorhizobium* (Lloret et al. 2007). The designation of *Agrobacterium* as *Rhizobium* (Young et al. 2001) was also criticized (Farrand et al 2003), and basically not accepted by researchers when publishing papers on *Agrobacterium*. More recently, the change of name of *Sinorhizobium* to *Ensifer* (Young et al. 2003) has elicited even more confusion and disagreement. The name *Ensifer* has priority over *Sinorhizobium* as it was described earlier and, therefore, when some *Ensifer* and *Sinorhizobium* strains were found to belong to the same genus, the name that took priority was *Ensifer* according to judicial rules. The latter rules appear to be quite inflexible and not always useful and accepted by researchers in the field. Indeed, owing to the arbitrary nature of these decisions, although all sinorhizobial species should, theoretically, be designated as *Ensifer*, most scientists have refused to change names and most published papers on the genus use the name *Sinorhizobium*. Another example of a potentially confusing name change in nodule bacteria comes from the β-proteobacterial genus, *Ralstonia*, the nodulating species of which, *R. taiwanensis* (Chen et al. 2001), was renamed twice in 1 year, first to *Wautersia* (Vaneechoutte et al. 2004) and then to *Cupriavidus* (Vandamme and Coenye 2004), which is the name currently used. A deeper knowledge of bacterial species and genera will probably emerge from genomic studies and will hopefully give taxonomy a new basis and perspective.

Table 8.3 lists the nodulating bacterial species as they are known in 2008. In the Table, tropical legume hosts are indicated in boldface.

Phylogenetic trees show nodule bacteria to be closely related to non-symbiotic bacteria isolated from non-nodule environments, and hence it appears that symbiotic genes, such as *nod* genes, are easily lost when they are not providing an advantage to the bacteria or in the absence of host plants. Non-symbiotic rhizobia such as *Rhizobium selenireducens*, *Rhizobium daejeonense*, *Rhizobium cellulosilyticus*, *Bradyrhizobium betae*, and bacteria of the genera *Agromonas*, *Aminobacter*, *Blastobacter*, *Brucella*, *Mycoplasma*, *Schinella* and others are closely related to nodulating bacteria. Another example is *Agrobacterium tumefaciens*, strains of which have been found in nodules from diverse legumes in many different locations (de Lajudie et al. 1999; Tan et al. 1999; Bala and Giller 2001; Chen et al. 2000; Khbaya et al. 1998; Man et al. 2008), but in most cases were not capable of nodulating when tested in single strain assays. In some cases nodulation was confirmed for *Agrobacterium* isolates (Rincón-Rosales et al. 2008), and those found in *Acaciella angustissima* showed very low levels of nitrogen fixation.

8.3 *Phaseolus vulgaris* Symbioses

The common bean has been introduced to many regions and has become the main legume food crop in some regions of Africa and Latin America. The improvement of its nitrogen-fixing capacity has been the goal of several projects (Hardarson and Atkins 2003; Snoeck et al. 2003). This is because bean has generally been considered to be amongst the crop legumes with the lowest levels of nitrogen fixation (Hardarson and Atkins 2003), although it should be noted that cultivars with a high capacity of fixing nitrogen are known (Peña-Cabriales and Castellanos 1993; Martínez-Romero et al. 1998). The symbiotic bacteria of *P. vulgaris* have been particularly well studied. The rhizobia associated with beans have been isolated and analyzed in many geographical regions, and different species have been found in the sites where beans originated as compared to those areas where beans have been

Table 8.3 Rhizobial and other legume-nodulating bacterial species

Genus	Species and type strain[a]	Original host[b]	Site of origin	Reference
α-Proteobacteria				
Azorhizobium	*A. caulinodans* ORS571^T	***Sesbania rostrata***	Senegal, Africa	Dreyfus et al. (1988)
	A. doebereinerae UFLA1-100	***S. virgata***	Southeast Brazil	Moreira et al. (2006)
Bradyrhizobium	*B. betae*[c] PL7HG1^T	*Beta vulgaris*[c]	Spain	Rivas et al. (2004)
	B. canariense BTA-1^T	*Chamaecytisus proliferus*	Tenerife, Canary Islands	Vinuesa et al. (2005)
	B. denitrificans			
	B. elkanii USDA76^T	*Glycine max*	Maryland, U.S.A.	Kuykendall et al. (1992)
	B. iriomotense EK05^T	*Entada koshunensis*	Okinawa, Japan	Islam et al. (2008)
	B. japonicum USDA6^T	*Glycine max*	Japan	Kuykendall et al. (1992)
	B. liaoningense USDA3622^T	*Glycine max*	Heilongjiang, China	Xu et al. (1995)
	B. yuanmingense CCBAU 1071^T	*Lespedeza cuneata*	Yuanmingyuan, Beijing, China	Yao et al. (2002)
Devosia	*D. neptuniae* J1^T	***Neptunia natans***	India	Rivas et al. (2003)
Methylobacterium	*M. nodulans* ORS2060^T	***Crotalaria pocarpa***	Senegal, Africa	Sy et al. (2001)
Mesorhizobium	*M. albiziae* CCBAU 61158^T	***Albizia kalkora***	Sichuan, China	Wang et al. (2007)
	M. amorphae LMG 18977^T	*Amorpha fruticosa*	China	Wang et al. (1999a, b)
	M. caraganae CCBAU 11299^T	*Caragana* spp.	Mongolia	Guan et al. (2008)
	M. chacoense PR5^T	*Prosopis alba*	Chancaní, Argentina	Velázquez et al. (2001)
	M. ciceri USDA 3383^T	*Cicer arietinum*	Spain	Jarvis et al. (1997); Nour et al. (1994)
	M. gobiense	*Astragalus filicaulis* and others	Xinjiang, China	Han et al. (2008a, b)
	M huakuii CCBAU 2609^T	*Astragalus sinicus*	Nanjing, China	Chen et al. (1991); Jarvis et al. (1997)
	M. loti USDA 3471^T	*Lotus tenuis*	New Zealand	Jarvis et al. (1997, 1982)
	M. mediterraneum USDA 3392, ^{T} *M. metallidurans*	*Cicer arietinum*	Spain	Jarvis et al. (1997); Nour et al. (1995)
	M. plurifarium ORS1032^T	***Acacia senegal***	Senegal, Africa	de Lajudie et al. (1998a)
	M. septentrionale SDW014 ^{T}	*Astragalus adsurgens*	North China	Gao et al. (2004)
	M. tarimense	*Lotus frondosus*	Xinjiang, China	Han et al. (2008a, b)
	M. temperatum SDW018	*Astragalus adsurgens*	North China	Gao et al. (2004)
	M. tianshanense CCBAU3306^T	*Glyzyrrhiza pallidiflora*	Xinjinang, China	Chen et al. (1995), Jarvis et al. (1997)
	M. thiogangeticum[c] SJT^T	Soil	Gangetic plains, India	Ghosh and Roy (2006)

Phyllobacterium	*P. trifolii* PETP02*T*	*Trifolium pratense*	Spain	Valverde et al. (2005)
Rhizobium	*R. cellulosilyticum* ALA10B2^{T}	*Populus alba*^{c}		Garcia-Fraile et al. (2007)
	R. daejeonense L61^{T}	cyanide degrading bioreactor	Korea	Quan et al. (2005)
	R. etli CFN 42^{T}	***Phaseolus vulgaris***,	Guanajuato, Mexico	Segovia et al. (1993)
	R. fabae	*V. faba*	China	Tian et al. (2008)
	R. galegae HAMBI 540	*Galega orientalis*	Finland	Lindström (1989)
	R. gallicum R602sp^{T}	*Phaseolus vulgaris*	Maine et Loire, France	Amarger et al. (1997)
	R. giardinii H152^{T}	*Phaseolus vulgaris*	Côte d´Or, France	Amarger et al. (1997)
	R. hainanense CCBAU 57015^{T}	*Desmodium sinuatum*	Hainan, China	Chen et al. (1997)
	R. huautlense LMG 18254^{T}	***Sesbania herbacea***	Sierra de Huautla, Morelos, México	Wang et al. (1998)
	R. indigoferae CCBAU 71042 ^{T}	*Indigofera* spp.	Loess, Northeast China	Wei et al. (2002)
	R. leguminosarum bv. *viciae* USDA 2370^{T}	*Vicia faba*		Frank (1879); Frank (1889); Jarvis et al. (1982)
	R. lusitanum P1-7^{T}(=LMG 22705^{T})	*Phaseolus vulgaris*	Portugal	Valverde et al. (2006)
	R. loessense CCBAU 7190B^{T}	*Astragalus complanatus*	Loess , China	Wei et al. (2003)
	R. miluonense CCBAU41251^{T}	*Lespedeza chinensis*	Hunan, China	Gu et al. (2008)
	R. mongolense USDA 1844^{T}	*Medicago ruthenica*	Mongolia	van Berkum et al. (1998)
	R. multihospitium CCBAU 83401^{T}	*Halimodendron halodendron*	Xinjiang, China	Han et al. (2008a,b)
	R. oryzae Alt 505^{T}	*Oryza sativa*	Guangzhou, China	Peng et al. (2008)
	R. pisi DSM 30132^{T}	*Pisum sativum*		Ramírez-Bahena et al. (2008)
	R. rhizogenes 163C	*Prunus persica*		Velázquez et al. (2005)
	R. selenireducens^{c} B1^{T}	*Bioreactor*		Hunter et al. (2007)
	R. tropici CIAT 899^{T}	***Phaseolus vulgaris***	Colombia	Martínez-Romero et al. (1991)
	R. undicola ORS 992 ^{T}	***Neptunia natans***	Senegal, Africa	de Lajudie et al. (1998b); Young et al. (2001)
	R. yanglingense CCBAU 71623^{T}	*Gueldenstaedtia multiflora*	Gansu, Northeast China	Tan et al. (2001)
Sinorhizobium	*S. adhaerens* ATCC 33212^{T}		Pennsylvania,U.S.A.	Casida (1982); Willems et al. (2003)

(*continued*)

Table 8.3 (continued)

Genus	Species and type strain[a]	Original host[b]	Site of origin	Reference
	S. americanum CFNEI156^{T}	***Acacia farnesiana***	Sierra de Huautla, Morelos, México	Toledo et al. (2003)
	S. arboris HAMBI 1552^{T}	***Prosopis chilensis***	Sudan, Africa	Nick et al. (1999)
	S. chiapanecum ITTG 570^{T}	*Acaciella angustissima*	Chiapas, Mexico	Rincón-Rosales et al. (2008)
	S. fredii[d] USDA 205^{T}	*Glycine max (soya)*	Honan, China	Chen et al. (1988); Scholla and Elkan (1984)
	S. kostiense HAMBI 1489^{T}	***Acacia senegal***	Sudan, Africa	Nick et al. (1999)
	S. kummerowiae CCBAU 71714 ^{T}	*Kummerowia stipulacea*	Loess , China	Wei et al. (2002)
	S. medicae USDA 1037^{T}	*Medicago truncatula*	France	Rome et al. (1996)
	S. meliloti USDA1002^{T}	*M. sativa*		Jordan (1984)
	S. mexicanum ITTG R7^{T}	***Acaciella angustissima***	Chiapas, Mexico	Lloret et al. (2007)
	S. morelense LMG 21331^{T}	***Leucaena leucocephala***	Sierra de Huautla, Morelos, Mexico	Wang et al. (2002)
	S. sahelense ORS 609^{T}	***Sesbania cannabina***	Senegal, Africa	de Lajudie et al. (1994)
	S. terangae ORS 1009^{T}	***Acacia laeta***	Senegal, Africa	de Lajudie et al. (1994)
Shinella	*S. kummerowiae* CCCBAU 25048^{T}	*Kummerowia stipulacea*	Chandong, China	Lin et al. (2008)
β-Proteobacteria				
Cupriavidus	*C. taiwanensis* LMG 19424^{T}	*Mimosa pudica*	Taiwan, China	Chen et al. (2001); Vaneechoutte et al. (2004)
Burkholderia	*B. mimosarum* PAS44^{T}	***Mimosa pigra***	Taiwan, China	Chen et al. (2006)
	B. nodosa Br 3437^{T}	***M. scabrella***		Chen et al. (2007)
	B. phymatum STM815^{T}	***Machaerium lunatum***[e]	French Guyana	Vandamme et al. (2002)
	B. sabiae	*Mimosa caesalpiniifolia*	Brazil	Chen et al. (2008)
	B. tuberum STM 678^{T}	*Cyclopia* spp.[f]	South Africa	Elliott et al. (2007)

[a]Designation of type strain is followed by a T superscript
[b]Tropical hosts are given in boldface
[c]Non-symbiotic
[d]*S. xinjiangense* is sister species of *S. fredii*
[e]Originally reported as *Machaerium lunatum* symbiont (Vandamme et al. 2002), but there is no evidence of its nodulation
[f]Originally reported as *Aspalathus carnosa* symbiont from South Africa(Vandamme et al. 2002), but there is no evidence of its nodulation

introduced (reviewed in Martínez-Romero 2003, also shown in Table 8.3). A considerable intraspecific diversity has been encountered within many species of bean-nodulating rhizobia (Piñero et al. 1988; Martínez-Romero et al. 1991; Segovia et al. 1991; Herrera-Cervera et al. 1999; Mhamdi et al. 1999; Rodriguez-Navarro et al. 2000; Mostasso et al. 2002; Silva et al. 2003, 2005; Aguilar et al. 2004). Furthermore, the promiscuity of beans and the large numbers of nodulating bacteria in soils often make it difficult to successfully introduce bacterial inoculants. However, the discovery of a new rhizobial species, *R. tropici*, represented a milestone in the history of bean inoculation, as it has an outstanding capacity to tolerate different stresses (Martínez-Romero et al. 1991). For example, in acid soils of Brazil and Kenya, *R. tropici* has an advantage over acid-sensitive bacteria. Nevertheless, numerous *Rhizobium etli* strains are encountered in Brazil (Grange and Hungria 2004). It was observed that soil liming, which increases the soil pH, has an effect on rhizobial diversity (Andrade et al. 2002). In the context of the present review, it is worth noting that *R. etli*, a species which is very competitive for bean nodulation, is used as inoculant not only for bean (Peralta et al. 2004), but also for maize in Los Tuxtlas, Mexico, and for maize and wheat in the USA (Smith, personal communication).

8.4 Rhizobial Inoculants

Inoculation of introduced crop legumes has resulted in yield increases in many sites. In some cases, the inoculated bacteria may face competition from native bacteria that are diverse and well adapted to the soil and environmental conditions. When this occurs, bacteria from the inoculum (all of which generally share the same genotype) do not necessarily replace all of the native strains in nodules, and often the inoculated strain represents only a very low proportion of the nodule rhizobial population (reviewed in Sessitsch et al. 2002; Burgos et al. 1999).

Along with inoculant technology, strategies to select high quality and competitive inoculant rhizobial strains have been reviewed by, amongst others, Sessitsch et al. (2002) and Hungria et al. (2005). Nodule bacteria have been used as inoculants in agriculture for more than 100 years and rhizobia are amongst the safest bacteria to be delivered to the environment, as no rhizobial species are known to be pathogenic to humans.

The bacteria that initially colonized the soybean nodules following soybean introduction in the USA were possibly those that survived their long transportation from Asia. The USA then pioneered soybean inoculation, although initially there was little knowledge about symbiotic nitrogen fixation, and hence the introduction of bradyrhizobial inoculants was not preceded by the selection of efficient strains. Many midwestern soybean growers faced the problem of growing soybeans in soils containing inefficient strains that were highly competitive (Schmidt et al. 1986). On the other hand, soybean inoculation in Brazil is the most outstanding example of the successful introduction of rhizobia in commercial agriculture

(Hungria et al. 2006; see Chaps. 4 and 11). In Brazil, a careful selection of efficient nitrogen-fixing bacteria is followed by an examination of the performance of selected strains in agricultural assays. Soybean inoculants are now produced commercially in the USA and in other countries, such as Brazil (Izaguirre-Mayoral et al. 2007). In Brazil, rhizobial inoculants are available for tropical legumes, such as *Pueraria*, *Vigna*, *Piptadenia*, *Prosopis*, *Lablab*, *Enterolobium*, and others, and the various bacteria used were classified by 16S rRNA gene sequencing (Menna et al. 2006).

Is there actually a need to inoculate tropical legumes? Some tropical legumes are considered to be promiscuous and hence the need to inoculate these is not obvious. However, we have shown that even those legumes that show low nodulation specificity have preferred symbionts, which seem to be more efficient for nitrogen fixation. For example, *A. angustissima* plants are found to be nodulated by different rhizobial species, but only *Sinorhizobium mexicanum* (Lloret et al. 2007) and *S. chiapanecum* are efficient and competitive species (Rincón-Rosales et al. 2008). Furthermore, whereas *Acaciella* and *Leucaena* plants do not respond to added chemical fertilizers (Rincón-Rosales et al. 2008), inoculation in nurseries is successful in Chiapas, Mexico. Similarly, inoculation increases the survival rate of acacias used in reforestation projects in Morelos, Mexico (Toledo unpublished).

8.5 Rhizobia-Legume Nodulation

There are several reviews on the basic aspects of nodulation and on nodule evolution (Sprent and James 2007), and these will not be covered here. The role of bacterial *nod* genes in determining the production of Nod factors has been well-studied, mainly in temperate rhizobia but also in some tropical species such as those nodulating *Sesbania rostrata* (D'Haeze and Holsters 2002; Suzuki et al. 2007) and *P. vulgaris* (Laeremans et al. 1996; Folch-Mallol et al. 1996; Poupot et al. 1993; Cardenas et al. 1995). A striking result was published in 2007 by Giraud et al. 2007, showing that *nod* genes and consequently Nod factors were not required for nodule formation in *Aeschynomene*, a tropical legume nodulated by *Bradyrhizobium*. This finding may be of relevance when trying to establish nitrogen fixation in non-legumes in the future, since it indicates that nodulation may occur without Nod factor recognition. Another candidate for the establishment of symbioses with non-legumes is *Azorhizobium caulinodans* (Lee et al. 2008), a bacterium that has been found to colonize wheat, rice and other plants and which, unlike *Rhizobium*, fixes nitrogen independently of the legume host. Some bradyrhizobial and nodulating *Burkholderia* strains (Elliott et al. 2007) also have the capacity to fix nitrogen in a free-living state.

It is possible that Nod factors were acquired late in the evolution of the interaction between legumes and rhizobia. The fact that many tropical legumes are nodulated by *Bradyrhizobium* led us to propose that *Bradyrhizobium* may be the oldest nodulating genus (Martínez-Romero 1994). In *A. caulinodans*, a small genomic island containing *nod* genes was probably acquired by lateral transfer

(Lee et al. 2008). *Bradyrhizobium* and *A. caulinodans nod* genes are located on the chromosome, whereas they reside in genomic islands or in plasmids in *Mesorhizobium* (Wang et al. 1998) and in plasmids in *Rhizobium* and *Sinorhizobium*. In *Cupriavidus*, *nod* and *nif* genes are in a plasmid of β-Proteobacterial descent; this plasmid harbors a compact symbiotic region that seemingly was acquired recently (Amadou et al. 2008).

Bacterial resistance to β-lactam antibiotics is considered to be originally chromosome-borne in *Klebsiella pneumoniae*. It was subsequently transferred to plasmids and then mobilized among bacteria by plasmids in the antibiotic era. Similarly, it may be that the presence of *nod* genes in readily transferred and recombined plasmids allowed bacteria to keep pace with the burst of diversification amongst legumes in evolutionary times.

Considerable funding has been obtained by groups studying the temperate legume model *Sinorhizobium meliloti*, a symbiont of alfalfa, and less support has been devoted to the study of tropical rhizobia. Whereas abundant funding and the proficiency of the groups working with *S. meliloti* have made this bacterium the best studied of all the rhizobial species (Fig. 8.1), there are still many genes with unknown function in *S. meliloti*, as in many other bacteria. Nodulation has also been thoroughly studied in model temperate legumes such as *Lotus japonicus* and *Medicago truncatula*. On the other hand, very few tropical symbionts have been studied with respect to the basic aspects of nodulation and nitrogen fixation, and

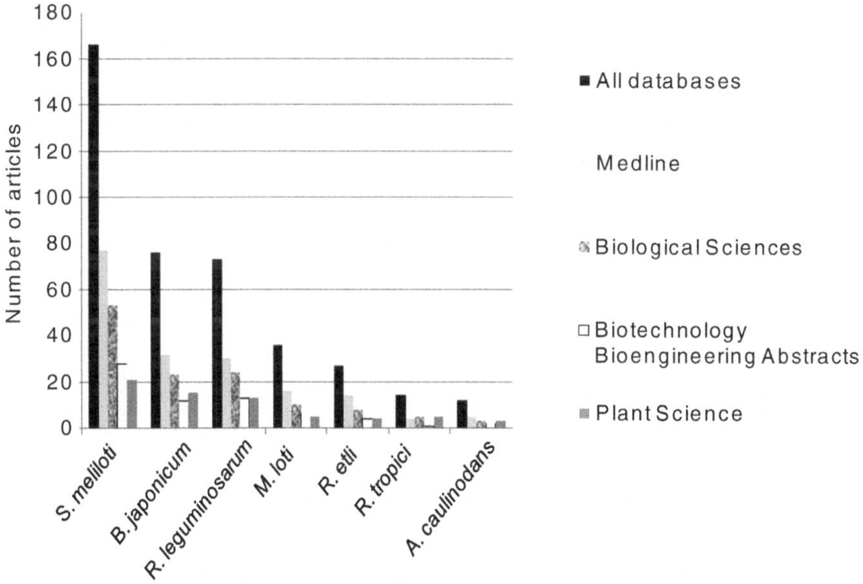

Fig. 8.1 Number of articles published in 2007–2009 with a particular species name in its title, as reported in different data bases. The *numbers along the ordinate* refer to the number of articles published between 2007 and 2009, with the corresponding species name (as given in the *abscissa*) in the title

there is little opportunity and little or no money for scientists in tropical countries to address these fundamental questions. The recent findings on the *Aeschynomene* symbiotic bacteria (see above) demonstrate that the study of tropical legumes may lead to key discoveries on basic symbiotic mechanisms. Indeed, the diversity of nodulation processes in legumes and of nodule morphology has long been addressed by Janet Sprent (Sprent 2001).

Although knowledge of the basic aspects of nodulation might not be required for the practical application of inoculants to field crops, knowledge of the identity of the bacteria to be inoculated is certainly desirable, especially so as to avoid introducing pathogens or potential pathogens to agricultural fields.

A common signaling pathway is involved in both the rhizobial and mycorrhizal symbioses (Catoira et al. 2000; Kistner et al. 2005) and this pathway could be conserved in many as yet unstudied legumes. Many plant genes ("nodulin genes") that participate in the nodule development process have been recognized. Will all this knowledge serve one day to transfer nodulation to non-legumes, such as cereals, which require large amounts of N fertilizer? Only the future will tell us.

More recently, interest in biofuels has led to increased support for research into legumes with a high oil seed content. One example is *Pongamia pinnata*, which is a member of the subfamily Papilionoideae and is native to Asia. Its seeds have a 27%–34% oil content, and it grows in marginal lands (Azam et al. 2005), most likely due to its symbiosis with nitrogen-fixing rhizobia.

8.6 Rhizobia as Plant Endophytes

It has long been known that rhizobia are very successful rhizospheric bacteria, and it is conceivable that during evolution rhizobia first colonized the plant root surface (as typical rhizospheric bacteria), and then invaded the interior of plants as bacterial endophytes. This colonization transition may have occurred early, and hence endophytism could be an ancestral characteristic of rhizobia.

There are probably no plants in nature devoid of bacterial endophytes (Rosenblueth and Martínez-Romero 2006). Endophytes inhabit plants without causing harm, and have been found to be beneficial to plants in different ways, e.g., by producing plant hormones and vitamins, by suppressing pathogens or, in some cases, by fixing nitrogen.

Endophytes may be seedborne (Holland and Polacco 1994; López-López et al. submitted), and be, therefore, transmitted vertically and distributed as far as plant seeds are spread. In addition, soil- and airborne bacteria may colonize the plants during their life cycle, and many bacterial species are known to become resident inside plants as endophytes (Rosenblueth and Martínez-Romero 2006). Endophytes constitute bacterial populations different from those encountered in the rhizosphere and the soil (reviewed in Rosenblueth and Martínez-Romero 2006), thus indicating that there is selection of the bacteria that may inhabit plants. Endophytes need to survive plant defense responses, and adequately use available plant nutrients without overgrowing or causing harm to the plant. By providing some benefit to their hosts, seedborne endophytes will favor seed production and consequently will facilitate their own spread, which suggests that natural selection might favor beneficial endophytes in seeds. For example, endophytes may stimulate seed germination, and accordingly it has been observed that seed germination may diminish if endophytic bacteria populations decrease (Holland and Polacco 1994). There appears to be functional redundancy in endophytes, as different species may be capable of producing similar plant hormones or vitamins (Phillips et al. 1999). It seems probable that bacterial communities, constituting biofilms or consortia, may exhibit complex interactions inside plants including bacteria–bacteria signaling and plant–bacteria signaling.

From a practical perspective, endophytes in plants constitute a problem when testing the effects of added bacteria in inoculation assays, as preexisting seedborne endophytes are difficult to eliminate. Hence, their presence and activity may interfere with the inoculation results.

It is not always easy to determine with certainty that bacteria isolated from surface-disinfected tissues of plants are bona fide endophytes, as they may rather be contaminants surviving an incomplete disinfection process (Reinhold-Hurek and Hurek 1998). As a consequence, many authors prefer to define isolates as plant-associated bacteria and not as endophytes. Many bacteria such as rhizobia, bacilli, and klebsiellas may be both endophytic and rhizospheric (Gutierrez-Zamora and Martinez-Romero 2001; Gardener 2004; Rosenblueth and Martinez-Romero 2004).

The role of bacteria in determining plant phenotypic traits needs to be reviewed. For example, inter-strain competition for alfalfa nodulation was conditioned by the presence of a seedborne *Erwinia* strain (Handelsman and Brill 1985). Bacilli, which have been found associated with many different plants (reviewed in Rosenblueth and Martínez-Romero 2006), produce hormones and volatiles that promote plant growth (Priest 1993; Ryu et al. 2003), reduce fungal infections (Pleban et al 1995), and suppress pathogens (Silo-Suh et al. 1994; Gardener 2004). Co-inoculation of rhizobia and bacilli stimulated nodulation in bean and soybean (Camacho et al. 2001; Halverson and Handelsman 1991; Lian et al. 2001; Estevez de Jensen et al. 2002). We recently found that the most common culturable endophytes in seeds of legumes such as *M. truncatula* and *P. vulgaris* are diverse species belonging to the genus *Bacillus*. Other Firmicutes, as well as actinobacteria that stimulated nodulation, were also obtained as endophytes from *P. vulgaris* (López-López et al. submitted).

Rhizobia have strong affinities for plants, not only in nodules, but also as rhizospheric and endophytic bacteria. Rhizobia have been reported to colonize the interior tissues of rice (see Sect. 7.4), poplar trees, maize and *Sesbania*. The closely related species, *R. etli* and *Rhizobium leguminosarum*, are natural endophytes in rice and maize (Yanni et al. 1997; Gutierrez-Zamora and Martinez-Romero 2001; Rosenblueth and Martinez-Romero 2004). Rhizobia probably originated as endophytes of legumes before nodules evolved, and nodulation was perhaps derived from an ancient endophytic process that had already been established with legume ancestors by the α-proteobacterial progenitor of rhizobia.

Less is known about endophyte–plant interactions than about other plant–bacterial interactions. However, the role of plant defense responses has been recognized. We recently described the role of lipopolysaccharide (LPS) in maize colonization by *R. tropici*. As compared to the wild type, three different mutants in LPS biosynthesis genes reached lower numbers in the rhizosphere and the root interior. The impaired colonization of maize could, in part, be explained by the sensitivity of the mutant *R. tropici* CIAT 899 to the maize antimicrobial 6-methoxy-2-benzoxazolinone (Ormeño-Orrillo et al 2008).

8.7 Practical Applications and Perspectives

Some legumes have become invasive when introduced as exotic species. For example, kudzu (*Pueraria lobata*) has invaded forests, fences, and large areas, self-propagating without the need of human participation. In contrast, other legumes have not been very successful: this is the case of *Mucuna pruriens*, a green manure introduced to provide nitrogen to some agricultural tropical areas. As it requires additional labor, interest in growing *M. pruriens* persisted in the rain forest of Los Tuxtlas in Veracruz, Mexico, only while the Rockefeller Foundation supported the project. Later mucuna was no longer cultivated, in spite of its proven advantages in preserving rain forest soil fertility.

Currently, projects aiming to increase nitrogen fixation may not be considered as high priorities, since humans have largely increased environmental nitrogen to limits such that even diazotrophs may no longer have an advantage (Martínez-Romero 2000). Even nitrogen from rainfall is changing plant communities (Pitcaim et al. 1991; Woodin and Farmer 1993). It could, therefore, be prudent to reuse human-fixed nitrogen instead of introducing higher quantities of nitrogen into the environment.

A global and local management of nitrogen is needed, and a more efficient use of nitrogen fixation by legumes and non-legumes would contribute both to the avoidance of nitrogen leakage into the environment from the inefficient use of fertilizers and to sustainability of tropical agriculture.

Acknowledgments We thank Marco A Rogel, Ivonne Toledo and Ernesto Ormeño-Orrillo for contributing to laboratory work, and Dr. Euan James and Dr. Michael Dunn for reading the manuscript. Partial financial support was from DGAPA PAPIIT IN200709, from GEF PNUMA,TSBF-CIAT and from Macroproyecto UNAM Manejo de Ecosistemas y Desarrollo Humano SDEI-ETID-02. E. M. is grateful to DGAPA UNAM for a postdoctoral fellowship during her sabbatical year at University of California, Davis.

References

Abaidoo RC, Keyser HH, Singleton PW, Borthakur D (2000) *Bradyrhizobium* spp. (TGx) isolates nodulating the new soybean cultivars in Africa are diverse and distinct from bradyrhizobia that nodulate North American soybeans. Int J Syst Evol Microbiol 50:225–234

Aguilar OM, Riva O, Peltzer E (2004) Analysis of *Rhizobium etli* and of its symbiosis with wild *Phaseolus vulgaris* supports coevolution in centers of host diversification. Proc Natl Acad Sci USA 101:13548–13553

Amadou C et al (2008) Genome sequence of the β-rhizobium *Cupriavidus taiwanensis* and comparative genomics of rhizobia. Genome Res 18:1472–1483

Amarger N, Macheret V, Laguerre G (1997) *Rhizobium gallicum* sp. nov. and *Rhizobium giardinii* sp. nov., from *Phaseolus vulgaris* nodules. Int J Syst Bacteriol 47:996–1006

Andrade DS, Murphy PJ, Giller KE (2002) The diversity of *Phaseolus*-nodulating rhizobial populations is altered by liming of acid soils planted with *Phaseolus vulgaris* L. in Brazil. Appl Environ Microbiol 68:4025–4034

Anonymous (1977) *Leucaena*: promising forage and tree crop for the tropics. The Philippine Council for Agriculture and Resources Research and the United States National Academy of Sciences, Washington

Anonymous (1979) Tropical legumes: resources for the future. National Academy of Sciences, Washington, DC

Azam MM, Waris A, Nahar NM (2005) Prospects and potential of fatty acid methyl esters of some non-traditional seed oils for use as biodiesel in India. Biomass Bioenergy 29:293–302

Bala A, Giller KE (2001) Symbiotic specificity of tropical tree rhizobia for host legumes. New Phytol 149:495–507

Bala A, Giller KE (2006) Relationships between rhizobial diversity and host legume nodulation and nitrogen fixation in tropical ecosystems. Nutr Cycling Agroecosyst 76:319–330

Bala A, Murphy P, Giller KE (2003) Distribution and diversity of rhizobia nodulating agroforestry legumes in soils from three continents in the tropics. Mol Ecol 12:917–929

Barrett CF, Parker MA (2006) Coexistence of *Burkholderia Cupriavidus* and *Rhizobium* sp. nodule bacteria on two *Mimosa* spp. in Costa Rica. Appl Environ Microbiol 72:1198–1206

Burgos PA, Castellanos J, Mora Y, Mora J (1999) Field inoculation of common bean (*Phaseolus vulgaris* L.) with high efficiency *Rhizobium* strains: nodule occupancy. In: Martínez E, Hernández G (eds) Highlights of nitrogen fixation research. Kluwer Academic, New York, pp 255–258

Camacho M, Santamaria C, Temprano F, Rodriguez-Navarro DN, Daza A (2001) Co-inoculation with *Bacillus* sp. CECT 450 improves nodulation in *Phaseolus vulgaris* L. Can J Microbiol 47:1058–1062

Cardenas L et al (1995) Isolation, chemical structures and biological activity of the lipo-chitin oligosaccharide nodulation signals from *Rhizobium etli*. Plant Mol Biol 29:453–464

Casas A, Otero-Arnaiz A, Pérez-Negrón E, Valiente-Banuet A (2007) *In situ* management and domestication of plants in Mesoamerica. Ann Bot 100:1101–1115

Casida LE Jr (1982) *Ensifer adherens* gen nov sp. nov.: a bacterial predator of bacteria in soil. Int J Syst Bacteriol 32:339–345

Catoira R et al (2000) Four genes of *Medicago truncatula* controlling components of a Nod factor transduction pathway. Plant Cell 12:1647–1666

Chacon SMI, Pickersgill B, Debouck DG (2005) Domestication patterns in common bean (*Phaseolus vulgaris* L.) and the origin of the Mesoamerican and Andean cultivated races. Theor Appl Genet 110:432–444

Chen WX, Yan GH, Li LJ (1988) Numerical taxonomic study of fast-growing soybean rhizobia and a proposal that *Rhizobium fredii* be assigned to *Sinorhizobium* gen. nov. Int J Syst Bacteriol 38:392–397

Chen WX, Li GS, Qi Y, Wang ET, Ruan HL, Li JL (1991) *Rhizobium huakuii* sp. nov., isolated from the root nodules of *Astragalus sinicus*. Int J Syst Bacteriol 41:275–280

Chen WX, Wang E, Wang S, Li Y, Chen X, Li Y (1995) Characteristics of *Rhizobium tianshanense* sp. nov., a moderately and slowly growing root nodule bacterium isolated from an arid saline environment in Xinjiang, People´s Republic of China. Int J Syst Bacteriol 45:153–159

Chen W-X, Tan Z-Y, Gao J-L, Li Y, Wang E-T (1997) *Rhizobium hainanense* sp. nov., isolated from tropical legumes. Int J Syst Bacteriol 47:870–873

Chen LS, Figueredo A, Pedrosa FO, Hungria M (2000) Genetic characterization of soybean rhizobia in Paraguay. Appl Environ Microbiol 66:5099–5103

Chen W-M, Laevens S, Lee T-M, Coenye T, De Vos P, Mergeay M, Vandamme P (2001) *Ralstonia taiwanensis* sp. nov., isolated from root nodules of *Mimosa* species and sputum of a cystic fibrosis patient. Int J Syst Evol Microbiol 51:1729–1735

Chen W-M et al (2006) *Burkholderia mimosarum* sp. nov., isolated from root nodules of *Mimosa* spp. from Taiwan and South America. Int J Syst Evol Microbiol 56:1847–1851

Chen W-M et al (2007) *Burkholderia nodosa* sp. nov., isolated from root nodules of the woody Brazilian legumes *Mimosa bimucronata* and *Mimosa scabrella*. Int J Syst Evol Microbiol 57:1055–1059

Chen W-M et al (2008) *Burkholderia sabiae* sp. nov., isolated from root nodules of *Mimosa caesalpiniifolia*. Int J Syst Evol Microbiol 58:2174–2179

Crews TE, Peoples MB (2004) Legume versus fertilizer sources of nitrogen: ecological tradeoffs and human needs. Agric Ecosyst Environ 102:279–297

Dance A (2008) What lies beneath. Nature 455:724–725

de Lajudie P et al (1994) Polyphasic taxonomy of rhizobia: emendation of the genus *Sinorhizobium* and description of *Sinorhizobium meliloti* comb. nov., *Sinorhizobium saheli* sp. nov., *Sinorhizobium teranga* sp. nov. Int J Syst Bacteriol 44:715–733

de Lajudie P et al (1998a) Characterization of tropical tree rhizobia and description of *Mesorhizobium plurifarium* sp. nov. Int J Syst Bacteriol 48:369–382

de Lajudie P et al (1998b) *Allorhizobium undicola* gen. nov., sp. nov., nitrogen-fixing bacteria that efficiently nodulate *Neptunia natans* in Senegal. Int J Syst Bacteriol 48:1277–1290

de Lajudie P et al (1999) *Agrobacterium* bv. 1 strains isolated from nodules of tropical legumes. Syst Appl Microbiol 22:119–132

D'Haeze W, Holsters M (2002) Nod factor structures responses and perception during initiation of nodule development. Glycobiology 12:79R–105R

Dreyfus B, García JL, Gillis M (1988) Characterization of *Azorhizobium caulinodans* gen. nov. sp. nov., a stem-nodulating nitrogen-fixing bacterium isolated from *Sesbania rostrata*. Int J Syst Bacteriol 38:89–98

Elliott GN et al (2007) Nodulation of *Cyclopia* spp. (Leguminosae Papilionoideae) by *Burkholderia tuberum*. Ann Bot 100:1403–1411

Estevez de Jensen C, Percich JA, Graham PH (2002) Integrated management strategies of bean root rot with *Bacillus subtilis* and *Rhizobium* in Minnesota. Field Crops Res 74:107–115

Farrand SK, van Berkum PB, Oger P (2003) *Agrobacterium* is a definable genus of the family Rhizobiaceae. Int J Syst Evol Microbiol 53:1681–1687

Folch-Mallol JL et al (1996) Characterization of *Rhizobium tropici* CIAT899 nodulation factors: the role of *nodH* and *nodPQ* genes in their sulfation. Mol Plant Microbe Interact 9:151–163

Frank B (1879) Ueber die Parasiten in den Wurzelanschwillingen der Papilionaceen. Bot Ztg 37 (376–387):394–399

Frank B (1889) Ueber die Pilzsymbiose der Leguminosen Ber. Dtsch Bot Ges 7:332–346

Fuller DQ (2007) Contrasting patterns in crop domestication and domestication rates: recent archaeobotanical insights from the old world. Ann Bot 100:903–924

Gao JL et al (2004) *Mesorhizobium septentrionale* sp. nov. and *Mesorhizobium temperatum* sp. nov., isolated from *Astragalus adsurgens* growing in the northern regions of China. Int J Syst Evol Microbiol 54:2003–2012

Garcia-Fraile P, Rivas R, Willems A, Peix A, Martens M, Martinez-Molina E, Mateos PF, Velazquez E (2007) *Rhizobium cellulosilyticum* sp. nov., isolated from sawdust of *Populus alba*. Int J Syst Evol Microbiol 57:844–848

Gardener BBM (2004) Ecology of *Bacillus* and *Paenibacillus* spp. in agricultural systems. Phytopathology 94:1252–1258

Gepts P (1998) Origin and evolution of common bean: past events and recent trends. Hort Sci 33:1124–1130

Gepts P, Papa R (2003) Evolution during domestication. In: Encyclopedia of life sciences, Wiley, Chichester, pp 1–7

Ghosh W, Roy P (2006) *Mesorhizobium thiogangeticum* sp. nov., a novel sulfur-oxidizing chemolithoautotroph from rhizosphere soil of an Indian tropical leguminous plant. Int J Syst Evol Microbiol 56:91–97

Giller KE (2001) Nitrogen fixation in tropical cropping systems. CAB International, Wallingford

Giraud E et al (2007) Legumes symbioses: absence of *nod* genes in photosynthetic bradyrhizobia. Science 316:1307–1312

Grange L, Hungria M (2004) Genetic diversity of indigenous common bean (*Phaseolus vulgaris*) rhizobia in two Brazilian ecosystems. Soil Biol Biochem 36:1389–1398

Gu CT et al (2008) *Rhizobium miluonense* sp. nov., a symbiotic bacterium isolated from *Lespedeza* root nodules. Int J Syst Evol Microbiol 58:1364–1368

Guan SH, Chen WF, Wang ET, Lu YL, Yan XR, Zhang XX, Chen WX (2008) *Mesorhizobium caraganae* sp. nov., a novel rhizobial species nodulated with *Caragana* spp. in China. Int J Syst Evol Microbiol 58:2646–2653

Gutiérrez-Zamora ML, Martínez-Romero E (2001) Natural endophytic association between *Rhizobium etli* and maize (*Zea mays* L.). J Biotechnol 91:117–126

Halverson LJ, Handelsman J (1991) Enhancement of soybean nodulation by *Bacillus cereus* UW85 in the field and in a growth chamber. Appl Environ Microbiol 57:2767–2770

Han TX et al (2008a) *Rhizobium multihospitium* sp. nov., isolated from multiple legume species native of Xinjiang, China. Int J Syst Evol Microbiol 58:1693–1699

Han TX et al (2008b) *Mesorhizobium gobiense* sp. nov. and *Mesorhizobium tarimense* sp. nov., isolated from wild legumes growing in desert soils of Xinjiang, China. Int J Syst Evol Microbiol 58:2610–2618

Handelsman J, Brill WJ (1985) *Erwinia herbicola* isolates from alfalfa plants may play a role in nodulation of alfalfa by *Rhizobium meliloti*. Appl Environ Microbiol 49:818–821

Hardarson G, Atkins C (2003) Optimising biological N_2 fixation by legumes in farming systems. Plant Soil 252:41–54

Haukka K, Lindstrom K, Young JPW (1998) Three phylogenetic groups of *nodA* and *nifH* genes in *Sinorhizobium* and *Mesorhizobium* isolates from leguminous trees growing in Africa and Latin America. Appl Environ Microbiol 64:419–426

Hernández Bermejo JE, León J (1994) Neglected crops: 1492 from a different perspective. Plant production and protection series No. 26. FAO, Rome, pp 273–288

Herrera-Cervera JA et al (1999) At least five rhizobial species nodulate *Phaseolus vulgaris* in a Spanish soil. FEMS Microbiol Ecol 30:87–97

Holland MA, Polacco JC (1994) PPFMs and other covert contaminants: is there more to plant physiology than just plant? Annu Rev Plant Physiol Plant Mol Biol 45:197–209

Hughes CE, Govindarajulu R, Robertson A, Filer DL, Harris SA, Bailey CD (2007) Serendipitous backyard hybridization and the origin of crops. Proc Natl Acad Sci USA 104:14389–14394

Hungria M, Loureiro MF, Mendes IC, Campo RJ, Graham PH (2005) Inoculant preparation production and application. In: Werner D, Newton WE (eds) Nitrogen fixation in agriculture, forestry, ecology and the environment. Springer, Dordrecht, pp 223–253

Hungria M, Campo RJ, Mendes IC, Graham PH (2006) Contribution of biological nitrogen fixation to the nitrogen nutrition of grain crops in the tropics: the success of soybean (*Glycine max* L Merr.) in South America. In: Singh RP, Shankar N, Jaiwal PK (eds) Nitrogen nutrition and plant productivity. Studium Press LLC, Houston, pp 43–93

Hunter WJ, Kuykendall LD, Manter DK (2007) *Rhizobium selenireducens* sp. nov.: a selenite-reducing α-Proteobacteria isolated from a bioreactor. Curr Microbiol 55:455–460

Islam MS, Kawasaki H, Muramatsu Y, Nakagawa Y, Seki T (2008) *Bradyrhizobium iriomotense* sp. nov., isolated from a tumor-like root of the legume *Entada koshunensis* from Iriomote island in Japan. Biosci Biotechnol Biochem 72:1416–1429

Izaguirre-Mayoral ML, Labandera C, Sanjuán J (2007) Biofertilizantes en Iberoamérica: visión técnica, científica y empresarial. Denad Internacional, Montevideo

Jarvis BDW et al (1997) Transfer of *Rhizobium loti, Rhizobium huakuii, Rhizobium ciceri, Rhizobium mediterraneum* and *Rhizobium tianshanense* to *Mesorhizobium* gen. nov. Int J Syst Bacteriol 47:895–898

Jarvis BDW, Pankhurst C, Patel JJ (1982) *Rhizobium loti*, a new species of legume root nodule bacteria. Int J Syst Bacteriol 32:378–380

Jordan DC (1984) Rhizobiaceae. In: Kreig NR, Holt JG (eds) Bergey's manual of systematic bacteriology, vol 1. Williams and Wilkins, Baltimore, pp 234–254

Kaga A, Isemura T, Tomooka N, Vaughan DA (2008) The genetics of domestication of the azuki bean (*Vigna angularis*). Genetics 178:1013–1036

Khbaya B, Neyra M, Normand P, Zerhari K, Filali-Maltouf A (1998) Genetic diversity and phylogeny of rhizobia that nodulate *Acacia* spp. in Morocco assessed by analysis of rRNA genes. Appl Environ Microbiol 64:4912–4917

Kistner C et al (2005) Seven *Lotus japonicus* genes required for transcriptional reprogramming of the root during fungal and bacterial symbiosis. Plant Cell 17:2217–2229

Kuykendall LD, Saxena B, Devine TE, Udell SE (1992) Genetic diversity in *Bradyrhizobium japonicum* Jordan 1982 and a proposal for *Bradyrhizobium elkanii* sp. nov. Can J Microbiol 38:501–505

Laeremans T, Caluwaerts I, Verreth C, Rogel MA, Vanderleyden J, Martínez-Romero E (1996) Isolation and characterization of the *Rhizobium tropici nod* factor sulfation genes. Mol Plant Microbe Interact 9:492–500

Lee K-B et al (2008) The genome of the versatile nitrogen fixer *Azorhizobium caulinodans* ORS571. BMC Genomics 9:271

Lewis G, Schrire B, Mackinder B, Lock M (eds) (2005) Legumes of the world. Kew Publishing, Kew

Lian B, Prithiviraj B, Souleimanov A, Smith DL (2001) Evidence for the production of chemical compounds analogous to Nod factor by the silicate bacterium *Bacillus circulans* GY92. Microbiol Res 156:289–292

Lie TA, Goktan D, Engin M, Pijnenborg J, Anlarsal E (1987) Co-evolution of the legume–*Rhizobium* association. Plant Soil 100:171–181

Lin DX et al (2008) *Shinella kummerowiae* sp. nov., a symbiotic bacterium isolated from root nodules of the herbal legume *Kummerowia stipulacea*. Int J Syst Evol Microbiol 58: 1409–1413

Lindström K (1989) *Rhizobium galegae*, a new species of legume root nodule bacteria. Int J Syst Bacteriol 39:365–367

Lloret L, Ormeno-Orrillo E, Rincón R, Martínez-Romero J, Rogel-Hernández MA, Martínez-Romero E (2007) *Ensifer mexicanus* sp. nov. a new species nodulating *Acacia angustissima* (Mill.) Kuntze in Mexico. Syst Appl Microbiol 30:280–290

Man CX, Wang H, Chen WF, Sui XH, Wang ET, Chen WX (2008) Diverse rhizobia associated with soybean grown in the subtropical and tropical regions of China. Plant Soil 310:77–87

Martínez-Romero E (1994) Recent developments in *Rhizobium* taxonomy. Plant Soil 161:11–20

Martínez-Romero E (2003) Diversity of *Rhizobium–Phaseolus vulgaris* symbiosis: overview and perspectives. Plant Soil 252:11–23

Martínez-Romero E (2000) Dinitrogen-fixing prokaryotes. In: Trüper HG, Dworkin M, Harder W, Schleifer KH (eds) The prokaryotes. Springer, Heidelberg, pp 1–12

Martínez-Romero E, Segovia L, Mercante FM, Franco AA, Graham P, Pardo MA (1991) *Rhizobium tropici*, a novel species nodulating *Phaseolus vulgaris* L. beans and *Leucaena* sp trees. Int J Syst Bacteriol 41:417–426

Martínez-Romero E, Hernández-Lucas I, Peña-Cabriales JJ, Castellanos JZ (1998) Symbiotic performance of some modified *Rhizobium etli* strains in assays with *Phaseolus vulgaris* beans that have a high capacity to fix N_2. Plant Soil 204:89–94

Menna P, Hungria M, Barcellos FG, Bangel EV, Hess PN, Martinez-Romero E (2006) Molecular phylogeny based on the 16S rRNA gene of elite rhizobial strains used in Brazilian commercial inoculants. Syst Appl Microbiol 29:315–332

Mhamdi R, Jebara M, Aouani ME, Ghrir R, Mars M (1999) Genotypic diversity and symbiotic effectiveness of rhizobia isolated from root nodules of *Phaseolus vulgaris* L. grown in Tunisian soils. Biol Fertility Soils 28:313–320

Moawad H, Bohlool BB (1984) Competition among *Rhizobium* spp. for nodulation of *Leucaena leucocephala* in two tropical soils. Appl Environ Microbiol 48:5–9

Moreira FMS et al (2006) *Azorhizobium doebereinerae* sp. nov. microsymbiont of *Sesbania virgata* (Caz.) Pers. Syst Appl Microbiol 29:197–206

Mostasso L, Mostasso FL, Dias BG, Vargas MAT, Hungria M (2002) Selection of bean (*Phaseolus vulgaris* L.) rhizobial strains for the Brazilian Cerrados. Field Crops Res 73:121–132

Moulin L, Munive A, Dreyfus B, Boivin-Masson C (2001) Nodulation of legumes by members of the β-subclass of Proteobacteria. Nature 411:948–950

Muller MH, Prosperi JM, Santoni S, Ronfort J (2003) Inferences from mitochondrial DNA patterns on the domestication history of alfalfa (*Medicago sativa*). Mol Ecol 12:2187–2199

Nick G et al (1999) *Sinorhizobium arboris* sp. nov. and *Sinorhizobium kostiense* sp. nov., isolated from leguminous trees in Sudan and Kenya. Int J Syst Bacteriol 49:1359–1368

Nour SM, Fernández MP, Normand P, Cleyet-Marel JC (1994) *Rhizobium ciceri* sp. nov., consisting of strains that nodulate chickpeas (*Cicer arietinum* L.). Int J Syst Bacteriol 44:511–522

Nour SM, Cleyet-Marel J-C, Normand P, Fernández MP (1995) Genomic heterogeneity of strains nodulating chickpeas (*Cicer arietinum* L.) and description of *Rhizobium mediterraneum* sp. nov. Int J Syst Bacteriol 45:640–648

Ormeño-Orrillo E, Rosenblueth M, Luyten E, Vanderleyden J, Martinez-Romero E (2008) Mutations in lipopolysaccharide biosynthetic genes impair maize rhizosphere and root colonization of *Rhizobium tropici* CIAT899. Environ Microbiol 10:1271–1284

Peña-Cabriales JJ, Castellanos JZ (1993) Effects of water stress on N_2 fixation and grain yield of *Phaseolus vulgaris* L. Plant Soil 152:151–155

Peng G, Yuan Q, Li H, Zhang W, Tan Z (2008) *Rhizobium oryzae* sp. nov., isolated from the wild rice *Oryza alta*. Int J Syst Evol Microbiol 58:2158–2163

Peoples MB, Herridge DF, Ladha JK (1995) Biological nitrogen fixation: an efficient source of nitrogen for sustainable agricultural production? Plant Soil 174:3–28

Peralta H, Mora Y, Salazar E, Encarnacion S, Palacios R, Mora J (2004) Engineering the *nifH* promoter region and abolishing poly-β-hydroxybutyrate accumulation in *Rhizobium etli* enhance nitrogen fixation in symbiosis with *Phaseolus vulgaris*. Appl Environ Microbiol 70:3272–3281

Phillips DA, Joseph CM, Yang G, Martinez-Romero E, Sanborn JR, Volpin H (1999) Identification of lumichrome as a *Sinorhizobium* enhancer of alfalfa root respiration and shoot growth. Proc Natl Acad Sci USA 96:12275–12280

Pickersgill B (2007) Domestication of plants in the Americas: insight from Mendelian and molecular genetics. Ann Bot 100:925–9490

Piñero D, Martínez E, Selander RK (1988) Genetic diversity and relationships among isolates of *Rhizobium leguminosarum* biovar *phaseoli*. Appl Environ Microbiol 54:2825–2932

Pitcairn CER, Fowler D, Grace J (1991) Changes in species composition of semi-natural vegetation associated with the increase in atmospheric inputs of nitrogen. Report to Nature Conservancy Council Institute of Terrestrial Ecology, Edinburgh

Pleban S, Ingel F, Chet I (1995) Control of *Rhizoctonia solani* and *Sclerotium rolfsii* in the greenhouse using endophytic *Bacillus* spp. Eur J Plant Pathol 101:665–672

Poupot R, Martinez-Romero E, Prome J-C (1993) Nodulation factors from *Rhizobium tropici* are sulfated or nonsulfated chitopentasaccharides containing an N-methyl-N-acetylglucosaminyl terminus. Biochemistry 32:10430–10435

Priest F (1993) Systematics and ecology of *Bacillus*. In: Sonnenshein AL, Hoch J, Losick R (eds) *Bacillus subtilis* and other Gram-Positive bacteria: Biochemistry, physiology and molecular genetics. American Society for Microbiology, Washington, DC, pp 3–16

Pueppke SG, Broughton WJ (1999) *Rhizobium* sp. strain NGR234 and *R. fredii* USDA257 share exceptionally broad nested host ranges. Mol Plant Microbe Interact 12:293–318

Qian J, Kwon S, Parker MA (2003) rRNA and *nifD* phylogeny of *Bradyrhizobium* from sites across the Pacific Basin. FEMS Microbiol Lett 219:159–165

Quan Z-X, Bae H-S, Baek J-H, Chen W-F, Im W-T, Lee S-T (2005) *Rhizobium daejeonense* sp. nov. isolated from a cyanide treatment bioreactor. Int J Syst Evol Microbiol 55:2543–2549

Ramírez-Bahena MH et al (2008) Revision of the taxonomic status of the species *Rhizobium leguminosarum* (Frank 1879) Frank 1889AL, *Rhizobium phaseoli* Dangeard 1926AL and *Rhizobium trifolii* Dangeard 1926AL. *R. trifolii* is a later synonym of *R. leguminosarum*. Reclassification of the strain *R. leguminosarum* DSM 30132 (=NCIMB 11478) as *Rhizobium pisi* sp. nov. Int J Syst Evol Microbiol 58:2484–2490

Reinhold-Hurek B, Hurek T (1998) Interactions of gramineous plants with *Azoarcus* spp. and other diazotrophs: Identification, localization and perspectives to study their function. Crit Rev Plant Sci 17:29–54

Rincón-Rosales R, Lloret L, Ponce E, Martínez-Romero E (2008) Rhizobia with different symbiotic efficiencies nodulate *Acaciella angustissima* in Mexico including *Sinorhizobium chiapanecum* sp. nov. that has common symbiotic genes with *S. mexicanum*. FEMS Microbiol Ecol 67:103–117

Rivas R et al (2003) Description of *Devosia neptuniae* sp. nov. that nodulates and fixes nitrogen in symbiosis with *Neptunia natans*, an aquatic legume from India. Syst Appl Microbiol 26:47–53

Rivas R et al (2004) *Bradyrhizobium betae* sp. nov. isolated from roots of *Beta vulgaris* affected by tumor-like deformations. Int J Syst Evol Microbiol 54:1271–1275

Rodriguez-Navarro DN, Buendia AM, Camacho M, Lucas MM, Santamaria C (2000) Characterization of *Rhizobium* spp. bean isolates from South-West Spain. Soil Biol Biochem 32:1601–1613

Rome S, Fernández MP, Brunel B, Normand P, Cleyet-Marel JC (1996) *Sinorhizobium medicae* sp. nov., isolated from annual *Medicago* spp. Int J Syst Bacteriol 46:972–980

Rosenblueth M, Martinez-Romero E (2004) *Rhizobium etli* maize populations and their competitiveness for root colonization. Arch Microbiol 181:337–344

Rosenblueth M, Martínez-Romero E (2006) Bacterial endophytes and their interactions with hosts. Mol Plant Microbe Interact 19:827–837

Ryu C-M et al (2003) Bacterial volatiles promote growth in *Arabidopsis*. Proc Natl Acad Sci USA 100:4927–4932

Sauer JD (1993) Historical geography of crop plants — a select roster. CRC Press, Boca Raton, FL

Schmidt EL, Zidwick MJ, Abebe HM (1986) *Bradyrhizobium japonicum* serocluster 123 and diversity among member isolates. Appl Environ Microbiol 51:1212–1215

Scholla MH, Elkan GH (1984) *Rhizobium fredii* sp. nov., a fast growing species that effectively nodulates soybeans. Int J Syst Bacteriol 34:484–486

Schrire BD, Lavin M, Lewis GP (2005) Global distribution patterns of the Leguminosae: insights from recent phylogenies. In: Friis I, Balslev H (eds) Plant diversity and complexity patterns: local, regional and global dimensions. Biologiske Skrifter 55:375–422

Segovia L, Pinero D, Palacios R, Martinez-Romero E (1991) Genetic structure of a soil population of nonsymbiotic *Rhizobium leguminosarum*. Appl Environ Microbiol 57:426–433

Segovia L, Young JPW, Martínez-Romero E (1993) Reclassification of American *Rhizobium leguminosarum* biovar phaseoli type I strains as *Rhizobium etli* sp. nov. Int J Syst Bacteriol 43:374–377

Sessitsch A, Howieson JG, Perret X, Antoun H, Martínez-Romero E (2002) Advances in *Rhizobium* research. Crit Rev Plant Sci 21:323–378

Silo-Suh LA, Lethbridge BJ, Raffel SJ, He H, Clardy J, Handelsman J (1994) Biological activities of two fungistatic antibiotics produced by *Bacillus cereus* UW85. Appl Environ Microbiol 60:2023–2030

Silva C, Vinuesa P, Eguiarte LE, Martinez-Romero E, Souza V (2003) *Rhizobium etli* and *Rhizobium gallicum* nodulate common bean (*Phaseolus vulgaris*) in a traditionally managed milpa plot in Mexico: population genetics and biogeographic implications. Appl Environ Microbiol 69:884–893

Silva C, Vinuesa P, Eguiarte LE, Souza V, Martínez-Romero E (2005) Evolutionary genetics and biogeographic structure of *Rhizobium gallicum sensu lato*, a widely distributed bacterial symbiont of diverse legumes. Mol Ecol 14:4033–4050

Smartt J (1984) Evolution of grain legumes. I. Mediterranean pulses. Expl Agric 20:275–296

Smartt J (1985a) Evolution of grain legumes. II. Old and new world pulses of lesser economic importance. Expl Agric 21:1–18

Smartt J (1985b) Evolution of grain legumes. III. Pulses in the genus Vigna. Expl Agric 21:87–100

Snoeck C, Vanderleyden J, Beebe S (2003) Strategies for genetic improvement of common bean and rhizobia towards efficient interactions. Plant Breeding Rev 23:21–72

Sprent JI (2001) Nodulation in legumes. Royal Botanic Gardens, Kew

Sprent JI, James EK (2007) Legume evolution: where do nodules and mycorrhizas fit in? Plant Physiol 144:575–581

Suzuki S et al (2007) Rhizobial factors required for stem nodule maturation and maintenance in *Sesbania rostrata-Azorhizobium caulinodans* ORS571 symbiosis. Appl Environ Microbiol 73:6650–6659

Sy A et al (2001) Methylotrophic *Methylobacterium* bacteria nodulate and fix nitrogen in symbiosis with legumes. J Bacteriol 183:214–220

Tan ZY, Wang ET, Peng GX, Zhu ME, Martinez-Romero E, Chen WX (1999) Characterization of bacteria isolated from wild legumes in the north-western regions of China. Int J Syst Bacteriol 49:1457–1469

Tan ZY, Kan FL, Peng GX, Wang ET, Reinhold-Hurek B, Chen WX (2001) *Rhizobium yanglingense* sp. nov., isolated from arid and semiarid regions in China. Int J Syst Evol Microbiol 51:909–914

Tian CF, Wang ET, Wu LJ, Han TX, Chen WF, Gu CT, Gu JG, Chen WX (2008) *Rhizobium fabae* sp. nov., a bacterium that nodulates *Vicia faba*. Int J Syst Evol Microbiol 58:2871–2875

Toledo I, Lloret L, Martínez-Romero E (2003) *Sinorhizobium americanus* sp. nov., a new *Sinorhizobium* species nodulating native *Acacia* spp. in Mexico. Syst Appl Microbiol 26: 54–64

Valverde A et al (2005) *Phyllobacterium trifolii* sp. nov., nodulating *Trifolium* and *Lupinus* in Spanish soils. Int J Syst Evol Microbiol 55:1985–1989

Valverde A, Igual JM, Peix A, Cervantes E, Velazquez E (2006) *Rhizobium lusitanum* sp. nov.a bacterium that nodulates *Phaseolus vulgaris*. Int J Syst Evol Microbiol 56:2631–2637

van Berkum P, Beyene D, Bao G, Campbell TA, Eardly BD (1998) *Rhizobium mongolense* sp. nov. is one of three rhizobial genotypes identified which nodulate and form nitrogen-fixing symbioses with *Medicago ruthenica* [(L.) Ledebour]. Int J Syst Bacteriol 48:13–22

Vandamme P, Coenye T (2004) Taxonomy of the genus *Cupriavidus*: a tale of lost and found. Int J Syst Evol Microbiol 54:2285–2289

Vandamme P, Goris J, Chen W-M, de Vos P, Willems A (2002) *Burkholderia tuberum* sp. nov. and *Burkholderia phymatum* sp. nov. nodulate the roots of tropical legumes. Syst Appl Microbiol 25:507–512

Vaneechoutte M, Kämpfer P, De Baere T, Falsen E, Verschraegen G (2004) *Wautersia* gen. nov., a novel genus accommodating the phylogenetic lineage including *Ralstonia eutropha* and related species and proposal of *Ralstonia* [*Pseudomonas*] *syzygii* (Roberts et al. 1990) comb. nov. Int J Syst Evol Microbiol 54:317–327

Velázquez E et al (2001) *Mesorhizobium chacoense* sp. nov., a novel species that nodulates *Prosopis alba* in the Chaco Arido region (Argentina). Int J Syst Evol Microbiol 51:1011–1021

Velázquez E et al (2005) The coexistence of symbiosis and pathogenicity-determining genes in *Rhizobium rhizogenes* strains enables them to induce nodules and tumors or hairy roots in plants. Mol Plant Microbe Interact 18:1325–1332

Vinuesa P et al (2005) *Bradyrhizobium canariense* sp. nov., an acid-tolerant endosymbiont isolated from the nodules of endemic genistoid legumes (Papilionoideae: Genisteae) growing in the Canary Islands. Int J Syst Evol Microbiol 55:569–575

Wang ET, van Berkum P, Beyene D, Sui XH, Dorado O, Chen WX, Martínez-Romero E (1998) *Rhizobium huautlense* sp. nov., a symbiont of *Sesbania herbacea* that has a close phylogenetic relationship with *Rhizobium galegae*. Int J Syst Bacteriol 48:687–699

Wang ET, Martínez-Romero J, Martínez-Romero E (1999a) Genetic diversity of rhizobia from *Leucaena leucocephala* nodules in Mexican soils. Mol Ecol 8:711–724

Wang ET, van Berkum P, Sui SH, Beyene D, Chen WX, Martínez-Romero E (1999b) Diversity of rhizobia associated with *Amorpha fructicosa* isolated from Chinese soils and description of *Mesorhizobium amorphae* sp. nov. Int J Syst Bacteriol 49:51–65

Wang ET, Tan ZY, Willems A, Fernández-López M, Reinhold-Hurek B, Martínez-Romero E (2002) *Sinorhizobium morelense* sp. nov., a *Leucaena leucocephala*-associated bacterium that is highly resistant to multiple antibiotics. Int J Syst Evol Microbiol 52:1687–1693

Wang FQ et al (2007) *Mesorhizobium albiziae* sp. nov., a novel bacterium that nodulates *Albizia kalkora* in a subtropical region of China. Int J Syst Evol Microbiol 57:1192–1199

Wei GH, Wang ET, Tan ZY, Zhu ME, Chen WX (2002) *Rhizobium indigoferae* sp. nov. and *Sinorhizobium kummerowiae* sp. nov., respectively isolated from *Indigofera* spp. and *Kummerowia stipulacea*. Int J Syst Evol Microbiol 52:2231–2239

Wei GH, Tan ZY, Zhu ME, Wang ET, Han SZ, Chen WX (2003) Characterization of rhizobia isolated from legume species within the genera *Astragalus* and *Lespedeza* grown in the Loess Plateau of China and description of *Rhizobium loessense* sp. nov. Int J Syst Evol Microbiol 53:1575–1583

Willems A et al (2003) Description of new *Ensifer* strains from nodules and proposal to transfer *Ensifer adhaerens* Casida 1982 to *Sinorhizobium* as *Sinorhizobium adhaerens* comb. nov. Request for an opinion. Int J Syst Evol Microbiol 53:1207–1217

Woodin SJ, Farmer AM (1993) Impacts of sulphur and nitrogen deposition on sites and species of nature conservation importance in Great Britain. Biol Conserv 63:23–30

Xu LM, Ge C, Cui Z, Li J, Fan H (1995) *Bradyrhizobium liaoningense* sp. nov., isolated from the root nodules of soybeans. Int J Syst Bacteriol 45:706–711

Yanni YG et al (1997) Natural endophytic association between *Rhizobium leguminosarum* bv trifolii and rice roots and assessment of potential to promote rice growth. Plant Soil 194:99–114

Yao ZY, Kan FL, Wang ET, Wei GH, Chen WX (2002) Characterization of rhizobia that nodulate legume species of the genus *Lespedeza* and description of *Bradyrhizobium yuanmingense* sp. nov. Int J Syst Evol Microbiol 52:2219–2230

Young JM, Kuykendall D, Martinez-Romero E, Kerr A, Sawada H (2001) A revision of *Rhizobium* Frank 1889 with an emended description of the genus and the inclusion of all species of *Agrobacterium* Conn 1942 and *Allorhizobium undicolade* Lajudie et al. 1998 as new combinations: *Rhizobium radiobacter, R. rhizogenes, R. rubi, R. undicola* and *R. vitis*. Int J Syst Evol Microbiol 51:89–103

Young JM et al (2003) The genus name Ensifer Casida 1982 takes priority over Sinorhizobium Chen 1988 and Sinorhizobium morelense Wang et al. 2002 is a later synonym of Ensifer adhaerens Casida 1982. Is the combination 'Sinorhizobium adhaerens' (Casida 1982) Willems et al. 2003 legitimate? Request for an Opinion. Int J Syst Evol Microbiol 53:2107–2110

Chapter 9
Mycorrhizal Associations in Agroforestry Systems

André Mundstock Xavier de Carvalho, Rodrigo de Castro Tavares, Irene Maria Cardoso, and Thomas W. Kuyper

9.1 Introduction

Since its first appearance 8,000 years BP, agriculture has caused large impacts on local to global scales (Ruddiman 2003), and it currently remains a major driver of human-induced environmental change. Agriculture impacts on biological, chemical, and physical properties of soils, leading to biodiversity losses, decreases in soil coverage, changes in element cycles and water balance of ecosystems, degradation of soil structure, erosion, and contamination of groundwater, amongst other consequences (Tilman et al. 2001; Foley et al. 2005). Many, if not all, of these unwanted side effects are due to the practices of agro-ecosystem simplification, where ecosystem services provided by the soil are increasingly bypassed. The problems that this simplification has generated have highlighted the need to utilize soil resources efficiently — not only water and nutrients, but also the biological resources of the soils. It is expected that increased ecological sustainability through alternative use of soil resources will feed back to increased social sustainability of agricultural systems. The perceived need to seek alternatives to current agricultural practices has resulted in an enhanced interest in agroforestry systems. Under agroforestry, the needs for ecological sustainability can be reconciled with the needs for sustainable food production (Young 1997). Agroforestry systems are based on a combination of tree with crop and/or animal species simultaneously or sequentially in the same area, and have as their major aim the optimization of beneficial ecological interactions among ecosystem components (Farrell and Altieri 2002).

A.M.X. de Carvalho, R. de Castro Tavares, and I.M. Cardoso (✉)
Department of Soil Science, Federal University of Viçosa, Viçosa, Minas Gerais, Brazil 36570-000
e-mail: irene@ufv.br

T.W. Kuyper
Department of Soil Quality, Wageningen University, P.O. Box 47, 6700 AA Wageningen, The Netherlands

P. Dion (ed.), *Soil Biology and Agriculture in the Tropics*, Soil Biology 21,
DOI 10.1007/978-3-642-05076-3_9,

Tropical agroforestry systems are planned with the objective of mimicking tropical forests. Agroforestry systems put much emphasis on species diversity, and on the interactions between these different species. It is generally assumed, on the basis of ecological principles, that agroforestry systems can be as productive as monocrop systems. Agroforestry systems also show larger resilience and resistance to attacks by pest species and diseases, in addition to maintaining organic matter, soil fertility, and biological activity at satisfactory levels (Souza 2006; Young 1997; Ewel 1999; Van Noordwijk and Ong 1999). However, we are confronted with a paradox. Whereas the path towards ecosystem simplification can successfully go along the path of methodological reductionism with the study of individual components in isolation, the route to sustainable agroforestry systems needs to take complexity and context-dependency as its point of departure. Therefore, knowledge on how to introduce and manage agroforestry systems is lagging behind, due to the specificity of each ecosystem and the great diversity and complexity of the interactions involved.

Among the interactions that occur in soils of agroforestry systems, we will focus in this chapter on interactions between plants and microbiota, especially those formed by roots and mycorrhizal fungi. However, mycorrhizal interactions cannot be studied in isolation, since further interactions with nitrogen-fixing bacteria and with other rhizosphere micro-organisms are an integral part of belowground complexity. The importance of mycorrhizal fungi and mycorrhizal associations is evidenced by the observation that they are ubiquitous in tropical soils. They have been present since the conquest of the terrestrial environment by higher plants (Brundrett 2002).

Mycorrhizas are mutualistic associations between the roots of plants and certain soil fungi, which can be grouped into four types: ectomycorrhiza, orchid mycorrhiza, ericoid mycorrhiza, and arbuscular mycorrhiza (Smith and Read 2008). In tropical agroforestry systems, the arbuscular mycorrhizal fungi (AMF) form the most important group. These fungi belong to the phylum Glomeromycota. At present, around 160–200 species have been described (INVAM 2008; Souza et al. 2008), and over 100 species of AMF have been recorded from Brazilian agroecosystems (Stürmer and Siqueira 2008).

A large part of the soil microbial biomass can be constituted of AMF material (Olsson et al. 1999). AMF can colonize virtually all plant types among Angiosperms and Gymnosperms. Some Pteridophytes and Bryophytes can also be colonized by AMF (Smith and Read 2008). Observations that the overwhelming majority of agroforestry tree species form arbuscular mycorrhiza confirm this overall picture. In a study of 101 tree species belonging to different families in southeastern Brazil, 93 of the evaluated species were colonized by AMF (Carneiro et al. 1998).

According to Siqueira et al. (2007) studies dealing with mycorrhizal associations in tropical tree species have been conducted in increasing numbers over recent decades, and about 500 noncrop species have been investigated for the occurrence, or the effect of mycorrhizal associations. Among the known mycorrhizal species, many are used by Atlantic Forest family farmers in agroforestry coffee systems (Cardoso et al. 2001; Siqueira 2008; Table 9.1). However, studies examining

Table 9.1 Mycorrhizal tree and shrub species used by smallholder farmers in agroforestry coffee systems, Zona da Mata of Minas Gerais, Atlantic Coastal Rainforest, Brazil

Family	Species	Reference
Anacardiaceae	*Mangifera indica*	Silveira and Gomes (2007)
	Miracroduon urundeuva	Siqueira et al. (2007)
	Schinus terebentifolium	Zangaro et al. (2003)
Annonaceae	*Annona muricata*[a]	Silveira and Gomes (2007)
Apocynaceae	*Aspidosperma polyneurum*	Zangaro et al. (2002)
Araucariaceae	*Araucaria angustifolia*	Siqueira et al. (2007)
Bignoniaceae	*Tabebuia impetiginosa*	Carneiro et al. (1998)
	T. schysotricha	Zangaro et al. (2003)
	T. serratifolia	Siqueira and Saggin-Júnior (2001)
Bixaceae	*Bixa orellana*[a]	Carneiro et al. (1998)
Bombacaceae	*Chorisia speciosa*	Zangaro et al. (2003)
Cannabaceae	*Trema micrantha*	Carneiro et al. (1996)
Caricaceae	*Carica papaya*[a]	Silveira and Gomes (2007)
Casuarinaceae	*Casuarina equisetifolia*	Carneiro et al. (1998)
Cecropiaceae	*Cecropia glaziovii*	Zangaro et al. (2002)
	C. pachystachya	Zangaro et al. (2003)
Ebenaceae	*Diospyrus kaki*	Silveira and Gomes (2007)
Euphorbiaceae	*Manihot esculenta*[a]	Silveira and Gomes (2007)
Lauraceae	*Persea americana*	Silveira and Gomes (2007)
Leguminosae–Caesalpinioideae	*Caesalpinea ferrea*	Siqueira et al. (2007)
	C. peltophoroides	Siqueira et al. (2007)
	Copaifera langsdorffii	Carneiro et al. (1998)
	Hymenaea courbaril	Zangaro et al. (2003)
	Pterogyne nitens	Zangaro et al. (2002)
	Schizolobium parahyba	Carneiro et al. (1998)
	Senna macranthera	Carneiro et al. (1998)
	S. multijuga	Carneiro et al. (1998)
Leguminosae–Mimosoideae	*Annadenanthera peregrina*	Siqueira et al. (2007)
	Enterolobium contortisiliquum	Zangaro et al. (2003)
	Inga sessilis	Zangaro et al. (2003)
	Leucaena leucocephala	Siqueira and Saggin-Júnior (2001)
	Piptadenia gonocantha	Carneiro et al. (1998)
	Mimosa caesalpiniaefolia	Siqueira et al. (2007)
Leguminosae–Papilionoideae	*Cajanus cajan*[a]	Siqueira et al. (2007)
	Machaerium nictitans	Carneiro et al. (1998)
	M. stipitatun	Zangaro et al. (2003)
Malpighiaceae	*Malpighia emarginata*[a]	Silveira and Gomes (2007)
Malvaceae	*Luehea divaricata*	Zangaro et al. (2002)
	L. grandiflora	Siqueira and Saggin-Júnior (2001)
Melastomataceae	*Tibouchina granulosa*	Siqueira and Saggin-Júnior (2001)
Meliaceae	*Azadirachta indica*	Siqueira et al. (2007)
	Cedrella fissilis	Carneiro et al. (1998)
	Melia azedarach	Carneiro et al. (1998)
Musaceae	*Musa* sp.[a]	Silveira and Gomes (2007)
Myrsinaceae	*Rapanea ferruginea*	Siqueira et al. (2007)

(*continued*)

Table 9.1 (continued)

Family	Species	Reference
Myrtaceae	*Campomanesia xanthocarpa*	Zangaro et al. (2002)
	Eugenia uniflora	Zangaro et al. (2003)
	Psidium guajava	Zangaro et al. (2002)
Palmae	*Euterpe edulis*	Zangaro et al. (2003)
	Syagrus romanzofianum	Zangaro et al. (2003)
Rhamnaceae	*Colubrina glandulosa*	Zangaro et al. (2003)
	Hovenia dulcis	Carneiro et al. (1998)
Rutaceae	*Citrus* sp.	Silveira and Gomes (2007)
Solanaceae	*Solanum argenteum*	Zangaro et al. (2002)
	S. granulosum	Siqueira and Saggin-Júnior (2001)
Verbenaceae	*Aegiphila sellowiana*	Zangaro et al. (2003)
	Cytharexyllum mirianthum	Zangaro et al. (2002)
	Vitex montevidensis	Zangaro et al. (2003)

[a]Shrub species

arbuscular mycorrhizal colonization in a complete agroforestry system (and not only with individual tree species) are scarce (Pande and Tarafdar 2004).

A search on the Scopus database with the words "agroforestry" and "mycorrhiza" yielded around 50 articles. However, fewer than 20 articles went beyond the evaluation of mycorrhizal responses of individual tree species. As a consequence of this meager database, many mycorrhizal aspects of agroforestry, related to relations between plant and fungal species diversity, ecological succession, effect of tree management such as pruning or shading, and interactions between plants under agroforestry systems, remain underinvestigated.

9.2 The Multi-Functionality of AMF in Agro-Ecosystems

The most obvious role of AMF in agro-ecosystems is to increase the soil volume exploited by the host plant. This leads to increased water and nutrient uptake, which in turn may enhance acquisition of other nutrients, for instance through associated N fixation (Bolan 1991; Garg et al. 2006; Smith and Read 2008). Other roles of AMF concern protection of the root system against pathogens (Pozo and Azcón-Aguilar 2007; Elsen et al. 2008), salinity (Al-Karaki 2000; Tavares 2007), phytotoxic elements such as Al (Rufyikiri et al. 2000), or heavy metals (Andrade et al. 2003). AMF are also involved in the formation and maintenance of soil structure (Rillig and Mummey 2006), and increase C input to soils (Rillig et al. 2001; Zhu and Miller 2003), both of these effects contributing to reduce erosion. Finally AMF also play a role in the maintenance of plant biodiversity (Van der Heijden et al. 1998). Mycorrhizal associations, therefore, are multifunctional in agro-ecosystems (Newsham et al. 1995). They have the potential to improve physical, chemical, and biological soil quality, including feedbacks between soil biota and plant communities.

A large body of literature on mycorrhizal associations has centered on a comparison between costs and benefits of the symbiosis, often expressed in the same carbon currency. Such studies have concentrated on conditions where nonmycorrhizal plants outperformed (in terms of biomass) mycorrhizal plants and have given rise to the concept of parasitic behavior of mycorrhizal fungi, or of a continuum between mutualism and parasitism (Bethlenfalvay et al. 1983; Johnson 1993; Johnson et al. 1997). Implicit in many of these studies is the argument that, considering the high costs for symbiosis which occur under conditions of nutrient, especially phosphorus, sufficiency, the mycorrhizal association represents a net carbon cost for the plant host in simplified agro-ecosystems, where plant nutrient uptake capacity almost matches plant needs (Bucher 2007). Bhadalung et al. (2005) observed that in soils where maize was grown for 27 years and with application levels of 180 kg ha^{-1} $year^{-1}$ of N and P_2O_5 (in the form of ammonium sulfate and triple superphosphate) there was a reduction of 70% in density and 40% in the diversity of AMF compared to the same soil without the application of fertilizers. Such conditions are hardly, if ever, found under agroforestry management, and the concept of the mutualism–parasitism continuum might be unimportant in such cases.

However, under conditions of severe nutrient limitation, the ability of the fungal mycelium to immobilize nutrients in its tissues, and hence to reduce the transfer of nutrients to the plant, could also result in lower performance of mycorrhizal plants compared to nonmycorrhizal plants. Such conditions could possibly occur in the most nutrient-depleted sites, hence diminishing the usefulness of agroforestry as a means for ecosystem rehabilitation.

The cost-benefit models of Johnson et al. (1997) and Schnepf and Roose (2006) propose that plants monitor the carbon-to-nutrient exchange rate, and that any deviation from the mean exchange rate is interpreted as a shift along the mutualism–parasitism continuum because of decreased coordination between plant and fungal responses. However, if carbon and nutrient fluxes are quantitatively unlinked, a larger context-dependency of mycorrhizal responses can be observed. The spatial and temporal heterogeneity of nutrient availability could result in larger variations in exchange rates between plants and fungus, without the need to invoke the parasitism terminology (Landis and Fraser 2008). Especially in pulsed environments, characterized by periods of prolonged drought alternated with periods of higher water and nutrient availability, adjustment of the mutualism–parasitism balance could then be context-dependent. Under such conditions, the dynamics of a mycorrhizal network should differ, depending on whether it is synchronized with perennial agroforestry trees or else with the growth cycle of annual crops.

9.2.1 *Implications of the AMF Mycelium Network*

Several functions performed by AMF are linked to the formation and maintenance of the mycelium network. In order for such a network to exist and be ecologically relevant, it is needed that at least some mycorrhizal species show low selectivity, and hence have the capacity to establish mycelial linkages between different plant species. Low fungal selectivity may or may not result in comparable responses of different plants to the same fungal species. The existence of such networks allows, at least from a theoretical perspective, the movement of carbon, water, and nutrients between plants belonging to different species, genera, or even families. The existence of such mycorrhizal networks (the wood-wide web) could therefore bring ecological consequences that seem to fit poorly with conventional theories on plant interactions (Fitter 2001; Simard and Durall 2004; Selosse et al. 2006). It has been proposed that such networks may reduce differences in competitive ability between plants. Plants benefiting from the network would include seedlings (Van der Heijden 2004) or plants whose photosynthetic performance is poorer because of shading. Conceivably, mycorrhizal fungi could enhance plant species richness, in the case where competitively inferior plant species would be most prone to mycorrhizal associations, or decrease it, when the competitively superior plant species would be most responsive (Urcelay and Diaz 2003).

With their very low levels of soil disturbance, agroforestry systems are particularly conducive to the build up and maintenance of mycorrhizal networks, as compared to the annual cropping systems characterized by regular disturbances and bare fallows. The existence of such networks can result in faster establishment of the mycorrhizal symbiosis in seedlings under agroforestry systems (Kuyper et al. 2004), because plant colonization tends to be faster from the mycelium than through spore germination (Brundrett and Abbott 1994). However, the possibility of further ecological consequences of such networks remains disputed. With regard to the transfer of carbon in arbuscular mycorrhizal networks, the position now held by almost all mycorrhizal researchers is that such carbon movements remain completely under the control of the fungus and therefore interplant carbon transfer is unimportant (Pfeffer et al. 2004) — except in the case of completely or partly mycoheterotrophic plants that can completely reverse carbon flow. Transfer of nutrients, especially transfer of nitrogen from a N-fixing plant to a neighboring plant in the same network, remains a more controversial issue (He et al. 2003). A recent study in agroforestry systems is consistent with a potentially large mycorrhizal role in this respect. Sierra and Nygren (2006) estimated N transfer from the N-fixing tree *Gliricidia sepium* to the grass *Dichanthium aristatum.* They claimed that around 25%–35% of the N in the grass adjacent to agroforestry plots was atmospheric in origin. They also observed a positive correlation between *Gliricidia* root density and amount of N in the grass derived from atmospheric sources, but no correlation between the grass N isotopic signature and the soil isotopic signature. Sierra and Nygren (2006) therefore suggested direct N transfer from trees to grass, for instance through root exudates (termed here "direct soil transfer") or via common mycorrhizal networks (corresponding to "direct transfer through the mycorrhiza"). Unfortunately, the mycorrhizal status of the plants examined in this study was not investigated. Lower N transfer levels were observed from leguminous trees to coffee plants in an agroforestry system, where 5% of fixed N was directly transferred; again, mycorrhizas were not studied (Snoeck et al. 2000).

In a simplified experimental design with nonmycorrhizal plants, Rao and Giller (1993) suggested that around 10%–15% of N found in the *Cenchrus ciliaris* grass was derived from N fixation by the legume *Leucaena diversifolia*.

Should urgently needed further research demonstrate the reality of direct N-transfer through the mycorrhizal network, then this recognition might lead to a re-evaluation of N-fixation studies based on ^{15}N natural abundance. Basically, such studies compare the differences in ^{15}N signal of legumes and reference plants, and make the implicit assumption that no N is directly transferred from legume to reference plants. Field data often show substantial variation in ^{15}N values of different reference plants, resulting in different estimates of the amounts of N fixed by the legume. Such variation in ^{15}N values of reference plants is poorly understood, and a re-evaluation of its possible mycorrhizal origin is a research priority.

Selosse et al. (2006) reported that phosphorus can be moved between plants (with unilateral transfer being more frequent than bidirectional transfer) and can end up in shoots of receiver plants. Because P is less mobile and required in lower amounts than N, the transfer of P is less than that of N (Johansen and Jensen 1996). Furthermore, the magnitude of P transfer is too small to significantly affect the nutrition of the recipient plant. However, more intense P transfer (and also N transfer) occurs from dying roots – with root death occurring, for instance, as a consequence of shoot pruning or root pruning, both regular processes in agroforestry management. How pruning would affect the carbon contribution of individual plants to the common mycorrhizal network, and how individual plants subsequently benefit from that network, remain to be determined.

The importance of common mycorrhizal networks for water transfer and redistribution has received remarkably little attention. However, the role of the mycorrhizal network in hydraulic redistribution could be of particular importance in agroforestry systems. After deeply-rooted plants have taken up water from profound soil layers, the activity of neighboring shallow-rooted plants could be sustained by nocturnal water efflux coupled to water uptake and transfer by mycorrhizal fungi in superficial soil layers. This process of mycorrhiza-mediated hydraulic redistribution has been demonstrated for ectomycorrhizal and arbuscular mycorrhizal systems (Allen 2007; Egerton-Warburton et al. 2007), but its importance for agroforestry still needs experimental proof.

9.2.2 Ecological Implications for Plants

Agroforestry systems are hypothesized to harbor a high AMF species richness and a large AMF biomass, due to increased productivity and species richness of host plants, whether cultivated or spontaneous, and to a greater extension of the fungal network. However, the relationship between plant species richness and AMF species richness remains to a large extent unexplained. Methodologically, addressing these questions is not easy. First, AMF species can be both drivers and passengers, determining or following changes in plant species richness (Hart et al. 2001). Second, with increasing numbers of plant species, the likelihood of including highly mycotrophic plants increases, resulting in a positive relation between plant and fungal species richness due to a sampling effect. However, from a management point of view, increasing AMF species richness in agroforestry systems is only relevant if the fungal species that associate with trees also associate with the agricultural crops in a common network. Ingleby et al. (2007) did in fact demonstrate the sharing of the mycorrhizal network between the agroforestry tree *Calliandra calothyrsus* and the crops maize and bean.

Methods for assessing AMF species richness may yield different outcomes. If a persistent fungal network, rather than an annual and labile mycorrhizal mycelium, characterizes agroforestry systems, then one would expect that the fungal population be preferentially comprised of K-selected rather than of r-selected species.

However, the more prolific spore formers may belong to the latter group, and in this case a diversity estimate based on spore data only would yield biased data. Molecular methods may yield better data, but application of such methods often lags behind in developing countries, due to the absence of well-equipped laboratories. Jefwa et al. (2006) observed lower species diversity in agroforestry systems with *Sesbania macrantha* and *S. sesban* than in maize monocrops, and suggested that the higher species diversity in the maize fields was due to the short maize cropping season, inducing rapid root dynamics and turnover, as compared to the much longer growth cycles of the agroforestry plots. Other studies yielded opposite data. In Ethiopia, Muleta et al. (2008) found higher abundance of AMF spores in agroforestry systems (especially when legumes served as shade crops) than in monocultural systems. Similar observations of higher spore abundance in agroforestry coffee systems than in monocultural systems were made in Brazil by Colozzi and Cardoso (2000) and Cardoso et al. (2003a).

Different plants show differential mycorrhizal dependence and responsiveness. This relation was first hypothesized by Baylis (1975) who argued that plants with a root system comprised of unbranched, relatively thick roots, with very few and short root hairs, will be more responsive to AMF than plant species with a finely branched root system, consisting of thin fine roots with numerous long root hairs. Janos (1980, 1996) further argued that mycorrhizal dependency increases during natural succession. In bare sites, colonizing plants tend to have a fine root system, allowing them to acquire nutrients even in the absence of AMF. Such plants are therefore often nonmycorrhizal or facultatively mycorrhizal. During succession, such plants are gradually replaced by plants that are obligatorily mycorrhizal. However, despite the conceptual clarity of the scheme, field data from Brazil do not fit with this pattern. Several studies demonstrated that responsiveness and susceptibility to mycorrhizal colonization were inversely related to succession (Siqueira et al. 1998; Siqueira and Saggin-Júnior 2001; Zangaro et al. 2003). Siqueira et al. (1998), working with a group of 28 native tree species belonging to different successional groups, observed that pioneer species responded relatively more to mycorrhizal inoculation than to P fertilizer, while climax species showed the opposite pattern. These findings support the hypothesis of a multifunctional nature of mycorrhizal associations. They also demonstrate, as already hypothesized by Newsham et al. (1995), that benefits of mycorrhiza other than P facilitation are important, particularly in plants with more effective root systems. However, the generality of this phenomenon and the implications for tree species choice in relation to agroforestry management deserve further study, because other plant traits besides root morphology, such as seed size and relative growth rate, are also correlated with susceptibility to mycorrhizal colonization and mycorrhizal responsiveness.

9.2.3 AMF and the Physical Quality of the Soil

AMF contribute in a direct way to the maintenance of soil structure, which is of fundamental importance for the stability of the agro-ecosystems (Rillig and Mummey 2006). The organization of soil particles determines the flow of water, gas, and nutrients in the soil (Rillig et al. 2002). In this context, the AMF are essential to the recovery of soil structure. Their actions cover a range of spatial scales, and include (1) directing clay particles around the hyphae, (2) producing polysaccharide secretions that connect clay particles, (3) performing a "packing" effect of particles by hyphae, leading to a new microstructure (Dorioz et al. 1993), and (4) directly transferring the carbon from plants to soil, which promotes aggregation.

Mycorrhizal effects do not depend only on the live mycelium. AMF produce and subsequently deposit on their hyphal walls a substance known as glomalin. For analytical reasons, this glycoprotein complex may be better referred to as "glomalin-related soil protein" (GRSP) rather than simply as "glomalin." The term GRSP points to the fact that other soil proteins may have similar properties and cannot always be separated from glomalin. Glomalin or GRSP is a very recalcitrant glycoprotein with high cementation capacity, which remains in the soil for longer time periods than the hyphae, thus contributing more persistently to the stabilization of aggregates (Driver et al. 2005). Hyphal residence time varies from days (for the smallest hyphae and the branched absorbing structures) to months (for runner hyphae of larger diameter) (Langley and Hungate 2003; Staddon et al. 2003) while residence time of GRSP varies from 6 to 42 years (Rillig et al. 2001).

GRSP is present in the soil in large quantities. In tropical forest soils of Costa Rica and Hawaii, GRSP levels reached values up to 12.5 and 60 mg of glomalin cm^{-3} respectively (Lovelock et al. 2004; Rillig et al. 2001). Lovelock et al. (2004) estimated that approximately 3.2% of soil total C and 5% of soil N in tropical forests was in the form of glomalin. Treseder and Turner (2007) suggested that as a rule of thumb around 10% of soil organic carbon is constituted by GRSP. These values suggest that the contribution of AMF to soil C sequestration is substantial. AMF also play a major role in the carbon dioxide fluxes through soil.

The mycorrhizal symbiosis implies carbon costs for the plants ranging from 10% to 20% of photoassimilates. If we scale up such levels (assuming that 15% of photosynthate are consumed by fungi) to the global scale, about 10 Pg

(10×10^{15} g) of C are annually used by fungi and (at equilibrium) returned to the atmosphere (Langley et al. 2006). Thus, estimates by Bago et al. (2000) that the annual C flux through AMF amounts to 10^{12} g, most certainly are far too low. Considering that tropical ecosystems are mainly composed of AMF-dominated vegetation, the role of AMF in the regulation of global carbon balance is substantial. By producing recalcitrant compounds such as GRSP, mycorrhizal agroforestry systems make a major contribution to the rehabilitation of degraded land, sequestration of carbon, and possibly reduction of emissions of other greenhouse gases such as N_2O (Lal 2005; Mutuo et al. 2005).

AMF do not act alone on soil structure. The interactions between soil fauna (especially soil ecosystem engineers such as earthworms and termites) and AMF in determining the balance between formation, stabilization, and breakdown of macro-aggregates and micro-aggregates could have a major impact. Synergistic effects between AMF and earthworms are probably very important in agroforestry systems, which provide an environment that is beneficial for both groups of organisms. However, such interactions have hardly been investigated.

Detrimental roles of mycorrhizal associations on soil carbon sequestration have also been reported. Chapela et al. (2001) observed that a reduction of soil organic carbon levels occurred over a 20-year period in Ecuador, following the massive introduction of non-native pines (*Pinus radiata*), which had been inoculated with the equally non-native ectomycorrhizal fungus *Suillus luteus*. This ecosystem was very productive in terms of edible fruitbody biomass, reaching levels up to 1,000 kg dry weight per hectare per year (Hedger 1986), thereby providing income for the rural population. However, this biomass production occurred at the expense of the soil organic carbon stock. While this study may well suggest a fundamental difference between ectomycorrhizal and arbuscular mycorrhizal associations in relation to soil carbon dynamics, it could also represent yet another example of the unexpected consequences of drastic ecosystem simplification, where one non-native tree and one non-native fungal species come to dominate the landscape. Presumably, most agroforestry systems would avoid this path towards excessive reduction of aboveground and belowground diversity, and the consequent reduction in the quality of ecosystem services.

9.2.4 AMF and the Chemical Quality of the Soil

9.2.4.1 AMF and Acquisition of Soil P

Agriculture under temperate climate conditions is often characterized by conditions of excess, while in tropical regions the problem is access (Van Noordwijk and Cadish 2002). This fundamental difference between both types of ecosystems translates into equally fundamental differences as regards the benefits to be expected from AMF, and the response of these fungi to agro-ecosystem management. In temperate agricultural systems, interest in mycorrhizal associations is increasing due to changes in the production system from conventional to organic (Mäder et al. 2002; Ryan and Graham 2002). It is still commonly held that mycorrhizal associations are unimportant under conventional agriculture, although this view neglects the multifunctionality of the arbuscular mycorrhizal symbiosis and overlooks the major contributions of AMF to soil structure (Ryan and Graham 2002). The situation in the tropics is very different. Most resource-poor tropical farmers are organic by default and under the force of circumstances, due to high agricultural input costs and low revenues. In agricultural systems under tropical climate, especially in agroforestry systems, AMF are essential for maximizing the utilization of P and other soil nutrients (Cardoso and Kuyper 2006).

Different nutrient inputs do not represent the only factor modulating the roles of mycorrhiza in temperate and tropical agro-ecosystems. Tropical soils are geologically older (on average) than soils in temperate regions, and this results in much stronger P deficiency in the tropics (see Chaps. 1 and 10). This P deficiency is not due to scarcity of phosphorus per se, but to the very low concentrations of P in the soil solution caused by the strong $H_2PO_4^-$ adsorption to iron (Fe) and aluminium (Al) oxides (Vance et al. 2003; Ticconi and Abel 2004). Because of this very strong adsorptive capacity of tropical soils, the greatest part of P occurs in chemical forms that are unavailable to plants (Novais and Smyth 1999). P fertilizer addition is also very inefficient, as most of the P fertilizer rapidly enters pools that are subsequently unavailable to plants.

In order to increase the efficiency of nutrient uptake and use by plants, enlarging the soil volume exploited by roots is imperative. The volume of exploited soil and the surface over which nutrients are taken up by hyphae of AMF (the total length of which is often in the range of 10–50 m $\times$ cm^{-3} soil) are generally substantially larger than those of plant root and root hairs (with a root length usually less than 5 cm $\times$ cm^{-3} soil). These measurements make it easy to grasp the crucial importance of mycorrhizal associations under conditions of low nutrient bioavailability, arising as a consequence of low water content and strong nutrient adsorption to oxides (Leake et al. 2004). Because the nutrient diffusion pathway is strongly dependent on soil water availability (both directly, and indirectly through soil moisture effects on tortuosity), water deficits contribute to limit the continuity of nutrient supply. In a pulsed climate, which is characterized by alternating wet and dry seasons, positive feedbacks between the mycorrhizal effects on water uptake and P uptake therefore occur (Augé 2001).

While enlarging the exploited soil volume is the most important mechanism for contribution of AMF to nutrient uptake, several other mechanisms have also been proposed, such as: (1) the kinetics of P absorption by hyphae differs from that of the roots, either because AMF have a higher affinity (lower K_m) or are less leaky (possess lower C_{min}) as compared to roots (Faquin et al. 1990; Silveira and Cardoso 2004); (2) roots and hyphae explore microsites differently, especially small amounts of organic matter and pores with small pore necks that can only be accessed by hyphae (Joner and Jakobsen 1995); (3) roots and hyphae of AMF have different means of solubilizing inorganic forms of P in the rhizosphere (Bolan 1991; Cardoso et al. 2006; Satter et al. 2006); (4) plants and mycorrhizal fungi make a differential contribution to the uptake of organic forms of P, because plants and fungi release to the soil different amounts of phosphatases and other enzymes (Hamel 2004).

There is disagreement in the literature with respect to possible differences in physiology of P uptake by roots and AMF. It is also unclear how such differences, where existent, would translate into significant differences in nutrient uptake. In a sensitivity analysis as part of a modeling study on nutrient uptake by plants, Williams and Yanai (1996) stated that changes in physiological parameters (K_m, C_{min}) of uptake models have little impact on uptake rates under nutrient-poor conditions. However, Schnepf and Roose (2006) reached the opposite conclusion using a mycorrhizal uptake model. Resolving this discrepancy would be a prerequisite for evaluating the potential for "mycorrhizal saturation" (O'Neill 1994), the condition where the density and spatial distribution of the mycorrhizal fungal mycelium is not optimal from the plant's perspective but seems to optimize the fungal fitness. Mycorrhizal saturation models, such as the model by Landis and Fraser (2008), do not automatically assume near-perfect coordination between plant and fungus.

Several studies have shown that plants with or without association with AMF use the same sources of P from soil (Bolan 1991; Hernández et al. 2000). However, other studies claimed that mycorrhizal plants obtained P from sources of inorganic P (P_i) (and possibly also from organic P, P_o) that are normally unavailable to nonmycorrhizal plants (Jayachandran et al. 1989; Feng et al. 2003; Shibata and Yano 2003). The latter studies were performed in an artificial medium or in soils enriched with artificial P compounds (such as phytates), raising doubts about the ability of AMF to use sources of P adsorbed or complexed to organic matter under field conditions. It should also be remembered that mycorrhizal symbiosis induces further changes in the rhizosphere, and that the occurrence and activity of phosphate-solubilizing bacteria (PSB) may be modified in the mycorrhizosphere. Synergistic and antagonistic interactions between AMF and PSB have been repeatedly described (Muthukumar et al. 2001), although, to our knowledge, studies under natural agroforestry conditions have not yet been published.

In order to examine whether plants associated with AMF gain differential access to P_i pools as compared to nonmycorrhizal plants, Cardoso et al. (2006) analyzed the different P pools through P fractionation before and after treatment with mycorrhiza in a natural soil. On the one hand, no change was noted in the fractions of P_i and P_o in the presence of nonmycorrhizal maize plants, and P balance analysis indicated that growth of the nonmycorrhizal plants was completely determined by seed P reserves. This complete inability of the maize cultivar to capture P in the nonmycorrhizal condition was surprising, because plant-available P (9 mg P kg^{-1} soil for Resin-P_i and $NaHCO_3$-P_i) in the soil exceeded the minimal concentration determined for plant growth (3 mg P kg^{-1} soil, according to Hayman (1983)).

On the other hand, the mycorrhizal maize plants completely used the pools of Resin-P_i and $NaHCO_3$-P_i (both of which are immediately accessible pools) and around 20% of the pool of NaOH-P_i (which is accessible in the medium term). These results remain to be confirmed under field conditions (Cardoso et al. 2006).

The various P_o pools remained constant or even increased in magnitude over the course of the above-mentioned experiment. This observation is in keeping with earlier suggestions that AMF have no saprotrophic ability. Under field conditions, the mycorrhiza may intercept and capture the newly mineralized nutrients before their fixation to soil particles. This strategy, amounting to a tightening of the P soil cycle, may be of great practical significance (Joner et al. 2000).

Cardoso et al. (2006) were unable to explain the uptake of P from pools (NaOH-P_i) that are considered to be unavailable in the short term. However, they noted that glomalin might have a role in P acquisition. Glomalin seems to interact with metal in the soil, because in analytical GRSP fractions, iron, aluminum and copper can be found in relatively high amounts (Rillig et al. 2001; Lovelock et al. 2004; Gonzales-Chavez et al. 2004). Iron levels varying from 0.8% to 8.8% have been reported in glomalin (Wright and Upadhyaya 1998). The metallic ions in GRSP seem to have a role in protecting and stabilizing this protein (Nichols and Wright 2005). Through this interaction, GRSP may affect not only the dynamics of the bound metals, but also that of other ions that bind to Fe or Al, such as P.

In order to explain P uptake by mycorrhizal plants from Fe-P sources, Bolan et al. (1987) suggested that AMF may break Fe-P links, but without proposing a specific mechanism for this. Possibly, one such mechanism may involve GRSP production. Assuming that for every molecule of Fe that ends up in GRSP, one molecule of P could become available for uptake, Cardoso et al. (2006) calculated that a soil with GRSP levels of 0.5 mg g^{-1} (with 1% of Fe being bound to GRSP) could deliver enough P to fully explain the changes in the NaOH-P_i pool. Further investigation is required to confirm this putative role of GRSP in P mobilization, and thoroughly understand the interrelated effect of mycorrhizal associations on biological, physical, and chemical properties of the soil.

9.2.4.2 AMF and Protection Against Heavy Metals and Al

Besides their effects on the chemical properties of the soil and the ability to supply nutrients to the plants, AMF may mitigate phytotoxic effects caused by elements such as heavy metals and aluminium. Heavy metal concentrations in host plant tissues may decrease as a result of fungal association, and reductions of Pb concentrations were noted in the aerial parts of soybean (Andrade et al. 2003) and *Brachiaria* (Silva et al. 2006). Hence, mycorrhizal plants gain improved heavy metal tolerance.

Similarly, the concentration of Al^{3+} is reduced in mycorrhizal banana plants as compared to nonmycorrhizal banana (Rufyikiri et al. 2000). Tolerance to Al may be conferred as a result of increased P acquisition or through other mechanisms. Again, a study of P uptake and Al resistance in relation to the dynamics of GRSP would be helpful. On the other hand, acting through mechanisms similar to those that are involved in enhanced P uptake, the mycorrhizal association can enhance Zn and Cu uptake when concentrations of these metals are limiting.

9.2.5 AMF and the Biological Quality of the Soil

AMF can reduce damage caused by pathogens such as fungi, bacteria, and nematodes (Azcón-Aguilar and Barea 1996; Cardoso and Kuyper 2006). It has been debated whether this is mainly an indirect effect, caused by an improved nutritional status of mycorrhizal plants as compared to nonmycorrhizal plants; or whether there are direct, non-nutritional effects. While improved nutrition is clearly involved in the observed protection, non-nutritional mechanisms are also acting, since mycorrhizal and nonmycorrhizal plants with a similar nutritional status may be affected differently by the same pathogen. Non-nutritional mechanisms include activation of the plant defense system, changes in exudate patterns resulting in concomitant changes in the microbial community of the mycorrhizosphere, increased lignification of the cell walls, and competition for space and infected sites (Elsen et al. 2008; Pozo and Azcón-Aguilar 2007). The latter mechanism may be of particular significance when AMF colonization takes place prior to pathogen infection, as would be expected to occur in agroforestry and other agro-ecosystems that help maintain the mycorrhizal network. However, direct evidence for this is currently lacking in the published literature.

Although the nutritional and non-nutritional pathogen control mechanisms almost certainly interact, their conceptual separation remains imperative for the rational application of mycorrhizal technologies. Should nutritional mechanisms prevail, then mineral or organic fertilizer application could, under certain circumstances at least, be cheaper and more effective than mycorrhizal management. However, to benefit from non-nutritional protection mechanisms calls for different management options (Cardoso and Kuyper 2006).

AMF can also play a role in antagonistic (parasitic) interactions between plants. In greenhouse experiments, mycorrhizal sorghum (*Sorghum bicolor*) plants suffered less damage caused by the root hemiparasite *Striga hermonthica* than non-mycorrhizal sorghum (Lendzemo and Kuyper 2001; Gworgwor and Weber 2003). The presence of AMF also suppressed emergence and abundance of *Striga*, both in greenhouse and field conditions (Lendzemo et al. 2005). Root exudates of mycorrhizal sorghum plants strongly inhibited the germination of *Striga* seeds (Lendzemo et al. 2007), most probably through downregulation of strigolactones, a class of substances that are important both in the AMF–plant molecular dialog and as *Striga* germination stimulants. *Striga* numbers are also reduced under the canopy of certain agroforestry trees (such as *Faidherbia albida*) and on termitaria, which are commonly found under such trees. Crushed material of *Cubitermes* termitaria improved the rate of sorghum mycorrhization, pointing to important synergistic effects between agroforestry trees, soil fauna, and mycorrhizal associations in the control of the world's most devastating parasitic plant (Andrianjaka et al. 2007).

AMF interact, not only with antagonistic organisms, but also with nitrogen-fixing bacteria. As a consequence of enhanced P uptake by the plant, AMF promote an increased legume nodulation by rhizobia: indeed, P is often the limiting factor for nodulation and nitrogen fixation. Direct interactions also occur between the two microsymbionts, such as the synthesis by AMF of proteins immunologically related to legume nodulins (Perotto et al. 1994). Several of these interactions reflect the fact that the early molecular dialog between legumes and rhizobia is derived from the molecular dialog between higher plants and AMF.

Field studies on the interactions between rhizobia and AMF in agroforestry systems are scarce. Most publications refer to studies where agroforestry trees were inoculated, individually or jointly, with rhizobia and AMF. Such studies demonstrated that the combination of both symbionts is effective in boosting the growth of tropical tree legumes (Weber et al. 2005; Lesueur and Sarr 2008). Dual colonization by AMF and rhizobia increased plant dry matter, concentration and content of N and P and nutrient uptake efficiency by *Sesbania virgata*, compared to the uninoculated control (Rodrigues et al. 2003a,b). It has often been mentioned that interactions between AMF and rhizobia lead to synergism. However, synergism has almost never been defined. We propose to call the effects of AMF plus rhizobia additive if an analysis of variance (ANOVA) indicates that the interaction term AMF * rhizobia is not significant, and synergistic if the interaction term is significant. Synergism includes instances of both negative interactions, as in the case of *Dalbergia nigra* (Santiago et al. 2002), and positive interactions (although we are not aware of publications on agroforestry legumes that show such positive synergism). In most cases the effect of AMF and rhizobia turns out to be additive — which is consistent with the theory that both root symbionts enhance plant availability for different limiting nutrients.

The beneficial effects from dual inoculation might be only temporary. Lesueur and Sarr (2008) showed that the benefit of the joint inoculation of the agroforestry legume *Calliandra calothyrsus* was significant until 12 months after transplanting to the field, but not after 24 months. Such changes, which again stress the need for field experiments with naturally occurring symbionts, is possibly due to the symbiotic capacity of native symbionts naturally present in the field. Inoculum potential, effectiveness, and competitive ability of the symbionts that naturally occur in the field might well differ from those of carefully selected strains used for experimentation under controlled conditions. The fact that many agro-ecosystems, unless very poorly managed, harbor indigenous AMF which might be competitively superior to commercial strains, suggests that for the rational use of the microbial potential, management is preferable to artificial inoculation (Kuyper et al. 2004). It should also be noted that commercial inoculants are often expensive and of variable quality. Recently, commercial formulations of AMF have been made available in the form of powders, pellets, tablets, granules, gel beads, and balls (Adholeya et al. 2005; see Chap. 11), but we are unaware of published evaluations of their effectiveness.

9.3 AMF and Agroforestry Systems: Beyond Compatibility

Plant monocultures, cultivation of nonmycorrhizal plants, prolonged bare fallow, erosion, compaction, use of excessive amounts of fertilizers, and the use of systemic pesticides are among the main factors that contribute to the reduction or elimination of the AMF (Siqueira et al. 2007; Abbott and Robson 1991). These adverse processes all derive from a shared (and implicit) principle of ecosystem simplification. A reversal of the simplification trend is therefore imperative, and agroforestry is one of the strategies that will allow full advantage to be taken of the ecosystem services provided by the soil biota.

Agroforestry is not a standard recipe fit for indiscriminate use under all agro-ecological conditions. Competition for resources (water, nutrients, and light) between annual crops and tree species has often been mentioned as a factor that constrains the success of agroforestry (Farrell and Altieri 2002). Because of the major impact that AMF have on uptake of water and nutrients, they could both mitigate and enhance competition. Whether mitigation or enhancement prevails depends on tree management, on properties of both the tree and crop components of the agroforestry system, and on the extent and implications of communication between these two components through a common mycorrhizal network. In such a network, maintenance of a mycorrhizal inoculum would insure a more rapid establishment on host plants and hence a better annual crop seedling performance. If nutrients and water are transferred between plants, and if the dominant flow is towards the competitively inferior plant, AMF have great potential to mitigate restrictions on plant growth (He et al. 2003; Selosse et al. 2006).

In a greenhouse study, Ingleby et al. (2007) observed that trees maintain active AMF propagules, hence increasing the density of AMF populations prone to annual crop colonization. This enhancement effect occurs both with indigenous propagules and after inoculum addition, and is related to the maintenance of spore banks and a stimulation of fungal mycelium spread in the soil. Shoot pruning of trees, which is a common practice in some agroforestry systems, did not restrict the mycorrhizal colonization of, or the spread of the mycelium network to, the annual crops. This effect may be of particular importance given that the mycelium network is responsible for the fast colonization of new roots, and thus for a growth stimulation and improved P absorption in young plants (Brundrett and Abbott 1994). However, the rate of spread of the AMF mycelium (1–3 mm day^{-1}) suggests that benefits from the network will develop slowly.

The common mycorrhizal network may further enhance the benefits of agroforestry through vertical niche expansion of AMF. The presence of perennial mycorrhizal tree species with deep roots increases the volume of soil to be exploited (Fig. 9.1) and thereby improves the efficiency of P cycling by AMF (Cardoso et al. 2003a; Muleta et al. 2008). The increase in the efficiency of P cycling is also linked to the maintenance of larger pools of P_0 and the greater abundance of roots, which reduce the possibility of fixation of the newly mineralized phosphate (Cardoso et al. 2003b). Such vertical niche expansion of AMF could also increase water use efficiency of the agroforestry system through mycorrhiza-mediated hydraulic distribution (Allen 2007; Egerton-Warburton et al. 2007).

Therefore, we propose that positive feedbacks exist between agroforestry practices and arbuscular mycorrhizal management. According to this hypothesis, agroforestry systems create conditions more conducive to arbuscular mycorrhizal functioning, while mycorrhizal management mitigates the competition between

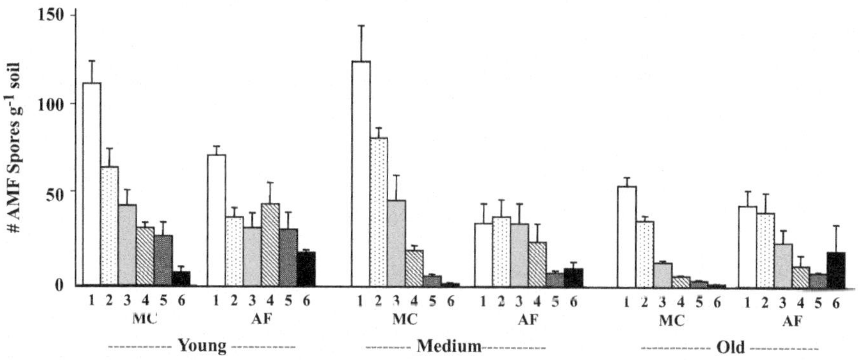

Fig. 9.1 Average (and standard error; n = 3) numbers (#) of arbuscular mycorrhizal fungi spores at different soil depths (1 = 0–1, 2 = 2–3, 3 = 5–7.5, 4 = 10–15, 5 = 20–30 and 6 = 40–60 cm depths) under monocultural coffee (MC) and agroforestry (AF) systems in the Atlantic Coastal Rainforest (Brazil). The agricultural systems were 5 years old (Young), 15–20 years old (Medium) and 20–24 years old (Old). Adapted from Cardoso et al. (2003a)

tree crops and annual crops. This beneficial impact between agroforestry management and mycorrhizal action may be depicted as a particular form of symbiosis. To confirm this hypothesis will require that mycorrhizal functions within agroforestry management be better understood.

9.4 Conclusions

Agroforestry systems can be a viable strategy for the preservation of natural resources while ensuring sustainable food production in the tropics. Agroforestry systems are designed with the objective of imitating tropical forests and optimizing beneficial ecological interactions among ecosystem components. In particular, the interactions between plants and soil as mediated by mycorrhizal fungi are of prime importance due to the wide range of functions that these fungi perform. Mycorrhizal fungi potentially improve physical, chemical, and biological soil quality. Several of these functions are linked to the formation of a common mycorrhizal network, which may mediate the transfer of water and nutrients between different plants. However, the ecological implications of these relationships remain poorly studied.

- Agroforestry systems are productive, in addition to containing a greater diversity of species than do simplified monoculture agro-ecosystems. However, the relationship of cause and effect between the diversity of AMF and diversity of plants, and the correlation of both with the productivity of agroforestry systems, remain largely undemonstrated.

- In tropical soils, which are commonly poor in nutrients and especially in available phosphorus, access to soil nutrients and the efficient use of these depend on the optimization of biogeochemical cycling and of the function of mycorrhizal fungi. The capacity and the access mechanisms to different fractions of soil P pools by AMF are not completely elucidated. The role of glomalin or GRSP in affecting P dynamics through GRSP interactions with Fe and Al needs to be further addressed.
- The practice of mycorrhizal inoculation remains very limited due to high inoculum costs, lack of quality standards, lack of technology adapted to resource-poor farmers, and especially the unresolved discrepancy between research results, obtained through short-term trials conducted under controlled conditions, and the phenomena occurring under long-term field conditions. In fact, management of native populations of mycorrhizal fungi might appear preferable to fungal inoculation.
- Agroforestry systems potentially maximize the benefits procured by AMF, which in turn could mitigate negative interactions between trees and annual crops. This positive mycorrhiza-agroforestry feedback, and the common mycorrhizal network which produces it, deserve closer attention.

Acknowledgments The authors thank the farmers and the Center of Alternative Technology of the Zona da Mata, Minas Gerais, partners in most of the research in agroforestry systems developed by the Brazilian authors. We also thank the Brazilian sponsors FAPEMIG (Fundação de Amparo à Pesquisa do Estado de Minas Gerais) for financial support, CAPES (Coordenação de Aperfeiçoamento de Pessoal de Nível Superior) and CNPq (Conselho Nacional de Desenvolvimento Científico e Tecnológico) for scholarships to the first and second authors and also for financial support.

References

Abbott LK, Robson AD (1991) Factors influencing the occurrence of vesicular–arbuscular mycorrhizas. Agr Ecosyst Environ 35:121–150

Adholeya A, Tiwari P, Singh R (2005) Large-scale inoculum production of arbuscular mycorrhizal fungi on root organs and inoculation strategies. In: Declerck S, Strullu D-G, Fortin JA (eds) In vitro culture of mycorrhizas. Springer, Berlin, pp 315–338

Al-Karaki GN (2000) Growth of mycorrhizal tomato and mineral acquisition under salt stress. Mycorrhiza 10:51–54

Allen MF (2007) Mycorrhizal fungi: highways for water and nutrients in arid soils. Vadose Zone J 6:291–297

Andrade SAL, Abreu CA, Abreu MF, Silveira APD (2003) Interação de chumbo, da saturação por bases do solo e de micorriza arbuscular no crescimento e nutrição mineral da soja. R Bras Ci Solo 27:945–954

Andrianjaka Z et al (2007) Biological control of *Striga hermonthica* by *Cubitermes* termite mound powder amendment in sorghum culture. Appl Soil Ecol 37:175–183

Augé RM (2001) Water relations, drought and vesicular–arbuscular mycorrhizal symbiosis. Mycorrhiza 11:3–42

Azcón-Aguilar C, Barea JM (1996) Arbuscular mycorrhizas and biological control of soil-borne plant pathogens — an overview of the mechanisms involved. Mycorrhiza 6:457–464

Bago B, Pfeffer PE, Shachar-Hill Y (2000) Carbon metabolism and transport in arbuscular mycorrhizas. Plant Physiol 124:949–957

Baylis GTS (1975) The magnolioid mycorrhiza and mycotrophy in root systems derived from it. In: Sanders FE, Mosse B, Tinker PB (eds) Endomycorrhizas. Academic, London, pp 373–389

Bethlenfalvay GJ, Bayne HG, Pacovsky RS (1983) Parasitic and mutualistic associations between a mycorrhizal fungus and soybean: the effect of phosphorus on host plant–endophyte interactions. Physiol Plantarum 57:543–548

Bhadalung N, Suwanarit A, Dell B, Nopamornbodi O, Thamchaipenet A, Rungchuang J (2005) Effects of long-term NP-fertilization on abundance and diversity of arbuscular mycorrhizal fungi under a maize cropping system. Plant Soil 270:371–382

Bolan NS (1991) A critical review on the role of mycorrhizal fungi in the uptake of phosphorus by plants. Plant Soil 134:189–207

Bolan NS, Robson AD, Barrow NJ (1987) Effects of vesicular–arbuscular mycorrhiza on the availability of iron phosphates to plants. Plant Soil 99:401–410

Brundrett MC (2002) Coevolution of roots and mycorrhizas of land plants. New Phytol 154: 275–304

Brundrett MC, Abbott LK (1994) Mycorrhizal fungus propagules in the jarrah forest. New Phytol 127:539–546

Bucher M (2007) Functional biology of plant phosphate uptake at root and mycorrhiza interfaces. New Phytol 173:11–26

Cardoso IM, Kuyper TW (2006) Mycorrhizas and tropical soil fertility. Agric Ecosyst Environ 116:72–84

Cardoso IM, Guijt I, Franco FS, Carvalho AF, Ferreira Neto PS (2001) Continual learning for agroforestry system design university, NGO and farmer partnership in Minas Gerais, Brazil. Agric Syst 69:235–257

Cardoso IM, Boddington C, Janssen BH, Oenema O, Kuyper TW (2003a) Distribution of mycorrhizal fungal spores in soils under agroforestry and monocultural coffee systems in Brazil. Agroforest Syst 58:33–43

Cardoso IM, Janssen BH, Oenema O, Kuyper TW (2003b) Phosphorus pools in Oxisols under shaded and unshaded coffee systems on farmers' fields in Brazil. Agroforest Syst 58:55–64

Cardoso IM, Boddington C, Janssen BH, Oenema O, Kuyper TW (2006) Differential access to phosphorus pools of an Oxisol by mycorrhizal and non-mycorrhizal maize. Commun Soil Sci Plant Anal 37:11–12

Carneiro MAC, Siqueira JO, Davide AC, Gomes LJ, Curi N, Vale FR (1996) Fungo micorrízico e superfosfato no crescimento de espécies arbóreas tropicais. Scient Forestalis 50:21–36

Carneiro MAC, Siqueira JO, Moreira FMS, Carvalho D, Botelho AS, Saggin-Júnior OJ (1998) Micorriza arbuscular em espécies arbóreas e arbustivas nativas de ocorrência no Sudoeste do Brasil. Cerne 41:129–145

Chapela IG, Osher LJ, Horton TR, Henn MR (2001) Ectomycorrhizal fungi introduced with exotic pine plantations induce soil carbon depletion. Soil Biol Biochem 33:1733–1740

Colozzi A, Cardoso EJBN (2000) Detection of arbuscular mycorrhizal fungi in roots of coffee plants and *Crotalaria* cultivated between rows. Pesqui Agropecu Bras 35:203–2042

Dorioz JM, Robert M, Chenu C (1993) The role of roots, fungi and bacteria on clay particle organization — an experimental approach. Geoderma 56:179–194

Driver JD, Holben WE, Rillig MC (2005) Characterization of glomalin as a hyphal wall component of arbuscular mycorrhizal fungi. Soil Biol Biochem 37:101–106

Egerton-Warburton LM, Querejeta JI, Allen MF (2007) Common mycorrhizal networks provide a potential pathway for the transfer of hydraulically lifted water between plants. J Exp Bot 58:1473–1483

Elsen A, Gervacio D, Swennen R, De Waele D (2008) AMF-induced biocontrol against plant parasitic nematodes in *Musa* sp.: a systemic effect. Mycorrhiza 18:251–256

Ewel JJ (1999) Natural systems as a model for the design of sustainable systems of land use. Agroforest Syst 45:1–21

Faquin V, Malavolta E, Muraoka T (1990) Cinética da absorção de fosfato em soja sob influência de micorriza vesículo–arbuscular. Rev Bras Cienc Solo 14:41–48
Farrell JG, Altieri MA (2002) Sistemas agroflorestais. In: Altieri MA (ed) Agroecologia: bases científicas para uma agricultura sustentável. Agropecuária, Guaíba, pp 413–440
Feng G, Song YC, Li XL, Christie P (2003) Contribution of arbuscular mycorrhizal fungi to utilization of organic sources of phosphorus by red clover in a calcareous soil. Appl Soil Ecol 22:139–148
Fitter AH (2001) Specificity, links and networks in the control of diversity in plant and microbial communities. In: Press MC, Huntly NJ, Levin S (eds) Ecology: achievement and challenge. Blackwell, Oxford, pp 95–114
Foley JA et al (2005) Global consequences of land use. Science 309:570–574
Garg N, Geetanjali K, Amandeep K (2006) Arbuscular mycorrhiza: nutritional aspects. Arch Agron Soil Sci 52:593–606
Gonzales-Chavez MC, Carillo-Gonzalez A, Wright AS, Nichols KA (2004) Glomalin: a mechanism for heavy-metal sequestration by arbuscular mycorrhizal fungi. Environ Pollut 130:317–323
Gworgwor NA, Weber HC (2003) Arbuscular mycorrhizal fungi–parasite–host interaction for control of *Striga hermonthica* (Del.) Benth. in sorghum [*Sorghum bicolor* (L.) Moench]. Mycorrhiza 13:27–281
Hamel C (2004) Impact of arbuscular mycorrhizal fungi on N and P cycling in the root zone. Can J Soil Sci 84:383–395
Hart MM, Reader RJ, Klironomos JN (2001) Life-history strategies of arbuscular mycorrhizal fungi in relation to their successional dynamics. Mycologia 93:1186–1194
Hayman DS (1983) The physiology of vesicular–arbuscular endomycorrhizal symbiosis. Can J Bot 61:944–963
He XH, Critchley C, Bledsoe C (2003) Nitrogen transfer within and between plants through common mycorrhizal networks (CMNs). Crit Rev Plant Sci 22:531–567
Hedger J (1986) *Suillus luteus* on the equator. Bull Br Mycol Soc 20:53–54
Hernández G, Cuenca G, García A (2000) Behaviour of arbuscular–mycorrhizal fungi on *Vigna luteola* growth and its effect on the exchangeable (^{32}P) phosphorus of soil. Biol Fertil Soils 31:232–236
Ingleby K, Wilson J, Munro RC (2007) Mycorrhizas in agroforestry: spread and sharing of arbuscular mycorrhizal fungi between trees and crops: complementary use of molecular and microscopic approaches. Plant Soil 294:125–136
International culture collection of vesicular arbuscular mycorrhizal fungi — INVAM (2008) http://invam.caf.wvu.edu
Janos DP (1980) Mycorrhizae influence tropical succession. Biotropica 12((Suppl.)):56–64
Janos DP (1996) Mycorrhizas, succession, and the rehabilitation of deforested lands in the humid tropics. In: Frankland JC, Magan N, Gadd GM (eds) Fungi and environmental change. Cambridge University Press, UK, pp 129–162
Jayachandran K, Schwab AP, Hetrick BAD (1989) Mycorrhizal mediation of phosphorus availability: synthetic iron chelate effects on phosphorus solubilization. Soil Sci Soc Am J 53:1701–1706
Jefwa JM, Sinclair R, Maghembe JA (2006) Diversity of glomale mycorrhizal fungi in maize/sesbania intercrops and maize monocrop systems in southern Malawi. Agroforestry Syst 67:107–114
Johansen A, Jensen ES (1996) Transfer of N and P from intact or decomposing roots of pea to barley interconnected by an arbuscular mycorrhizal fungus. Soil Biol Biochem 28:73–81
Johnson NC (1993) Can fertilization of soil select less mutualistic mycorrhizae? Ecol Appl 3:749–757
Johnson NC, Graham JH, Smith FA (1997) Functioning of mycorrhizal associations along the mutualism–parasitism continuum. New Phytol 135:575–585
Joner EJ, Jakobsen I (1995) Growth and extracellular phosphatase activity of arbuscular mycorrhizal hyphae as influenced by soil organic matter. Soil Biol Biochem 27:1153–1159
Joner EJ, Van Aarle IM, Vosatka M (2000) Phosphatase activity of extra-radical arbuscular mycorrhizal hyphae. Plant Soil 226:199–210

Kuyper TW, Cardoso IM, Onguene NA, Van Noordwijk M, Van Noordwijk M (2004) Managing mycorrhiza in tropical multispecies agroecosystems. In: Van Noordwijk M, Cadish G, Ong CK (eds) Below-ground interactions in tropical agroecosystems. CABI, Wallingford, pp 243–261

Lal R (2005) Soil carbon sequestration in natural and managed tropical forest ecosystems. J Sustain Forest 21:1–30

Landis FC, Fraser LH (2008) A new model of carbon and phosphorus transfers in arbuscular mycorrhizas. New Phytol 177:466–479

Langley JA, Hungate BA (2003) Mycorrhizal controls on belowground litter quality. Ecology 84:2302–2312

Langley JA, Chapman SK, Hungate BA (2006) Ectomycorrhizal colonization slows root decomposition: the *post-mortem* fungal legacy. Ecol Lett 9:955–959

Leake J, Johnson D, Donnelly D, Muche G, Boddy L, Read D (2004) Networks of power and influence: the role of mycorrhizal mycelium in controlling plant communities and agroecosystem functioning. Can J Bot 82:1016–1045

Lendzemo VW, Kuyper TW (2001) Effects of arbuscular mycorrhizal fungi on damage by *Striga hermonthica* on two contrasting cultivars of sorghum, *Sorghum bicolor*. Agric Ecosyst Environ 87:29–35

Lendzemo VW, Kuyper TW, Kropff MJ, Van Ast A (2005) Field inoculation with arbuscular mycorrhizal fungi reduces *Striga hermonthica* performance on cereal crops and has the potential to contribute to integrated *Striga* management. Field Crops Res 91:51–61

Lendzemo VW, Kuyper TW, Matusova R, Bouwmeester HJ, Van Ast A (2007) Colonization by arbuscular mycorrhiza fungi of sorghum leads to reduced germination and subsequent attachment and emergence of *Striga hermonthica*. Plant Sign Behav 2:58–62

Lesueur D, Sarr A (2008) Effects of single and dual inoculation with selected microsymbionts (rhizobia and arbuscular mycorrhizal fungi) on field growth and nitrogen fixation of *Calliandra calothyrsus* Meissn. Agroforestry Syst 73:37–45

Lovelock CE, Wright SF, Clark DA, Ruess RW (2004) Soil stocks of glomalin produced by arbuscular mycorrhizal fungi across a tropical rain forest landscape. J Ecol 92:278–287

Mäder P, Fliessbach A, Dubois D, Gunst L, Fried P, Niggli U (2002) Soil fertility and biodiversity in organic farming. Science 196:1694–1697

Muleta D, Assefa F, Nemomissa S, Granhall U (2008) Distribution of arbuscular mycorrhizal fungi spores in soils of smallholder agroforestry and monocultural coffee systems in southwestern Ethiopia. Biol Fertil Soils 44:653–659

Muthukumar T, Udaiyan K, Rajeshkannan V (2001) Response of neem (*Azadirachta indica* A. Juss.) to indigenous arbuscular mycorrhizal fungi, phosphate-solubilizing and asymbiotic nitrogen-fixing bacteria under tropical nursery conditions. Biol Fert Soils 34:417–426

Mutuo PK, Cadisch G, Albrecht A, Palm CA, Verchot L (2005) Potential of agroforestry for carbon sequestration and mitigation of greenhouse gas emissions from soils in the tropics. Nutr Cycl Agroecosyst 71:43–54

Newsham KK, Fitter AH, Watkinson AR (1995) Multifunctionality and biodiversity in arbuscular mycorrhizas. Trends Ecol Evol 10:407–411

Nichols KA, Wright SF (2005) Comparison of glomalin and humic acid in eight native US soils. Soil Sci 170:985–997

Novais RF, Smyth TJ (1999) Fósforo em solo e planta em condições tropicais. MG, Universidad Federal de Viçosa, Viçosa

Olsson PA, Thingstrup I, Jakobsen I, Baath F (1999) Estimation of the biomass of arbuscular mycorrhizal fungi in a linseed field. Soil Biol Biochem 31:1879–1887

O'Neill EG (1994) Responses of soil biota to elevated atmospheric carbon dioxide. Plant Soil 165:55–65

Pande M, Tarafdar JC (2004) Arbuscular mycorrhizal fungal diversity in neem-based agroforestry systems in Rajasthan. Appl Soil Ecol 26:233–241

Perotto S, Brewin NJ, Bonfante P (1994) Colonization of pea roots by the mycorrhizal fungus *Glomus versiforme* and by *Rhizobium* bacteria: immunological comparison using monoclonal antibodies as probes for plant cell surface components. Mol Plant Microbe Interact 7:91–98

Pfeffer PE, Douds DD Jr, Bucking H, Schwartz DP, Shachar-Hill Y (2004) The fungus does not transfer carbon to or between roots in an arbuscular mycorrhizal symbiosis. New Phytol 163:617–627

Pozo MJ, Azcón-Aguilar C (2007) Unraveling mycorrhiza-induced resistance. Curr Opin Plant Biol 10:393–398

Rao AV, Giller KE (1993) Nitrogen fixation and its transfer from *Leucaena* to grass using ^{15}N. For Ecol Manage 61:221–227

Rillig MC, Mummey DL (2006) Mycorrhizas and soil structure. New Phytol 171:41–53

Rillig MC, Wright SF, Nichols KA, Schmidt WF, Torn MS (2001) Large contribution of arbuscular mycorrhizal fungi to soil carbon pools in tropical forest soils. Plant Soil 233:167–177

Rillig MC, Wright SF, Eviner VT (2002) The role of arbuscular mycorrhizal fungi and glomalin in soil aggregation: comparing effects of five plant species. Plant Soil 238:325–333

Rodrigues LA, Martins MA, Salomão MSMB (2003a) Uso de micorrizas e rizóbios em cultivo consorciado de eucalipto e sesbânia. I – Crescimento, absorção e transferência de nitrogênio entre plantas. Rev Bras Cienc Solo 27:583–591

Rodrigues LA, Martins MA, Salomão MSMB (2003b) Uso de micorrizas e rizóbios em cultivo consorciado de eucalipto e sesbânia. II — absorção e eficiência de utilização de fósforo e frações fosfatadas. Rev Bras Cienc Solo 27:593–599

Ruddiman WF (2003) The anthropogenic greenhouse era began thousands of years ago. Clim Change 61:261–293

Rufyikiri GS, Declerck S, Dufey JE, Delvaux B (2000) Arbuscular mycorrhizal fungi might alleviate aluminium toxicity in banana plants. New Phytol 148:343–352

Ryan MH, Graham JH (2002) Is there a role for arbuscular mycorrhizal fungi in production agriculture? Plant Soil 244:263–271

Santiago GM, Garcia Q, Scotti MR (2002) Effect of post-planting inoculation with *Bradyrhizobium* sp and mycorrhizal fungi on the growth of Brazilian rosewood, *Dalbergia nigra* Allem. ex Benth., in two tropical soils. New Forests 24:15–25

Satter MA, Hanafi MM, Mahmud TMM, Azizah H (2006) Influence of arbuscular mycorrhiza and phosphate rock on uptake of major nutrients by *Acacia mangium* seedlings on degraded soil. Biol Fertil Soils 42:345–349

Schnepf A, Roose T (2006) Modelling the contribution of arbuscular mycorrhizal fungi to plant phosphate uptake. New Phytol 171:669–682

Selosse MA, Richard F, He X, Simard SW (2006) Mycorrhizal networks: des liaisons dangereuses? Trends Ecol Evol 21:621–628

Shibata R, Yano K (2003) Phosphorus acquisition from non-labile sources in peanut and pigeonpea with mycorrhizal interaction. Appl Soil Ecol 24:133–141

Sierra J, Nygren P (2006) Transfer of N fixed by a legume tree to the associated grass in a tropical silvopastoral system. Soil Biol Biochem 38:1893–1903

Silva S, Siqueira JO, Soares CRFS (2006) Fungos micorrízicos no crescimento e na extração de metais pesados pela braquiária em solo contaminado. Pesqui Agropecu Bras 41:1749–1757

Silveira APD, Cardoso EJBN (2004) Arbuscular mycorrhiza and kinetic parameters of phosphorus absorption by bean plants. Sci Agric 61:203–209

Silveira APD, Gomes VFF (2007) Micorrizas em plantas frutíferas tropicais. In: Silveira APD, Freitas SS (eds) Microbiota do solo e qualidade ambiental. Instituto Agronômico, Campinas, pp 57–77

Simard SW, Durall DM (2004) Mycorrhizal networks: a review of their extent, function, and importance. Can J Bot 82:1140–1165

Siqueira LC (2008) Levantamento florístico e etnobotânico do estrato arbóreo em sistemas naturais e agroflorestais, Araponga. Magister Scientiae dissertation, Universidade Federal de Viçosa, Viçosa, Minas Gerais

Siqueira JO, Saggin O Jr (2001) Dependency on arbuscular mycorrhizal fungi and responsiveness of some Brazilian native woody species. Mycorrhiza 11:245–255

Siqueira JO, Carneiro MAC, Curi N, Rosado SCS, Davide AC (1998) Mycorrhizal colonization and mycotrophic growth of native woody species as related to successional groups in Southeastern Brazil. For Ecol Manage 107:241–252

Siqueira JO, Soares CRFS, Santos JGD, Schneider J, Carneiro MAC (2007) Micorrizas e degradação do solo: caracterização, efeitos e ação recuperadora. In: Ceretta CA, Silva LS, Reichert JM (eds) Tópicos em ciência do solo, vol 5. Sociedade Brasileira de Ciência do Solo, Viçosa, pp 219–305

Smith SE, Read DJ (2008) Mycorrhizal Symbiosis, 3rd edn. Academic, London

Snoeck D, Zapata F, Domenach A-M (2000) Isotopic evidence of the transfer of nitrogen fixed by legumes to coffee trees. Biotechnol Agron Soc Environ 4:95–100

Souza HN (2006) Sistematização da experiência participativa com sistemas agroflorestais: rumo à sustentabilidade da agricultura familiar na Zona da Mata mineira. Magister Scientiae dissertation, Universidade Federal de Viçosa, Viçosa

Souza FA, Lima da Silva IC, Berbara RLL (2008) Fungos micorrízicos arbusculares: Muito mais diversos do que se imaginava. In: Moreira FMS, Siqueira JO, Brussaard L (eds) Biodiversidade do solo em ecossistemas brasileiros. Editora UFLA, Lavras, pp 483–536

Staddon PL, Bronk-Ramsey C, Ostle N, Ineson P, Fitter AH (2003) Rapid turnover of hyphae of mycorrhizal fungi determined by AMS microanalysis of ^{14}C. Science 300:1138–1140

Stürmer SL, Siqueira JO (2008) Diversidade de fungos micorrízicos arbusculares em ecossistemas brasileiros. In: Moreira FMS, Siqueira JO, Brussaard L (eds) Biodiversidade do solo em ecossistemas brasileiros. Editora UFLA, Lavras, pp 537–583

Tavares RC (2007) Efeito da inoculação com fungo micorrízico arbuscular e da adubação orgânica no desenvolvimento de mudas de sabiá (*Mimosa caesalpiniaefolia* Benth.), sob estresse salino. Magister Scientiae dissertation, Universidade Federal do Ceará, Fortaleza

Ticconi CA, Abel S (2004) Short on phosphate: plant surveillance and countermeasures. Trends Plant Sci 9:548–555

Tilman D et al (2001) Forecasting agriculturally driven global environmental change. Science 292:281–284

Treseder KK, Turner KM (2007) Glomalin in ecosystems. Soil Sci Soc Am J 71:1257–1266

Urcelay C, Diaz S (2003) The mycorrhizal dependence of subordinates determines the effect of arbuscular mycorrhizal fungi on plant diversity. Ecol Lett 6:388–391

Van der Heijden MGA (2004) Arbuscular mycorrhizal fungi as support systems for seedling establishment in grassland. Ecol Lett 7:293–303

Van der Heijden MGA et al (1998) Mycorrhizal fungal diversity determines plant biodiversity, ecosystem variability and productivity. Nature 396:69–72

Van Noordwijk M, Cadish G (2002) Access and excess problems in plant nutrition. Plant Soil 247:25–40

Van Noordwijk M, Ong CK (1999) Can the ecosystem mimic hypotheses be applied to farms in African savannahs? Agroforest Syst 45:131–158

Vance CP, Uhde-Stone C, Allan DL (2003) Phosphorus acquisition and use: critical adaptations by plants for securing a nonrenewable resource. New Phytol 157:423–447

Weber J et al (2005) Co-inoculation of *Acacia mangium* with *Glomus intraradices* and *Bradyrhizobium* sp. in aeroponic culture. Biol Fertil Soils 41:233–239

Williams M, Yanai RD (1996) Multi-dimensional sensitivity analysis and ecological implications of a nutrient uptake model. Plant Soil 180:311–324

Wright SF, Upadhyaya A (1998) A survey of soils for aggregate stability and glomalin, a glycoprotein produced by hyphae of arbuscular mycorrhizal fungi. Plant Soil 198:97–107

Young A (1997) Agroforestry for soil management, 2nd edn. ICRAF and CAB International, Wallingford

Zangaro W, Nisizaki SMA, Domingos JCB, Nakano EM (2002) Micorriza arbuscular em espécies arbóreas nativas da bacia do rio Tibagi, Paraná. Cerne 8:77–87

Zangaro W, Nisizaki SMA, Domingos JCB, Nakano EM (2003) Mycorrhizal response and successional status in 80 woody species from south Brazil. J Trop Ecol 19:315–324

Zhu YG, Miller RM (2003) Carbon cycling by arbuscular mycorrhizal fungi in soil-plant systems. Trends Plant Sci 8:407–409

Chapter 10
Agricultural Development in Tropical Acidic Soils: Potential and Limits of Phosphate-Solubilizing Bacteria

Luis Andrés Yarzábal

10.1 Introduction

Freedom from hunger is a fundamental human right. As such, it is intrinsically linked to, as well as essential for, the full enjoyment of other rights such as health, education, and work, and everything that emanates from them. Nevertheless, according to recent estimates from the Food and Agriculture Organization of the United Nations (FAO), more than 923 million people are chronically hungry, most of them in rural areas of poor countries (FAO 2005). The number of hungry people is steadily rising, having increased by about 50 million in 2007 alone. If world population continues to grow at the mean estimated annual rate of 1.1%, it will reach 7.4 billion by 2017. In 2050, it will head towards 9.3 billion people (FAO 2005). Therefore, the world urgently needs a second Green Revolution – sometimes referred to as the "Gene Revolution" – in order to increase food production by at least 50% in the next 20 years.

The challenge we are facing is enormous. Two of the requirements for producing additional food are of paramount importance. First of all, agriculture productivity must significantly increase, particularly in the lesser developed, tropical regions of the world. The second condition is that the agricultural frontier be expanded into regions considered as marginal lands, consisting of fragile soils with a lower productive capacity and a higher risk of degradation (Zapata and Roy 2004).

L.A. Yarzábal
Laboratorio de Microbiología Molecular y Biotecnología, Departamento de Biología, Facultad de Ciencias, Núcleo de La Hechicera, Universidad de Los Andes, Mérida 5101, Estado Mérida, Venezuela
e-mail: yluis@ula.ve

P. Dion (ed.), *Soil Biology and Agriculture in the Tropics*, Soil Biology 21,
DOI 10.1007/978-3-642-05076-3_10,

10.2 Acidic Soils and Agriculture in the Tropics

Acid soils occupy about 3.95 billion ha and account for 30% of the world's ice-free land area (von Uexküll and Mutert 1995). Approximately 43% of the world's tropical land area is classified as acidic, comprising about 68% of tropical America, 38% of tropical Asia, and 27% of tropical Africa (reviewed in Herrera Estrella 1999). Thus, acid soils represent the largest potential land area for future agricultural development. These soils are highly weathered because of warm temperatures and high rainfall. They are also inherently infertile. Many reasons account for this infertility: a strongly acid (5.5–4.5) to extremely acid (<4.5) pH; a low cation-exchange capacity; a low base saturation; a high phosphorus-fixing (P-fixing) capacity; and high to toxic levels of Al^{3+} cations (reviewed in He et al. 2003; see Chap. 1). Intensive weathering of tropical soils removes nutrient cations, leaving behind more stable materials rich in Fe and Al oxides. Human intervention on soils through fertilization, irrigation, mining, and long-term legume cultivation also contributes to land degradation and acidification.

The dominant acid soils in the tropics and subtropics include Ultisols and Oxisols (termed Ferralsols in the FAO taxonomy scheme). These soils are widely cultivated, and are most important to world agriculture. However, they generate low crop yields due to a combination of factors that includes their extreme nitrogen (N) and phosphorus (P) deficiency (reviewed by He et al. 2003). Since not all acid soils can be used for agriculture (the tropical forest soils, for example, should be left unexploited), efforts to develop acid soils for agriculture should be directed to the acid savannas of the world, which have an enormous potential for food production and cover an area of over 700 million ha (which represents approximately 50% of the global area that is currently under cultivation) (Herrera Estrella 1999). Even though there are some good examples of successful conversion of acid savannas into productive lands for agriculture, new, efficient, and environmentally friendly technologies should be developed for this productivity to be increased in a sustainable way and without causing major harm to the environment.

10.3 Phosphorus and Acid Soils

After nitrogen, phosphorus is the most important nutrient for plant growth and development. Unfortunately, while most mineral nutrients in soil solution are present in millimolar amounts, the free soluble P concentration is in the range of micromolar or lesser quantities even at pH 6.5, where P is most soluble (He et al. 2003). This is paradoxical, since tropical soils contain total P at concentrations which are orders of magnitude greater than those of plant-available P. A large proportion of this total P corresponds to organic forms, traditionally considered as relatively unavailable to plants, although this view has been challenged (reviewed by Turner et al. 2006). On the other hand, inorganic P is either "occluded" within, or else strongly fixed at the surface of soil minerals, mainly Fe and Al hydrous oxides (sesquioxides). Indeed, free Al and Fe cations react readily with phosphate, leading to relatively insoluble precipitates. This reversible process is known as "P fixation" (Johnson and Loeppert 2006).

To cope with such an extreme deficiency, most developing countries import chemical fertilizers, which are often in limited supply and represent a major expenditure for resource-poor farmers. In addition, intensification of agricultural production in these regions necessitates the addition of large P inputs, not only to increase crop production but also to improve soil P status in order to avoid further soil degradation. However, a significant proportion of this soluble P is either rapidly fixed by soil components or else is leached in the short term, causing eutrophication of surface waterbodies (Zapata and Roy 2004). While there are some efficient ways of improving soil N status (by applying crop residues and other organic sources or by inoculating crops with N-fixing bacteria, NFB), soil P status is more difficult to ameliorate. Therefore, it is imperative to explore alternative methods for improving the status of P in acidic soils. Very attractive in this regard is the possibility of enhancing P uptake of crops by the inoculation of P-solubilizing bacteria (PSB).

10.4 The Use of Soil Bacteria as Biofertilizers

Soil micro-organisms play a significant role in the major global biogeochemical cycles, regulating the dynamics of organic matter decomposition and the availability of plant nutrients such as N, P and S. As such, soil micro-organisms play a fundamental role in promoting plant growth. In the rhizosphere, (i.e. the zone surrounding and influenced by roots) micro-organisms interact with plants in several ways. Bacteria that act positively on plant growth and development, through direct or indirect mechanisms, are collectively known as "plant growth-promoting rhizobacteria (PGPR). While some PGPR act by inhibiting or antagonizing plant pathogens (these are termed "bioprotectants" or "biocontrollers") others promote plant growth by releasing phytostimulators (in the case of the "biostimulants") or by providing nutrients, such as N and P (for "biofertilizers") (reviewed by Zahir et al. 2004). Today, the use of N-fixing bacteria (NFB) to enhance crop yield, particularly through the improvement of nodulation and biological N_2 fixation in legumes, is widespread. By comparison, the use of PSB as biofertilizers is much less common. As we will review in this chapter, even though literally thousands of PSB have been isolated, identified, characterized, and proposed as potential biofertilizers, their effectiveness in the soil-plant system is still unclear, particularly under field conditions.

10.4.1 The Phosphate-Solubilizing Bacteria

Rhizospheric micro-organisms may affect P supply to plants in different ways: (1) by immobilizing P in microbial biomass, (2) by enzymatic decomposition of organic P compounds, and/or (3) by promoting the solubilization of insoluble phosphates. The last two mechanisms release soluble inorganic orthophosphate ($H_2PO_4^-$) into soil solution, from which it can be absorbed by plant roots. The ability of some soil-borne bacteria to dissolve mineral phosphates was noticed by researchers more than a century ago (Goldstein and Krishnaraj 2007). Most of the early studies were centered on the isolation of P-solubilizing micro-organisms (PSM) from the rhizospheric soil and the characterization of their phosphate-solubilizing activity under in vitro conditions. Research into solubilization of P under field conditions and its uptake by plants began much later (Gerretsen 1948).

Even though fungi are among the best studied PSM, this chapter will focus on the activity and potential use of P-solubilizing bacteria (PSB) for agricultural development. In his pioneer work, Gerretsen concluded that the "solvent action of bacteria" (today called mineral phosphate solubilizing ability, MPS) is a general character of several types of micro-organisms. Since then, the so-called MPS^+ phenotype has been identified in numerous bacterial isolates (for review see Rodriguez and Fraga 1999). Strikingly, PSB can also enhance growth and development of plants by mechanisms other than P solubilization. Indeed, they can release phytohormones that promote root growth, enhance availability of other nutrients (e.g. Mn and Fe), and control plant pathogens (reviewed by Vassilev et al. 2006, Nehl and Knox 2006, Avis et al. 2008). Additionally, some PSB may protect plants against the detrimental effects of heavy metals (Ma et al. 2008) and induce plant systemic resistance (Han et al 2008).

According to some authors, PSB occur in most soils and may represent up to 40% of the culturable population (Richardson 2001), even though this proportion has been said to be overestimated (see below). PSB started to be massively applied to soils during the 1950s in the former Soviet Union, with rather spectacular results. However, serious concerns about the reliability and reproducibility of these results arose very soon (Mishustin and Naumova 1962).

Although thousands of PSB have been isolated in the past, these belong to only a few bacterial genera, namely: *Bacillus*, *Paenibacillus*, *Pseudomonas*, *Rahnella*, *Escherichia*, *Enterobacter*, *Burkholderia*, *Serratia*, *Pantoea*, and some other members of the *Enterobacteriaceae* family. Species belonging to the genus *Rhizobium* and related organisms have also been identified as PSB (Kämpfer 2007).

Considering the diverse mechanisms which are involved in the solubilization of P (see below), it is no surprise that knowledge of the genetics of the MPS^+ phenotype remains scanty (Goldstein 1995; Rodriguez and Fraga 1999; Rodríguez et al. 2007). Only a few genes, mainly encoding proteins related to the production of organic acids, have been isolated and characterized. Almost nothing is known about the regulation of the expression of these genes, and results of efforts at improving P-solubilizing ability through genetic modification of PSB remain modest.

10.4.2 Mechanisms of Inorganic Phosphate Mobilization by PSB

As we will review in Sect. 10.5 of this chapter, it is widely agreed that soil bacteria participate actively in P solubilization from sparingly soluble P-containing minerals by producing and excreting organic acids (e.g. see review by Rodriguez and Fraga 1999). However, other microbially based mechanisms have also been proposed as contributing to soil P mobilization, often on the basis of an observed lack of correlation between acidification of PSM culture supernatants and P solubilization (Sperber 1958b; Halvorson et al. 1990; Illmer and Schinner 1992 and 1995a). These alternative mechanisms of P solubilization are briefly summarized below.

10.4.2.1 Sink Theory

The "sink theory," proposed by Halvorson et al. (1990), considers that the very effective P uptake systems of rhizospheric micro-organisms would enable the assimilation of large amounts of P from the soil solution, thus disturbing the equilibrium between insoluble and soluble P. Sparingly soluble phosphates would then be dissolved indirectly by the continuous removal of P from the soil solution.

10.4.2.2 Acidification by H^+ Excretion Theory

This mechanism was proposed to explain P solubilization, dependent on the production of microbial biomass and accompanied by a decrease in pH, *in the absence* of organic acid or chelating agent production (Illmer and Schinner 1995a). The H^+ release is thought to be associated, under these circumstances, with ammonium ion (NH_4^+) assimilation. Protons may then act as agents for P solubilization. Plants evolved a similar strategy to release P from sparingly soluble minerals (reviewed in Hiradate et al. 2007).

10.4.2.3 The Bacterial Cell Wall and P Solubilization

Due to their negative charge, bacterial surfaces can adsorb a wide range of metal cations. Therefore, bacteria are commonly adsorbed on mineral surfaces in soils, sediments and groundwaters, mainly by hydrogen bonding and/or by electrostatic interactions (Young and Crawford 2004; Rong et al. 2008). This non-metabolic cell wall adsorption is significantly affected by pH, with larger amounts of bacteria adsorbed at lower pHs. The interaction of bacteria with minerals can cause both an increase in the extent of mineral dissolution and an inhibition of secondary mineral formation (Wightman and Fein 2004; Rong et al. 2008). The magnitude of this bacterial effect is proportional to the abundance of cells in a geologic system, and to the magnitude of the stability constants of the principal metal-bacterial surface complexes (Wightman and Fein 2004). Since minerals containing Fe^{3+} or Al^{3+} cations form highly stable surface complexes with bacteria, the solubility of these minerals should be strongly affected by bacterial surface adsorption (Fein et al. 1997).

Generally, bacterial adsorption to mineral surfaces is the first step towards the establishment of a biofilm. Cells in biofilms are embedded within an extracellular matrix composed of self-produced polymeric substances. Biofilm formation has been proposed to play an important role in altering the solubility of minerals (Welch et al. 2002) and also in contributing to P solubilization in soils (Jayasinghearachchi and Seneviratne 2006). One possible explanation for this phenomenon is that the production of both organic acids and exopolysaccharides (EPS) increases in mixed biofilms (Bandara et al. 2006). In keeping with these observations, the role of bacterial EPS in the microbially enhanced dissolution of P has been recently demonstrated (Yi et al. 2008). The increase of P solubilization brought about by EPS is attributed to its capacity for holding free P in the medium, therefore modifying the solubility equilibrium to accelerate the release of soluble P. On the other hand, Liu et al. (2006) showed that bacterial EPS strongly adsorb organic acids, which results in high concentrations of organic acids near the mineral.

10.4.2.4 Inorganic Acid Production and P Mobilization

The production of soluble P fertilizers is usually accomplished by sulfuric acid treatment of rock phosphate (RP). Therefore it is no surprise that acidophilic bacteria (e.g. *Acidithiobacillus* sp.), able to oxidize reduced sulfur compounds (such as pyrite) to sulfuric acid (H_2SO_4), may participate in the solubilization of P in soils. For instance, addition of *Acidithiobacillus thiooxidans* (formerly *Thiobacillus thiooxidans*) and low-grade phosphatic rock (PR) to an Oxisol caused a rapid drop in soil pH to levels below 4.0 and a concurrent increase in soluble-P level (Muchovej et al. 1989). The amount of P thus solubilized was sufficient for sustaining sorghum (*Sorghum bicolor*) growth at levels similar to those obtained using superphosphate. Similar results were obtained by Stamford et al. (2007) with the use of *A. ferrooxidans* and yam bean (*Pachyrhizus erosus*), in acid soils with low available P.

10.4.2.5 Bacterial Chelators and P Solubilization

As we will see in Sect. 10.5, organic acids both sequester cations and acidify the soil solution, thus causing solubilization of P. The chelating property of the organic anions is therefore as important as the proton effect. Indeed, it has been shown that the addition to soil of well-known chelators, such as EDTA, may enhance both mineral dissolution and P solubilization (Campbell and Eick 2002). Some bacterial metabolites, different from organic acid anions, may also contribute to P solubilization through chelation. Siderophores, for example, are biogenic chelators with a high affinity and specificity for iron, which can alter the solubility of iron oxides over a wide pH range (reviewed by Kraemer 2004). This can release P adsorbed on the surface of the Fe oxides. Recently, Hamdali et al. (2008) found that P solubilization from RP correlated well with the production of siderophores – but not with that of organic acids – by actinobacteria (*Streptomyces* sp. and *Micromonospora* sp.).

10.4.2.6 Reductive Dissolution of P-Containing Minerals

The reduction of hydrous ferric oxides under anaerobic conditions, characteristic of flooded soils, may release occluded P and ferrous iron (Fe^{2+}) to the soil solution (Stemmler and Berthelin 2003). This phenomenon is well documented in rice paddy fields. If soils are highly saturated with inorganic P, relatively large concentrations of soluble P can be released to the soil solution by reductive dissolution of minerals (Young and Ross 2001). Conversely, almost no soluble P will be mobilized to floodwater if the soil initial P content is low. The presence of redox-inactive Al oxides on the surface of ferric oxyhydroxides may negatively affect reductive dissolution rates of PO_4 and Fe^{2+} (Murray and Hesterberg 2006).

10.4.3 Organic P Mineralization

There is growing recognition of the importance of organic forms of P (P_o) for the development of sustainable agriculture practices, especially in tropical soils (Turner et al. 2006). A large proportion of soil P occurs in organic forms (phosphate esters, phosphonates, and anhydrides), which must be hydrolyzed by specific enzymes to release PO_4. Numerous soil micro-organisms are able to mineralize P_o by synthesizing phosphatases, and some are able to perform simultaneously inorganic P solubilization and P_o mineralization. For example, *Enterobacter agglomerans* and *Burkholderia cepacia* exhibited significant abilities to both solubilize hydroxyapatite and hydrolyze P_o (Kim et al. 1997, 1998; Rodriguez and Fraga 1999). Other bacterial strains with high P-solubilizing abilities have also been found to exhibit substantial acid or/and alkaline phosphatase activities (de Freitas et al. 1997), and it was found recently that a considerable proportion of soil micro-organisms harbor these combined properties (Oliveira et al. 2008).

10.5 Role of Organic Acids in P Mobilization from Acidic Soils

Both plants and soil micro-organisms exude or excrete low-molecular weight carboxylates (organic acid anions), able to efficiently chelate metal ions. Microbial production of organic acid P ligands may occur either constitutively or in response to P deficiency and is enhanced in the rhizosphere. Indeed, a large proportion of bacteria thriving in the rhizosphere may solubilize P by transforming sugars exuded through the roots into their respective sugar acids, whose amount and nature would depend on the type of sugars available (Deubel et al. 2000).

Two primary mechanisms have been proposed to explain the release of P from Fe-oxide surfaces in the presence of organic ligands and in the context of acidic soils: (1) ligand exchange and (2) ligand-enhanced dissolution of the Fe-oxides (reviewed in Johnson and Loeppert 2006). In ligand exchange processes, organic ligand exchanges for inorganic P at a mineral surface site, thus releasing P into the soil solution. During ligand-enhanced dissolution, the organic ligand is adsorbed first at a surface structural Fe site, and then the Fe-oxide surface is slowly dissolved, releasing adsorbed P to the soil solution. The relative effectiveness of specific organic acids in releasing P by either mechanism is related to the number and arrangement of organic acid carboxyl- and hydroxyl-groups (reviewed in Jones et al. 2003, and in Arcand and Schneider 2006). In general, the relative capacity to desorb P in soils varies in the following order: tricarboxylic- > dicarboxylic- > monocarboxylic-acid.

Some of the organic acids that have been shown to mobilize P from soils are citric, glutamic, succinic, lactic, oxalic, glyoxalic, maleic, fumaric, tartaric, and α-ketobutyric acids (Rodriguez and Fraga 1999; Khan et al. 2006). In addition, gluconic and 2-ketogluconic acids may be the most effective, particularly because of their extremely low pK_a(s) (~3.4 and ~2.6 respectively) (Goldstein 1995; Rodríguez Rodriguez and Fraga 1999), although their ability to solubilize P adsorbed to Fe- and/or Al-oxides has been debated (Whitelaw et al. 1999). These acids are produced through direct oxidation (or nonphosphorylating oxidation) of glucose, an alternative aldose utilization pathway which is expressed in a number of rhizobacteria (Goldstein 2007).

Citric acid is perhaps the best studied P-mobilizing agent in the context of acidic soils. P solubilization mediated by citric acid is not only due to acidification of the soil solution, but also to formation of Al and Fe complexes (reviewed by Jones 1998). For instance, citrate was among the most effective acids in releasing P from two Fe oxides, ferrihydrite and goethite (Johnson and Loeppert 2006), and from P-loaded synthetic goethite (Geelhoed et al. 1998). Input of citrate to acid soils not only results in release of soluble P (reviewed by Hocking 2001) but also reduces the sorption of newly applied fertilizer P (Jones and Darrah 1994). Addition of citric acid to soils may increase the amount of dissolved molybdate-reactive P (Drouillon and Merckx 2003; Hutchison and Hesterberg 2004). Citric acid mobilized and released more P from an upland clay loam Ultisol than did tartaric and oxalic acids (Wang et al. 2008).

A number of bacterial species are efficient producers of citric acid and therefore have been tested in P solubilization experiments. In one of these studies, Hoberg et al. (2005) demonstrated that *Pseudomonas fluorescens* released up to 60 mM citrate in the presence of P-loaded goethite.

In some cases, however, addition of organic acids (e.g. citric, oxalic, tartaric, gluconic, succinic, lactic, and acetic acid) to acid soils amended with RP failed in releasing P, even though the acid concentration was sufficient to decrease the pH to less than pH 4.0 (Srivastava et al. 2007). Indeed, besides the identity and the concentration of the organic acids, a number of other factors, such as Fe-oxide crystallinity, soil solution pH, and initial P content, may affect P release from natural soils (Johnson and Loeppert 2006; Hiradate et al 2007).

Organic acids may also affect P mobilization in soils indirectly, particularly by improving the ability of crops to obtain P from organic P compounds. For instance, organic acids can promote the growth of micro-organisms that mineralize organic forms of P (Richardson 1994). There is also evidence that organic acids can liberate adsorbed and complexed phytate, rendering it susceptible to breakdown by extra-cellular phytases (Jones 1998). In addition to their P-mobilizing abilities, organic acids also mediate detoxification of metals, particularly Al^{3+}, allowing plants to grow (Hocking 2001; Jones et al. 2003).

When considering the role of organic acids on P availability and uptake, it should be taken into account that research has often been conducted under laboratory conditions, which are vastly different from those that exist in vivo. Besides, organic acids are not only produced, but also consumed by soil micro-organisms (Jones et al. 2003). Organic acids are metabolized two to three times faster in the rhizosphere than in bulk soil, typically with 60% of the molecules present being mineralized and the remainder being incorporated into microbial biomass (Jones 1998). Nevertheless, organic acids may be protected from rapid biodegradation through an interaction with the soil solid phase (occurring by adsorption or fixation) (Jones et al 2003). Plants may also reduce the rate of organic acid microbial degradation by secreting phenolic compounds, mainly isoflavonoids, through their roots (Weisskopf et al. 2006).

10.6 Field Experiences with Phosphate-Solubilizing Bacteria

Massive application of PSB to increase the fertility of natural soils was carried out in the former Soviet Union during the 1950s (Mishustin and Naumova 1962). A large proportion of agricultural soils were inoculated with a fertilizer consisting of kaolin-impregnated *Bacillus megaterium* var. *phosphaticum* spores. The biofertilizer was called phosphobacterin. In general, the results were astonishing: addition of phosphobacterin to Soviet soils allowed crop yield increases of up to 70%! This was attributed to the mobilization of soil P as a result of microbial production of organic acids. In India, where almost 46% of all soils are classified as P-deficient, researchers also reported positive responses — although not as spectacular — to phosphobacterin applications (reviewed in Sundara et al. 2002). In contrast, experiments conducted in the United States during the 1960s did not confirm the value of phosphobacterin as a universal biofertilizer (Smith et al. 1961).

Significant areas of cultivated soils in Korea and China are P-deficient and have low crop productivity (Xie 1998). Therefore, scientists from these countries contributed numerous articles dealing with the use of PSB for agricultural purposes.

The enormous interest in this area is better reflected in the numerous patents issued to researchers, R&D institutions, and private companies. A number of PSB are even commercially produced as biofertilizers (Lucy et al. 2004; see Chap. 11). Application of PSB under field conditions proved beneficial to plant growth, yield, P uptake, and overall quality of various crops and legumes in different regions of the world (recently reviewed by Khan et al. 2006 and by Osorio Vega 2007). Field tests have been performed as single, dual- or multi-species inoculations alone or in combination with the application of PR. Although most experiments were conducted in non-acidic soils, various observations are relevant in the context of the present review and will be summarized here. For a more detailed account, the reader is referred to the excellent reviews from Rodriguez and Fraga (1999), Gyaneshwar et al. (2002), Khan et al. (2006) and Osorio Vega (2007).

10.6.1 Single-Species Inoculations

Numerous authors have reported on the beneficial effects of single-PSB species inoculation on growth and development of different crops, in many cases showing consistent increases in crop yield over several years. These published accounts were compiled by Lucy et al. (2004), and some will be presented here. Following a 3-year study, Sundara et al. (2002) concluded that application of *B. megaterium* var. *phosphaticum* to sugarcane cultures resulted in enhanced tillering, stalk population, and weight. The authors also reported increases in cane and sugar yields and an improvement in juice quality. The same study showed that a 25% reduction in the amount of P applied to sugarcane was possible when the P fertilizer was used in combination with PSB. Similarly, a 2-year study conducted under different field conditions showed that inoculation of sugar beet seeds with *Paenibacillus polymyxa* and *Pseudomonas putida* significantly increased leaf yield, when compared to the uninoculated controls (Çakmakçi et al. 2006). However, in many cases, inoculation of plants with PSB did not result in an increase of the crop parameters monitored (Gyaneshwar et al. 2002). Indeed, as recently pointed out by Goldstein and Krishnaraj (2007) and by Richardson (2007), when considering the hundreds of papers published on this subject, the number of failures equals the number of successful trials. The reasons explaining such discouraging results are discussed in Sect. 10.8 of this chapter.

A large number of legume-nodulating *Rhizobium* strains are able to solubilize inorganic phosphates (Chabot et al. 1996; Sridevi et al. 2007). Apart from their dual beneficial nutritional effect resulting from both P mobilization and N_2 fixation (Peix et al. 2001), rhizobia also establish synergistic interactions with arbuscular mycorrizal fungi (AMF) (Barea et al. 2002). Further advantages of rhizobial use as PGPR include their integration in well-established crop rotation systems and the ready availability of technologies for inoculation and inoculant production. Furthermore, since they have been used with legumes for many years without causing harm to the environment or to farmers (see Chaps. 8 and 11), rhizobia are perceived as environmentally friendly. In a pioneering work, Chabot et al. (1996) assayed P-solubilizing rhizobia in field trials, observing growth stimulations for lettuce (*Lactuca sativa*) and maize (*Zea mays*) similar in magnitude to those obtained with other PSM. Sorghum growth and P uptake were also significantly increased by rhizobial inoculation (Matiru and Dakora 2004). Kumar et al. (2001) showed that wheat inoculation with another N-fixing bacterium, *Azotobacter chroococcum*, increased P uptake and plant growth.

Unfortunately, few inoculation studies have been conducted in acid soils. In one such study, acid-resistant PSB were isolated from samples of various soil classes and cropping histories in the Himalayan regions of India by enrichment culture techniques (Pal 1998). The acid-tolerant PSB were further tested in field experiments conducted in a typical Inceptisol (pH 4.8), with and without added P sources. Seed inoculation with one of these strains, which exhibited the highest P-solubilization activity and acid tolerance, resulted in significant increases in grain and vegetative yield of finger millet (*Elosine coracana*), maize, amaranth (*Amaranthus hypochondriacus*), buckwheat (*Fagopyrium esculentum*), and French bean (*Phaseolus vulgaris*). Similarly, when tested in a pot experiment conducted in an Indonesian Ultisol (pH 4.9), a P-solubilizing *Pseudomonas* strain significantly improved sorghum growth, height, root P content, shoot dry weight, and root dry weight when compared to the uninoculated control (Widada et al. 2007). Recently, Fankem et al. (2008) reported that inoculation of either one of three PSB (*P. fluorescens* CB501, CD511 and CE509), isolated from acid soils in Cameroon and able to mobilize P from Al- and Fe-phosphates, increased grain yield and P uptake of maize when tested in the greenhouse (see Sect. 5.3.1.4).

10.6.2 Dual Inoculation Assays

Field effects of PSB may be enhanced upon their mixed inoculation with either other PSB, P-solubilizing fungi (PSF), NFB or AMF.

10.6.2.1 Co-Inoculation of PSB and NFB

Since N and P are the two major plant nutrients, combined inoculation of NFB and PSB may benefit the plant better than inoculation with either group of organisms alone (Gull et al. 2004; Khan et al. 2006). NFB and PSB are able to interact synergistically by providing nutrients, removing inhibitory products, or otherwise. In particular, it is well known that N_2-fixation is highly dependent on P- availability (Barea et al. 2005).

Kopler et al. (1988) found that inoculation of legumes with *Pseudomonas* spp. enhanced rhizobial-induced nodulation. Similar results were obtained by others in field assays conducted with different combinations of PSB, NFB and test crops (reviewed by Khan et al. 2006). Synergistic interactions on plant growth and yield have also been observed when co-inoculating PSB with other well-known NFB such as *Azospirillum* or *Azotobacter*. Furthermore, PSB exert a beneficial influence on the rhizosphere survival of *Azotobacter* (Ocampo et al. 1975). In some cases, results from dual inoculations involving two different PSB have been better than those obtained with each strain used separately (Kundu and Gaur 1980; Tiwari et al. 1989; Çakmakçi et al. 1999; Han et al. 2006).

10.6.2.2 Co-Inoculation of PSB and Nonmycorrhizal Fungi

Phosphate-solubilizing fungi (PSF) are among the best known and well-studied PSM (Whitelaw 2000; Bandara et al. 2006). As in the case of PSB, their P-solubilizing ability is related to the production of important amounts of organic acids. Not surprisingly, combined inoculation of PSF and NFB resulted in enhanced growth, nutrient uptake, and yield in several crops. For example, inoculation of mungbean with either *Bacillus subtilis* or *Bacillus circulans* and *Aspergillus niger* increased nodulation of plant roots by indigenous rhizobia, root and shoot biomass, as well as straw and grain yield (Gaind and Gaur 1991). Similarly, combined inoculation of either *B. megaterium* or *Pseudomonas striata* with *Aspergillus awamori* significantly increased productivity of potato (*Solanum tuberosum* L.) (Dubey and Billore 1992). Mixed populations of soil bacteria and fungi are currently prepared as commercial biofertilizers for improving P nutrition of plants (Richardson 2007).

10.6.2.3 Co-Inoculation of PSB and AMF

Arbuscular mycorrhizal fungi increase plant P uptake by: (1) increasing the mycorrhizal root absorptive area, (2) improving nutrient transfer efficiency and utilization of P within the host plant, and/or (3) enhancing the solubility of P in the rhizosphere through pH alteration of the surrounding soil (reviewed in Barea et al. 2005; see Chap. 9). Numerous reports of greenhouse and field trials have highlighted the beneficial effects of dual inoculations with PSB and AMF (reviewed in Barea et al. 2005; Khan et al. 2006; Osorio Vega 2007). Basically, a synergistic microbial interaction is established between AMF and PSB, thus improving P acquisition by the plant (Barea et al. 2005). For instance, mycorrhizal plants can release higher amounts of organic compounds into their rhizosphere than do non-mycorrhizal plants (Linderman 1988). PSB survive longer when associated to mycorrhizal roots rather than to non-mycorrhizal roots (Singh 1990; Toro et al. 1997; Barea et al. 2002), thus increasing the possibility of delivering P into the soil solution. Some AMF also produce and excrete organic acids, which contribute to the solubilization of mineral phosphates, particularly Fe-P (Bolan 1991).

The effect of AMF-PSB dual inoculations on the growth and nutrient uptake of sorghum was studied in pot experiments using an Indonesian Ultisol (pH 4.9) with a low P content (Widada et al. 2007). Inoculation of the AMF *Glomus manihotis* and the PSB *Pseudomonas* sp. increased plant dry weight 112 times with respect to that of the uninoculated plant. Plant dry weight and uptake of N, P, Fe and Zn increased more upon dual inoculation than with the PSB alone.

There is a certain degree of specificity between PSB, AMF, and crops. When AMF (*Glomus mosseae* or *Glomus fasciculatum*) were co-inoculated with different PSB isolated from an Oxisol, there was an increase in kudzu (*Pueraria phaseoloides*) growth, yield, and nutritional status with some bacteria, but not with others (Toro et al. 1996).

10.6.3 Multi-Species Inoculations

The higher plant P concentration obtained through co-inoculation of AMF and PSB might benefit resident or inoculated NFB and the functioning of their nitrogenases, leading to increased N fixation and positive feedback on root and mycorrhizal development (Barea et al. 2005). The effectiveness of multispecies inoculations was examined in a number of studies, with encouraging results (reviewed by Khan et al. 2006 and by Osorio Vega 2007). Again, few of these studies dealt with acidic soils.

10.6.4 Combined Use of PR and PSB

To enhance P nutrition of crops cultured on acidic soils, direct application of phosphatic rock (PR) has been recommended (Zapata and Roy 2004; Arcand and Schneider 2006). PRs are rich in calcium phosphate (Ca-P) complexes and, as such, are weakly soluble in acidic soils. PRs are cheaper than inorganic P fertilizer, create less environmental pollution as they require minimum processing, and their dissolution results in the slow release of P in the soil. However, the agricultural effectiveness of PR may differ with soil properties, climatic conditions, and the nature of the crop (Nahas 1996; Rajan et al. 1996). Because PR minerals are released slowly and their use as fertilizer often causes insignificant increases in crop yield (Zapata and Roy 2004), organic acid-secreting PSB, alone or in combination with AMF or PSF, have been used to improve PR agronomic value (Goenadi et al. 2000; Barea et al. 2002; Lin et al. 2002, Şahin et al. 2004). The combined use of PR and *Bacillus megaterium* var. *phosphaticum* for sugarcane cultures in India permitted a 50% decrease in superphosphate use without sacrificing yields (Sundara et al. 2002).

An acidophilic, sulfuric acid-producing *Acidithiobacillus* sp. has been inoculated together with elemental sulfur and PR. This resulted in increased total N and P in yam bean (*P. erosus*) when grown in a typical Brazilian Fragiudult soil (pH 5.8) with low available P (Stamford et al. 2007).

In spite of some successful trials, it is not clear if PSB can actually increase the effectiveness of PR under standard field conditions to such a magnitude that it can be used as an alternative fertilizer. For example, in a recent study, three well-characterized PSB (*Enterobacter asburiae*, *Bacillus coagulans* and *Citrobacter koseri*) failed to release P from an Indian Alfisol (pH 6.5) amended or not with PR (Srivastava et al. 2007).

A partially acidulated phosphatic rock (PAPR) is normally obtained by the partial chemical acidulation of PR to convert only a part of its P into water-soluble form. PSB have also been used to prepare PAPR (Rodriguez and Fraga 1999; Whitelaw 2000; Vassilev et al. 2001). This biologically-treated phosphatic rock fertilizer increased P nutrition of plants in acid soils (Zapata and Roy 2004). As compared to the chemical process, the PSB-mediated acidulation process presents the advantages of occuring at low temperatures and of being more selective for P extraction (Whitelaw 2000; Vassilev and Vassileva 2003).

10.7 How to Isolate Acid-Proficient Phosphate Solubilizing Bacteria?

For several decades, Pikovskaya (PVK) medium (Pikovskaya 1948) or Sperber medium (Sperber 1958a) were routinely used to screen PSB from soil samples. Both media contain some form of calcium phosphate, added as the sole P source, and the identification of PSB relies on observing a clear solubilization halo surrounding bacterial colonies. To improve the clarity and visibility of the results, bromophenol blue may be added to the medium (Gupta et al. 1994), yielding yellow-colored halos. In 1999, Nautiyal formulated a new chemically-defined growth medium (NBRIP medium) for rapid screening of PSM, in which each of the components was tested separately to increase the efficiency of P solubilization and, consequently, the sensitivity of the screening technique. Being chemically defined, NBRIP medium facilitates interpretation of the experimental outcome. However, the formation of a clear halo, which remained the only criterion for identification of PSM, still depended on a series of uncontrolled factors, such as the varying diffusion rates of different organic acids. Furthermore, the reliability of these halo-based techniques was questioned because many isolates not producing any visible halo in the plates were indeed able to solubilize inorganic phosphates in liquid media. Hence, a broth assay with bromophenol blue was developed, which resulted in more reliable outcomes (Mehta and Nautiyal 2001).

The media mentioned above contain slightly soluble P sources (calcium phosphate, tricalcium phosphate, hydroxyapatite, fluorapatite) and are designed for neutrophiles. Accordingly, few of the bacteria recovered on these media can effectively solubilize the more insoluble Fe- and Al phosphates, which are the main forms of P occurring in acidic, tropical soils (Gyaneshwar et al. 2002; He et al. 2003). Furthermore, with only a few exceptions, colonies grown on screening agar are mainly representatives of r-strategists, with a preferential ability to rapidly metabolize simple substrates (Brimecombe et al. 2001).

To isolate PSB useful as biofertilizers in acidic soils (here called acid-proficient PSB), the source of P in screening media was changed to more insoluble forms of P, typical of these soils. Such media contained $AlPO_4$, $FePO_4$ or zinc phosphate as the sole source of P (Illmer et al. 1995; Srivastava et al. 2007; Fankem et al. 2008).

Instead of pure mineral phosphates, test media used to isolate PSM or monitor their activity may contain a series of variable-charge minerals (artificially synthesized goethite, natural kaolinite, and/or montmorillonite) surface-loaded with inorganic P (He and Zhu 1998; Hoberg et al. 2005).

In addition to containing Al-P, Fe-P or other insoluble P-containing minerals, media used to isolate acid-proficient PSB should also reproduce the buffering properties of natural soils. Indeed, the buffering capacity of soils may diminish the efficiency of microbial acid production in P solubilization (Cunningham and Kuiack 1992). The addition of 100 mM Tris-HCl (pH 8.0) to a screening medium negatively affected the ability of two PSB to solubilize di-calcium phosphate and PR (Gyaneshwar et al. 1998).

The nature of the carbon source also influences the outcome of PSB isolation trials. Different carbon sources will favor various carbon utilization pathways, and thus influence the identity and amount of excreted organic acids.

Not only is the presence of a particular combination of sugars (glucose plus sucrose) necessary for solubilizing Al-P, but an increased concentration of total C (from 4 g l^{-1} to 16 g l^{-1}) allows this to happen more efficiently (Illmer et al. 1995). Furthermore, glucose concentration is a key factor for P solubilization in NBRIP liquid medium (Nautiyal 1999). Thus, the concentration of sugars added to different screening and/or test media may be considered one of the factors responsible for the isolation of PSB that *did not* perform well under field conditions (Richardson 2001). Indeed, most screening media contain unrealistically high concentrations of a single carbon source, glucose, which is only found in very low concentrations in most root exudates. In fact, rhizodepositions are composed of a mixture of sugars and other C sources, which can be further metabolized or transformed by rhizobacteria (Deubel et al. 2000). Consequently, PSB isolated using glucose as the sole C source may not be typical rhizosphere inhabitants or be capable of solubilizing P under rhizosphere conditions.

With a view to developing a screening method that would more realistically reflect nutrient levels in the rhizosphere, Harris et al. (2006) added lower amounts of glucose and sucrose (0.05%), plus exudates from wheat roots, to their medium (modified MIS medium). They were then able to isolate PSB that were not only good P solubilizers in laboratory tests, but resulted in increased grain yield and grain phosphorus content in test plants grown in non-sterile soil in the greenhouse.

Another important factor to be considered when screening PSB for biofertilization purposes is the pH of the medium, since this factor may drastically influence microbial utilization of C sources (Yao et al. 2000). Therefore, when isolating strains from acid soils, it appears advisable to utilize a medium with an acid, rather than a neutral, pH (Pal 1998).

The N source also influences the in vitro P-solubilizing ability of many PSB. In general, greater solubilization activities are observed in the presence of ammonium salts (NH_4^+), as compared to nitrate (NO_3^-) (Halder et al. 1992; Illmer et al. 1995). This has been attributed to the extrusion of protons to compensate for NH_4^+ uptake.

Culture-independent methods may be used both to detect and identify PSB. Since the direct oxidation of glucose has been proposed as the metabolic basis for a superior MPS^+ phenotype in Gram-negative bacteria, molecular probes targeting direct oxidation pathway genes may be used for detecting these bacterial populations in situ (Goldstein 2007). Indeed, this approach allowed the detection of populations of PSB in two alkaline desert soils (Goldstein et al. 1999). However, as was mentioned earlier, other bacterial-mediated mechanisms for mobilizing P in soils exist (see Sect. 10.4.2), which makes it difficult to define universal criteria for detection of all potentially beneficial PSB by molecular biology protocols.

In addition to optimizing the formulation of screening media, other factors should be considered in the selection of acid-proficient PSB. For example, as Sperber (1958a) first noticed, many isolates rapidly and irreversibly lose their ability to dissolve apatite on sub-culturing (Halder et al. 1990; Illmer and Schinner 1992). This phenomenon has not received much attention, although it may give rise to contradictory results. Such instability might arise from the segregation of genes located in plasmids, in the absence of a strong selection force under in vitro conditions. In line with this argument, genes related with the direct oxidation pathway of glucose are indeed localized in plasmids in certain bacterial species (Gupta et al 1997).

10.8 Why Do PSB Fail When Tested in the Field?

As mentioned above, comparisons of the effect of PSB-based biofertilizers in laboratory, greenhouse, and field studies lead frequently to very contrasting results. The factors involved in these inconsistencies, some of which were reviewed by Lucy et al. (2004), include the following:

1) *Lack of fundamental knowledge concerning the microbial mechanisms involved in P mobilization.* Early biofertilization trials were conducted with very limited information concerning the bacterial mechanisms involved in P solubilization from sparingly soluble sources. Even though this gap began to close during the 1990s, the basic solubilization mechanisms still remain incompletely understood (Illmer and Schinner 1995a; Rodriguez and Fraga 1999).
2) *Lack of adequate experimental design and analysis of results.* Illmer and Schinner (1995b) were the first to point out the frequent methodological mistake of comparing growth of plants in sterile soils with growth in non-sterile soils inoculated with PSM. Obviously, the extrapolation of these results

to outdoor experiments is impossible because of the non-sterile conditions and climatic variability in the field.

3) *Variable effectiveness of PSB as influenced by different soil types*. In many cases, the P-solubilization ability of PSB in natural soils is different from that observed under in vitro conditions. Indeed, most PSB have been isolated using neutral and unbuffered media, although it is well known that both the acidity and the buffering capacity of soils could limit microbially mediated P solubilization (Gyaneshwar et al. 1998). For example, inoculation with *E. asburiae* PSI3 was effective in enhancing plant growth in a Vertisol (pH 8.2) but not in an Alfisol (pH 6.6) (Srivastava et al. 2007). Therefore, PSB survival in a microcosm of the field environment should be examined during the initial stages of laboratory testing (Tang et al. 1995).
4) *Climatic variability*. Temperature fluctuations have been shown to significantly affect the P-solubilizing activity of micro-organisms (Dwivedi et al. 2004; Stamford et al. 2007).
5) *Antagonism and competition with other indigenous micro-organisms in the rhizosphere*. Many PSB have proven to be too vulnerable and unreliable to be used in agriculture (Illmer and Schinner 1995b). This has been attributed to their reduced ability to (a) survive and multiply, (b) colonize the rhizosphere, (c) compete for resources with native micro-organisms, and (d) avoid predation. All these factors may determine the rapid decline in the population size or density of exogenous PSM upon introduction into soils (Jacoud et al. 1998). Hence, it is highly recommended to select bacteria intended to be used as biofertilizers from naturally occurring, endemic populations, in order to take advantage of their ability to colonize a particular niche (Barea et al. 2005). It is also advisable to monitor the dynamics of rhizosphere colonization by the introduced bacteria, with a particular emphasis on their effects on native microbial populations (Castro-Sowinski et al. 2007).
6) *The frequent use of inappropriate screening media to identify and select PSB*. As discussed in Sect. 10.7, very often the media used for screening PSB favor the isolation of PSB with a better efficiency under laboratory conditions than in the soil.
7) *The pre-existent nutrient level of the soils and their moisture content*. The number, diversity, and metabolic activity of soil micro-organisms are influenced both by the identity and the availability of organic and inorganic nutrients (Welbaum et al. 2004). Field trials have shown that the effectiveness of PSB in promoting plant growth is highly dependent on the nutrient status of the soil (de Freitas and Germida 1990; Çakmakçi et al. 2006). Furthermore, in some field trials the effectiveness of PSB varied according to the amount of P previously added to the soil (Chabot et al. 1996).
8) *Bacteria-plant specificity*. Some PSB are only able to produce organic acids when grown in the presence of root exudates of the host plant (Hwangbo et al. 2003), which points to the existence of plant–PSB compatibility effects.
9) *Inoculum size*. Researchers have applied high PSB inoculant doses (reviewed by Lucy et al. 2004), even though the ability to colonize plant roots in high

numbers is not always necessary for a positive effect (Harris et al. 2006; Jacoud et al. 1998). In fact, excessively large numbers of bacteria may sometimes be detrimental to the germination and growth of certain seeds or plants (Chanway 1997). On the other hand, survival of the inoculant in large numbers in rhizospheric soil depends on the initial inoculum density (Jjemba and Alexander 1999).

10) *The nature of the carrier employed to apply bacteria to soils.* Even though no universal formulation for PSB inocula is presently available, a good carrier should have the capacity to deliver the right number of viable cells in appropriate physiological condition at the right time (Bashan 1998).

While keeping in mind the various factors mentioned above, it should also be stressed that, in order to lead to useful results, the field trials must be conducted under the full range of current and future conditions that will be experienced by the farmer (Hameeda et al. 2008).

10.9 Conclusions

Inoculation of crops with PSB may be a cost-effective alternative to inorganic P fertilizer application, and use of these bacteria may lead to improved crop yield and quality, particularly in tropical acidic soils with high P-fixation capacities. In the last two decades, significant contributions were made towards understanding the mechanisms underlying bacterially mediated P solubilization. The interactions taking place between rhizospheric micro-organisms, and between those micro-organisms and the plant, are also better known (Nehl and Knox 2006). Nevertheless, even though a few commercial products are already available and marketed in some countries, we are far from a general use of PSB. As Richardson (2007) stated recently, nearly 60 years after Gerretsen (1948) first demonstrated the potential of PSB as biofertilizers, not much has been delivered… besides promises. This long delay in establishing the use of PSB as a mature biotechnology is due to a combination of obstacles, some of which are biological (see above), while others are of a different nature.

For instance, a majority of farmers in developing countries are small-scale and resource-poor, and therefore rely primarily on low-input agriculture practices. Hence, farmers' access to bank loans with the aim of improving their agricultural techniques is extremely limited. This can be considered one major constraint to enhance crop productivity, even by means of traditional fertilization practices. Furthermore, even in countries where funds can be relatively easily obtained, the cost of using patented technologies and/or commercial biofertilizers (where available) might represent a major obstacle.

A second problem is logistic: how to bring these technologies within the reach of the small- and mid-scale farmers, who are usually settled in remote areas, far away from cities and ports? Also, how can artificial inoculation with PSB be promoted when the infrastructure required to store and transport biological products in large quantities into rural areas is not available?

A third obstacle, which is possibly the most difficult to overcome, is cultural in nature. Indeed, with the exception of rhizobia, the use of beneficial micro-organisms in many regions of the world is almost unknown. Consequently, when farmers are asked to replace well-known chemical fertilizers with bacterial inoculants, they are reluctant to do so. These hesitations result from fear of reducing soil fertility or because of the frequent association of microbes with human or animal diseases (Bashan 1998).

To overcome these constraints, it will be necessary to establish mechanisms of technology transfer, from developed countries to the developing world, from academic institutions to government agencies, and from private companies to the public sector. To promote this transfer, research centers with the capability of acquiring new technologies should be created in lesser developed countries. These centers should be able to adapt foreign technologies to local crops and to develop their own technologies (Herrera Estrella 1999). Moreover, national public programs aimed towards diffusing these new technologies among farmers should be launched, together with information and education campaigns. Altogether, this implies that the public sector should direct more resources towards agricultural research and technology diffusion.

Fortunately, there are encouraging signs coming from certain countries. For example, a year ago the Venezuelan government launched a project oriented towards mass-producing biofertilizers (PSB, NFB) and biological control agents to be delivered at low or no cost to farmers. In the particular case of PSB, two indigenous PSB, isolated from a Venezuelan savanna soil, are currently being produced in small-scale facilities throughout the country.

One of the most important messages that emerged from the State of Food and Agriculture report (FAO 2004) is that biotechnology — i.e. "the use of living (micro) organisms to make or modify a product for practical purposes" — is indeed capable of benefiting small, resource-poor farmers. However, as we have briefly discussed, biotechnology itself cannot overcome the problems that arise as a consequence of underdevelopment.

References

Arcand M, Schneider KD (2006) Plant- and microbial-based mechanisms to improve the agronomic effectiveness of phosphate rock: a review. An Acad Bras Cienc 78:791–807

Avis TJ, Gravel V, Antoun H, Tweddell RJ (2008) Multifaceted beneficial effects of rhizosphere microorganisms on plant health and productivity. Soil Biol Biochem 40:1733–1740

Bandara WMMS, Seneviratne G, Kulasooriy SA (2006) Interactions among endophytic bacteria and fungi: effects and potentials. J Biosci 31:645–650

Barea JM, Azcón R, Azcón-Aguilar C (2002) Mycorrhizosphere interactions to improve plant fitness and soil quality. Antonie Van Leeuwenhoek 81:343–351

Barea JM, Pozo MJ, Azcón R, Azcón-Aguilar C (2005) Microbial co-operation in the rhizosphere. J Exp Bot 56:1761–1778

Bashan Y (1998) Inoculants of plant growth-promoting bacteria for use in agriculture. Biotechnol Adv 16:729–770

Bolan NS (1991) A critical review on the role of mycorrhizal fungi in the uptake of phosphorus by plants. Plant Soil 134:189–207

Brimecombe MJ, De Leij FA, Lynch JM (2001) The effect of root exudates on rhizosphere microbial populations. In: Pinton R, Varanini Z, Nannipieri P (eds) The rhizosphere: biochemistry and organic substances at the soil–plant interface. Marcel Dekker, New York, pp 95–140

Çakmakçi R, Kantar F, Algur ÖF (1999) Sugar beet and barley yields in relation to *Bacillus polymyxa* and *Bacillus megaterium* var. *phosphaticum* inoculation. J Plant Nutr Soil Sci 162:437–442

Çakmakçi R, Dönmez F, Aydn A, Sahin F (2006) Growth promotion of plants by plant growth-promoting rhizobacteria under greenhouse and two different field soil conditions. Soil Biol Biochem 38:1482–1487

Campbell JL, Eick MJ (2002) Effects of oxyanions on the EDTA-promoted dissolution of goethite. Clays Clay Miner 50:336–341

Castro-Sowinski S, Herschkovitz Y, Okon Y, Jurkevitch E (2007) Effects of inoculation with plant growth-promoting rhizobacteria on resident rhizosphere microorganisms. FEMS Microbiol Lett 276:1–11

Chabot R, Antoun H, Cescas MP (1996) Growth promotion of maize and lettuce by phosphate-solubilizing *Rhizobium leguminosarum* biovar. *phaseoli*. Plant Soil 184:311–321

Chanway CP (1997) Inoculation of tree roots with plant growth promoting soil bacteria: an emerging technology for reforestation. Forest Sci 43:99–112

Cunningham JE, Kuiack C (1992) Production of citric and oxalic acids and solubilization of calcium phosphate by *Penicillium bilaii*. Appl Environ Microbiol 58:1451–1458

De Freitas JR, Germida JJ (1990) Plant growth promoting Rhizobacteria for winter wheat. Can J Microbiol 36:265–272

De Freitas JR, Banerjee MR, Germida JJ (1997) Phosphate-solubilizing rhizobacteria enhance the growth and yield but not phosphorus uptake of canola (*Brassica napus* L.). Biol Fertil Soils 24:358–364

Deubel A, Gransee A, Merbach W (2000) Transformation of organic rhizodepositions by rhizosphere bacteria and its influence on the availability of tertiary calcium phosphate. J Plant Nutr Soil Sci 163:387–392

Drouillon M, Merckx R (2003) The role of citric acid as a phosphorus mobilization mechanism in highly P-fixing soils. Gayana Bot 60:55–62

Dubey SK, Billore SD (1992) Phosphate solubilizing microorganisms (PSM) as inoculant and their role in augmenting crop productivity in India. Crop Res 5:11–24

Dwivedi BS, Singh VK, Dwivedi V (2004) Application of phosphate rock with or without *Aspergillus awamori* inoculation to meet phosphorus demands of rice-wheat systems in the Indo-Gangetic plains of India. Aust J Exp Agr 44:1041–1050

Fankem H, Ngo Nkot L, Deubel A, Quinn J, Merbach W, Etoa FX, Nwaga D (2008) Solubilization of inorganic phosphates and plant growth promotion by strains of *Pseudomonas fluorescens* isolated from acidic soils of Cameroon. Afr J Microbiol Res 2:171–178

Food and Agriculture Organization of the United Nations (FAO) (2004) The state of food and agriculture 2003–2004. Agricultural biotechnology: meeting the needs of the poor? In: Agriculture Series No. 35. FAO Corporate Document Repository. http://www.fao.org/docrep/006/Y5160E/Y5160E00.HTM. Accessed 20 Nov 2008

Food and Agriculture Organization of the United Nations (FAO) (2005) The state of food insecurity in the world: eradicating world hunger – key to achieving the Millenium development goals. FAO Corporate Document Repository. http://www.fao.org/docrep/008/a0200e/a0200e00.htm. Accessed 20 Nov 2008

Fein JB, Daughney CJ, Yee N, Davis T (1997) A chemical equilibrium model for metal adsorption onto bacterial surfaces. Geochim Cosmochim Acta 61:3319–3328

Gaind S, Gaur AC (1991) Thermotolerant phosphate solubilizing microorganisms and their interaction with mungbean. Plant Soil 133:141–149

Geelhoed JS, Hiemstra T, van Riemsdijk JH (1998) Competitive adsorption between phosphate and citrate on goethite. Environ Sci Technol 32:2119–2123

Gerretsen FC (1948) The influence of microorganisms on the phosphate intake by the plant. Plant Soil 1:51–81

Goenadi DH, Siswanto SY (2000) Bioactivation of poorly soluble phosphate rocks with a phosphorus-solubilizing fungus. Soil Sci Soc Am J 64:927–932

Goldstein AH (1995) Recent progress in understanding the molecular genetics and biochemistry of calcium phosphate solubilization by gram negative bacteria. Biol Agr Hort 12:185–193

Goldstein AH (2007) Future trends in research on microbial phosphate solubilization: One hundred years of insolubility. In: Velázquez E, Rodríguez-Barrueco C (eds) First international meeting on microbial phosphate solubilization. Springer, Dordrecht, pp 91–96

Goldstein AH, Krishnaraj PU (2007) Phosphate solubilizing microorganisms vs. phosphate mobilizing microorganisms: what separates a phenotype from a trait? In: Velázquez E, Rodríguez-Barrueco C (eds) First international meeting on microbial phosphate solubilization. Springer, Dordrecht, pp 203–213

Goldstein AH, Braverman KE, Osorio N (1999) Evidence for mutualism between a plant growing in a phosphate-limited desert environment and a mineral phosphate solubilizing (MPS) rhizobacterium. FEMS Microbiol Ecol 30:295–300

Gull FY, Hafeez I, Saleem M, Malik KA (2004) Phosphorus uptake and growth promotion of chickpea by co-inoculation of mineral phosphate solubilizing bacteria and a mixed rhizobial culture. Aust J Exp Agric 44:623–628

Gupta RR, Singal R, Shanker A, Kuhad RC, Saxena RK (1994) A modified plate assay for screening phosphate solubilizing microorganisms. J Gen Appl Microbiol 40:255–260

Gupta A, Verma V, Qazi GN (1997) Transposon induced mutation in *Gluconobacter oxydans* with special reference to its direct-glucose oxidation metabolism. FEMS Microbiol Lett 147: 181–188

Gyaneshwar P, Naresh Kumar G, Parekh LJ (1998) Effect of buffering on the phosphate solubilising ability of microorganisms. World J Microbiol Biotechnol 14:669–673

Gyaneshwar P, Naresh Kumar G, Parekh LJ, Poole PS (2002) Role of soil microorganisms in improving P nutrition of plants. Plant Soil 245:83–93

Halder AK, Mishra AK, Bhattacharyya P, Chakrabartty PK (1990) Solubilization of rock phosphate by *Rhizobium* and *Bradyrhizobium*. J Gen Appl Microbiol 36:81–92

Halder AK, Banerjee A, Mishra AK, Chakrabartty PK (1992) Role of NH_4 or NO_3 on release of soluble phosphate from hydroxyapatite by *Rhizobium* and *Bradyrhizobium*. J Basic Microbiol 32:325–330

Halvorson HO, Keynan A, Kornberg HL (1990) Utilization of calcium phosphates for microbial growth at alkaline pH. Soil Biol Biochem 22:887–890

Hamdali H, Bouizgarnea B, Hafidid M, Lebrihic A, Virolleb MJ, Ouhdouch Y (2008) Screening for rock phosphate solubilizing actinomycetes from Moroccan phosphate mines. Appl Soil Ecol 38:12–19

Hameeda B, Harini G, Rupela OP, Wani SP, Reddy G (2008) Growth promotion of maize by phosphate-solubilizing bacteria isolated from composts and macrofauna. Microbiol Res 163:234–242

Han HS, Supanjani LKD (2006) Effect of co-inoculation with phosphate and potassium solubilizing bacteria on mineral uptake and growth of pepper and cucumber. Plant Soil Environ 52: 130–136

Han SH et al (2008) Inactivation of pqq genes of Enterobacter intermedium 60–2G reduces antifungal activity and induction of systemic resistance. FEMS Microbiol Lett 282:140–146

Harris JN, New PB, Martin PM (2006) Laboratory tests can predict beneficial effects of phosphate-solubilising bacteria on plants. Soil Biol Biochem 38:1521–1526

He ZL, Zhu J (1998) Microbial utilization and transformation of phosphate adsorbed by variable charge minerals. Soil Biol Biochem 30:917–923

He ZL, Yang XE, Baligar BC, Calvert DV (2003) Microbiological and biochemical indexing systems for assessing quality of acid soils. Adv Agron 78:89–138

Herrera Estrella L (1999) Transgenic plants for tropical regions: some considerations about their development and their transfer to the small farmer. Proc Natl Acad Sci USA 96:5978–5981

Hiradate S, Ma JF, Matsumoto H (2007) Strategies of plants to adapt to mineral stresses in problem soils. Adv Agron 96:65–132

Hoberg E, Marschner P, Lieberei R (2005) Organic acid exudation and pH changes by *Gordonia* sp and *Pseudomonas fluorescens* grown with P adsorbed to goethite. Microbiol Res 160: 177–187

Hocking J (2001) Organic acids exuded from roots in phosphorus uptake and aluminum tolerance of plants in acid soils. Adv Agron 74:63–97

Hutchison KJ, Hesterberg D (2004) Dissolution of phosphate in a phosphorus-enriched ultisol as affected by microbial reduction. J Environ Qual 33:1793–1802

Hwangbo H et al (2003) 2-Ketogluconic acid production and phosphate solubilization by *Enterobacter intermedium*. Curr Microbiol 47:87–92

Illmer P, Schinner F (1992) Solubilization of inorganic phosphates by microorganisms isolated from forest soils. Soil Biol Biochem 24:389–395

Illmer P, Schinner F (1995a) Solubilization of inorganic calcium phosphates — solubilization mechanisms. Soil Biol Biochem 27:257–263

Illmer P, Schinner F (1995b) Phosphate solubilizing microorganisms under non-sterile conditions. Bodenkultur 46:197–204

Illmer P, Barbato A, Schinner F (1995) Solubilization of hardly-soluble $AlPO_4$ with P-solubilizing microorganisms. Soil Biol Biochem 27:265–270

Jacoud C, Faure D, Wadoux P, Bally R (1998) Development of strain specific probe to follow inoculated *Azospirillum lipoferum* CRT1 under field conditions and enhancement of maize root development by inoculation. FEMS Microbiol Ecol 27:43–51

Jayasinghearachchi HS, Seneviratne G (2006) Fungal solubilization of rock phosphate is enhanced by forming fungal–rhizobial biofilms. Soil Biol Biochem 38:405–408

Jjemba PK, Alexander M (1999) Possible determinants of rhizosphere competence of bacteria. Soil Biol Biochem 31:623–632

Johnson SE, Loeppert RH (2006) Role of organic acids in phosphate mobilization from iron oxide. Soil Sci Soc Am J 70:222–234

Jones DL (1998) Organic acids in the rhizosphere — a critical review. Plant Soil 205:25–44

Jones DL, Darrah P (1994) Role of root derived organic acids in the mobilization of nutrients from the rhizosphere. Plant Soil 166:247–257

Jones DL, Dennis PG, Owen AG, Van Hees PAW (2003) Organic acid behavior in soils — misconceptions and knowledge gaps. Plant Soil 248:31–41

Kämpfer P (2007) Taxonomy of phosphate solubilizing bacteria. In: Velázquez E, Rodríguez-Barrueco C (eds) First international meeting on microbial phosphate solubilization. Springer, Dordrecht, pp 101–106

Khan MS, Zaidi A, Wani PA (2006) Role of phosphate-solubilizing microorganisms in sustainable agriculture. A review. Agron Sustain Dev 26:1–15

Kim KY, Jordan D, Mcdonald GA (1998) Effect of phosphate-solubilizing bacteria and vesicular–arbuscular mycorrhizae on tomato growth and soil microbial activity. Biol Fertil Soils 26: 79–87

Kim KY, McDonald GA, Jordan D (1997) Solubilization of hydroxyapatite by *Enterobacter agglomerans* and cloned *Escherichia coli* in culture medium. Biol Fertil Soils 24:347–352

Kraemer S (2004) Iron oxide dissolution and solubility in the presence of siderophores. Aquat Sci 66:3–18

Kumar V, Behl RK, Narula N (2001) Establishment of phosphate solubilizing strains of *Azotobacter chroococcum* in the rhizosphere and their effect on wheat cultivars under greenhouse conditions. Microbiol Res 156:87–93

Kundu BS, Gaur AC (1980) Effect of phosphobacteria on the yield and phosphate uptake of potato crop. Curr Sci 49:159

Lin QM, Rao ZH, Sun YX, Yao J, Xing LJ (2002) Identification and practical application of silicate dissolving bacteria. Agr Sci China 1:81–85

Linderman RG (1988) Mycorrhizal interaction with the rhizosphere microflora: the mycorrhizosphere effect. Phytopathology 78:366–371

Liu W, Xu X, Wu X, Yang Q, Luo Y, Christie P (2006) Decomposition of silicate minerals by *Bacillus mucilaginosus* in liquid culture. Environ Geochem Health 28:133–140

Lucy M, Reed E, Glick BR (2004) Applications of free living plant growth-promoting rhizobacteria. Antonie Van Leeuwenhoek 86:1–25

Ma Y, Rajkumar M, Freitas H (2008) Inoculation of plant growth promoting bacterium *Achromobacter xylosoxidans* strain *Ax10* for the improvement of copper phytoextraction by *Brassica juncea*. J Environ Manage 20:1–7

Matiru VN, Dakora FD (2004) Potential use of rhizobial bacteria as promoters of plant growth for increased yield in landraces of African cereal crops. Afr J Biotechnol 3:1–7

Mehta S, Nautiyal CS (2001) An efficient method for qualitative screening of phosphate-solubilizing bacteria. Curr Microbiol 43:51–56

Mishustin EM, Naumova AN (1962) Bacterial fertilizers, their effectiveness and mode of action. Mikrobiologiya 31:442–452

Muchovej RMC, Muchovej JJ, Alvarez VH (1989) Temporal relations of phosphorus fractions in an oxisol amended with rock phosphate and *Thiobacillus thiooxidans*. Soil Sci Soc Am J 53:1096–1100

Murray GC, Hesterberg D (2006) Iron and phosphate dissolution during abiotic reduction of ferrihydrite–boehmite mixtures. Soil Sci Soc Am J 70:1318–1327

Nahas E (1996) Factors determining rock phosphate solubilization by microorganisms isolated from soil. World J Microbiol Biotechnol 12:567–572

Nautiyal CS (1999) An efficient microbiological growth medium for screening of phosphate solubilizing microorganisms. FEMS Microbiol Lett 170:265–270

Nehl DB, Knox OGG (2006) Significance of bacteria in the rhizosphere. In: Mukerji KG, Manoharachary C, Singh J (eds) Microbial activity in the rhizosphere. Springer, Berlin, pp 89–119

Ocampo JA, Barea JM, Montoya EM (1975) Interaction between *Azotobacter* and "phosphobacteria" and their establishment in the rhizosphere as effected by soil fertility. Can J Microbiol 21:1160–1165

Oliveira CA, et al. (2008) Phosphate solubilizing microorganisms isolated from rhizosphere of maize cultivated in an oxisol of the Brazilian Cerrado biome. Soil Biol Biochem. doi:10.1016/j.soilbio.2008.01.012

Osorio Vega NW (2007) A review on beneficial effects of rhizosphere bacteria on soil nutrient availability and plant nutrient uptake. Rev Fac Nal Agr Medellín 60:3621–3643

Pal SS (1998) Interactions of an acid tolerant strain of phosphate solubilizing bacteria with a few acid tolerant crops. Plant Soil 198:169–177

Peix A, Rivas-Boyero A, Mateos PF, Rodríguez-Barrueco C, Martínez-Molina E, Velázquez E (2001) Growth promotion of chickpea and barley by a phosphate solubilizing strain of *Mesorhizobium mediterraneum* under growth chamber conditions. Soil Biol Biochem 33:103–110

Pikovskaya RI (1948) Mobilization of phosphorus in soil in connection with vital activity of some microbial species. Microbiology 17:362–370

Rajan SSS, Watkinson JH, Sinclair AG (1996) Phosphate rock for direct application to soils. Adv Agron 57:77–159

Richardson AE (1994) Soil microorganisms and phosphorus availability. In: Pankhurst CE, Doube BM, Gupta VVSR (eds) Soil biota: Management in sustainable farming systems. CSIRO Publishing, Collingwood, pp 50–62

Richardson AE (2001) Prospects for using soil microorganisms to improve the acquisition of phosphorus by plants. Aust J Plant Physiol 28:897–906

Richardson AE (2007) Making microorganisms mobilize soil phosphate. In: Velazquez E, Rodriguez-Barrueco C (eds) Microbial phosphate solubilization: Developments in plant and soil sciences. Kluwer Academic Publishers, Dordrecht, pp 85–90

Rodriguez H, Fraga R (1999) Phosphate solubilizing bacteria and their role in plant growth promotion. Biotechnol Adv 17:319–339

Rodríguez H, Fraga R, Gonzalez T, Bashan Y (2007) Genetics of phosphate solubilization and its potential applications for improving plant growth-promoting bacteria. In: Velázquez E, Rodríguez-Barrueco C (eds) First international meeting on microbial phosphate solubilization. Springer, Dordrecht, pp 43–53

Rong X, Huang Q, He X, Chen H, Cai P, Liang W (2008) Interaction of *Pseudomonas putida* with kaolinite and montmorillonite: a combination study by equilibrium adsorption, ITC, SEM and FTIR. Colloids Surf B 64:49–55

Şahin F, Çakmakçi R, Kantar F (2004) Sugar beet and barley yields in relation to inoculation with N_2-fixing and phosphate solubilizing bacteria. Plant Soil 265:123–129

Singh HP (1990) Response of dual inoculation with *Bradyrhizobium* and VAM mycorrhiza or phosphate solubilizer on soybean in mollisol. In: Jalali BL, Chand H (eds) Trends in mycorrhizae. Research proceedings of the national conference on mycorrhiza. Hisar, India, pp 14–16

Smith JH, Allison FE, Soulides DA (1961) Evaluation of Phosphobacterin as a soil inoculant. Soil Sci Soc Am J 25:109–111

Sperber JI (1958a) The incidence of apatite-solubilizing organisms in the rhizosphere and soil. J Agric Res 9:778–781

Sperber JI (1958b) Solution of apatite by soil microorganisms producing organic acids. Aust J Agric Res 9:782–787

Sridevi M, Mallaiah KV, Yadav NCS (2007) Phosphate solublization by *Rhizobium* isolates from *Crotalaria* species. J Plant Sci 2:635–639

Srivastava S, Kausalya MT, Archana G, Rupela OP, Naresh-Kumar G (2007) Efficacy of organic acid secreting bacteria in solubilization of rock phosphate in acidic alfisols. In: Velazquez E, Rodriguez-Barrueco C (eds) Microbial phosphate solubilization: developments in plant and soil sciences. Kluwer Academic Publishers, Dordrecht, pp 117–124

Stamford NP, Santos PR, Santos CES, Freitas ADS, Dias SHL, Lira MA Jr (2007) Agronomic effectiveness of biofertilizers with phosphate rock, sulphur and *Acidithiobacillus* for yam bean grown on a Brazilian tableland acidic soil. Biores Technol 98:1311–1318

Stemmler SJ, Berthelin J (2003) Microbial activity as a major factor in the mobilization of iron in the humid tropics. Eur J Soil Sci 54:725–733

Sundara B, Natarajan V, Hari K (2002) Influence of phosphorus solubilizing bacteria on the changes in soil available phosphorus and sugarcane and sugar yields. Field Crops Res 77:43–49

Tang WZ, Pasternak JJ, Glick BR (1995) Persistence in soil of the plant growth promoting rhizobacterium *Pseudomonas putida* GR12–2 and genetically manipulated derived strains. Can J Microbiol 4:445–451

Tiwari VN, Lehri LK, Pathak AN (1989) Effect of inoculation crops with phospho-microbes. Exp Agric 25:47–50

Toro M, Azcón R, Herrera R (1996) Effects on yield and nutrition of mycorrhizal and nodulated *Pueraria phaseoloides* exerted by P-solubilizing rhizobacteria. Biol Fertil Soils 21:23–29

Toro M, Azcón R, Barea JM (1997) Improvement of arbuscular mycorrhiza development by inoculation of soil with phosphate-solubilizing rhizobacteria to improve rock phosphate bioavailability (^{32}P) and nutrient cycling. Appl Environ Microbiol 63:4408–4412

Turner BL, Frossard E, Oberson A (2006) Enhancing phosphorus availability in low-fertility soils. In: Uphoff NT et al (eds) Biological approaches to sustainable soil systems. CRC Press, Boca Raton, FL, pp 191–205

Vassilev N, Vassileva M (2003) Biotechnological solubilization of rock phosphate on media containing agro-industrial wastes. Appl Microbiol Biotechnol 61:435–440

Vassilev N, Vassileva M, Fenice M, Federici F (2001) Immobilized cell technology applied in solubilization of insoluble inorganic rock phosphates and P plant acquisition. Bioresour Technol 79:263–271

Vassilev N, Vassileva M, Nikolaeva I (2006) Simultaneous P-solubilizing and biocontrol activity of microorganisms: potentials and future trends. Appl Microbiol Biotechnol 71:137–144

von Uexküll HR, Mutert E (1995) Global extent, development and economic impact of acid soils. Plant Soil 171:1–15

Wang Y, He Y, Zhang H, Schroder J, Li C, Zhou D (2008) Phosphate mobilization by citric, tartaric, and oxalic acids in a clay loam ultisol. Soil Sci Soc Am J 72:1263–1268

Weisskopf L, Tomasi N, Santelia D, Martinoia E, Langlade NB, Tabacchi R, Abou-Mansour E (2006) Isoflavonoid exudation from white lupin roots is influenced by phosphate supply, root type and cluster-root stage. New Phytol 171:657–668

Welbaum GE, Sturz AV, Dong Z, Nowak J (2004) Managing soil microorganisms to improve productivity of agro-ecosystems. Crit Rev Plant Sci 23:175–193

Welch SA, Taunton AE, Banfield JF (2002) Effect of microorganisms and microbial metabolites on apatite dissolution. Geomicrobiol J 19:343–367

Wightman PG, Fein JB (2004) The effect of bacterial cell wall adsorption on mineral solubilities. Chem Geol 212:247–254

Whitelaw MA (2000) Growth promotion of plants inoculated with phosphate-solubilizing fungi. Adv Agron 69:99–151

Whitelaw MA, Harden TJ, Helyar KR (1999) Phosphate solubilization in solution culture by the fungus *Penicillium radicum*. Soil Biol Biochem 31:655–665

Widada J, Damarjaya DI, Kabirun S (2007) The interactive effects of arbuscular mycorrhizal fungi and rhizobacteria on the growth and nutrients uptake of sorghum in acid soil. In: Velázquez E, Rodríguez-Barrueco C (eds) First international meeting on microbial phosphate solubilization. Springer, Dordrecht, pp 173–177

Xie JC (1998) Present situation and prospects for the world's fertilizer use. Plant Nutr Fertil Sci 4:321–330

Yao H, He ZL, Wilson MJ, Campbell DC (2000) Microbial biomass and community structure in a sequence of soils with increasing fertility and changing land use. Microbial Ecol 40:223–237

Yi Y, Huang W, Ge YJ (2008) Exopolysaccharide: a novel important factor in the microbial dissolution of tricalcium phosphate. J Microbiol Biotechnol 24:1059–1065

Young EO, Ross DS (2001) Phosphate release from seasonally flooded soils: a laboratory microcosm study. J Environ Qual 30:91–101

Young IM, Crawford JW (2004) Interactions and self-organization in the soil–microbe complex. Science 1304:1634–1637

Zahir ZA, Arshad M, Frankenberger WT (2004) Plant growth promoting rhizobacteria: applications and perspectives in agriculture. Adv Agron 8:97–167

Zapata F, Roy RN (2004) Use of phosphate rock for sustainable agriculture. In: Fertilizer and Plant Nutrition Bulletin 13, Food and Agriculture Organization of the United Nations, FAO Corporate Document Repository. http://www.fao.org/docrep/007/y5053e/y5053e00.htm. Accessed 20 Nov 2008

Chapter 11
Role of Microbial Biofertilizers in the Development of a Sustainable Agriculture in the Tropics

Daniel Uribe, Jimena Sánchez-Nieves, and Javier Vanegas

11.1 Introduction

Modern agriculture is currently following a path that few had predicted. Countries all around the world are facing historically high agricultural commodities prices, due to a changing social and economic world order. This situation is characterized by unstable oil prices, changing diets, urbanization, economic growth, and expanding populations (OECD–FAO 2008). Climate change is also contributing to the high prices of agricultural commodities. This situation presents us with new incentives for the development of a sustainable agriculture, which will boost agricultural production without putting the future at risk. The need for such improvements is greatest in the tropics where most of the least developed and poorest countries are located.

The agricultural outlook for the period 2008–2017, prepared by the Organization for Economic Co-operation and Development (OECD) and the Food and Agriculture Organization of the United Nations (FAO), indicates that additional measures, besides an increase in the land area used for agriculture, will be required to meet new food demands. Therefore, and in view of the high prices of mineral nitrogen and phosphorus fertilizers, agronomic action should be taken so as to improve plant nutrient uptake through the use of microbial inoculants.

Unfortunately, a large number of tropical agricultural systems are currently unsustainable (Boddey et al. 1997). This is primarily due to the soil physical degradation, widespread erosion, and the lack of nutrient replenishment caused by crop harvest, heavy rains, and rapid organic matter decomposition (Esilaba et al. 2005; Boddey et al. 1997; Döbereiner 1997). Between 1960 and 1990, an average of

D. Uribe (✉) and J. Vanegas
Instituto de Biotecnologia, Universidad Nacional de Colombia Bogotá, Colombia A.A 14-490
e-mail: duribev@unal.edu.co

J. Sánchez-Nieves
Departamento de Biología, Universidad Nacional de Colombia Bogotá, Colombia A.A 14-490

P. Dion (ed.), *Soil Biology and Agriculture in the Tropics*, Soil Biology 21,
DOI 10.1007/978-3-642-05076-3_11,

660 kg ha^{-1} of nitrogen (N), 75 kg ha^{-1} of phosphorus (P), and 450 kg ha^{-1} of potassium (K) were lost from 200 million ha of cultivated land in 37 African countries, and current annual rates of total nutrient losses are estimated to be 4.4 million tons of N, 0.5 million tons of P, and 3 million tons of K (Omamo et al. 2002).

Many of the soils of the humid and sub-humid tropics are low in clay content, and have low cation exchange capacity, water-holding capacity, and aggregate stability (see Chap. 1). These characteristics, which are responsible for poor soil fertility, are derived mainly from the low organic matter content of tropical soils, and therefore practices have been adopted which maintain or increase soil organic matter (SOM) reserves (Boddey et al. 1997). However, in resource-poor farming systems, organic inputs alone cannot sustain crop production due to limitations in their quality and availability. Also, organic resources can stimulate pests and diseases (Vanlauwe and Guiller 2006; Boddey et al. 1997). Therefore, other strategies, such as the improvement of nutrient uptake by plants through the utilization of microbial inoculants, should be evaluated (Esilaba et al. 2005).

In tropical countries, low- and high-income farmers adopt different strategies for soil fertility management. The smallholder farmers are more oriented towards a subsistence production pattern, where development limitations are related to socio-economic issues, such as the high cost of transportation of agricultural commodities from and to the farm (Perry et al. 2004; Omamo et al. 2002). In spite of this situation, several efforts have been directed towards technical improvement of soil fertility management by smallholder farmers. Nongovernmental organizations (NGOs), public and private universities, and governments are developing collaborative projects along those lines, in some cases with the support of European countries. For example, in 1992, Holland established an international cooperative program in four tropical countries, namely India, Kenya, Zimbabwe, and Colombia. In Colombia, several objectives were defined, including clean seed production, the development of disease-resistant cultivars, and the production of microbial biofertilizers. Currently, biofertilizers based on mycorrhizas, compost derived from earthworms, and microbial inoculants are successfully produced by smallholder farmers in the Caribbean coast of Colombia for application to cassava, yam, and banana plants (Perry et al. 2004).

An Indian and Swiss collaborative project used a similar approach. Its main goal was to develop new biotechnologies such as the use of inoculants to improve plant growth and soil health in marginal rain-fed regions of India (Roesti et al. 2006). As part of yet another biofertilizer project supported by the Forum for Nuclear Cooperation in Asia (FNCA), experts from eight different countries (China, Japan, Vietnam, Korea, Indonesia, Malaysia, The Philippines, and Thailand), met to prepare a biofertilizer manual. Through the improvement and dissemination of biofertilizer technology, this project aims to increase the yield of grain legumes and other crops that are important food and animal feed sources in Asia, and to enhance environmental sustainability of farming practices by reducing mineral fertilizer applications (FNCA 2006). Another example of soil fertility management project can be found in tropical Africa, where continuous cropping in the absence of external inputs to soils has generated poorly productive soil patches (Place et al. 2003; Woomer et al. 2003). As a result, in some countries such as Kenya, a fertilization product called PREP-PAC® (Table 11.1) was developed by researchers at public universities in a joint project with NGOs. This product is applied at a cost of US$ 0.56/25 m^2, meeting the needs of small farmers who are not reached by retail fertilizer networks (Woomer et al. 2003).

In tropical countries, basic research oriented towards the isolation, identification and functional characterization of biofertilizers and their development as inoculants has led to important industrial developments for the agricultural sector (Oliveira et al. 2008; Moreno-Sarmiento et al. 2007; Yadav and Tarafdar 2003; Gyaneshwar et al. 2002; Rodriguez and Fraga 1999; see Sect. 4.3.1). In the following review we will examine the contributions of the inoculant industry to the development of sustainable agriculture in the tropics, with an emphasis on tropical Latin American and Asian countries. We will pay special attention to the enhancement of N and P uptake through two general phenomena. These are, first, symbiotic and nonsymbiotic biological nitrogen fixation (BNF), and, second, phosphate solubilization (PS) and translocation to plants (see Chaps. 4, 5, 8, 9).

11.2 Biological Nitrogen Fixation

Agriculture might be expected to be more dependent on nitrogen fertilizers in the tropics than in temperate regions, because heavy rains and rapid organic matter decomposition lead to leaching and rapid loss of the applied fertilizer (Döbereiner 1997). High nitrogen fertilizer prices, particularly in tropical countries where subsidies are not available, plus the related and sometime elevated costs of transportation, which have to be met entirely by the farmers, create a situation that stimulates the development of microbial inoculants with BNF activity (Matiru and Dakora 2004; Omamo et al. 2002; Döbereiner 1997). BNF microbial inoculants may be symbiotic or nonsymbiotic.

11.2.1 Use and Application of Symbiotic Diazotrophs

The symbiotic inoculants marketed as commercial products are based mostly on strains of nitrogen-fixing rhizobia. Such rhizobia-based inoculants are heavily used in Brazil and Argentina, which are two of the three largest producers of soybean in the world. In Brazil, where soybean were first introduced in the 1960s, massive use of BNF bacteria was immediately and successfully implemented, placing the country in a privileged situation as compared to others where N mineral fertilizers were in use (Hungria and Campo 2007; Döbereiner 1997). Of the 22.2 million ha devoted to soybean cultivation in Brazil during the 2005–2006 season, 60%–75% were inoculated with *Bradyrhizobium* sp. (Castro-Sowinski et al. 2007;

Table 11.1 Some commercial microbial biofertilizer products manufactured in tropical countries

Product	Microorganism/active principle	Crop	Country	Reference
Arbuscular mycorrhizal fungi				
EcoMic®	*Glomus fasciculatum*	Rice, cotton, soybean, corn, coffee, sorghum, grass, sunflower, citrus trees, flowers	Cuba	Taset (2004)
Micofert UV®	*G. intraradices, G. etunicatum, Gigaspora* sp.	Coffee , papaya, corn, onion, avocado, yam bean *(Pachyrrhizus erosus)*, chilli, strawberry and flowers	Mexico	Cobos (2005)
Mycobiol®	*Glomus* sp., *Entrophospora colombiana, Acaulospora mellea*	Banana, cassava, lettuce, tomato, cape gooseberry *(Physalis peruviana)*	Colombia	CORPOICA (2008)
BuRIZE®	*Glomus intraradices*	Alfalfa, apple, tomato, orange, avocado, cotton, onion, strawberry	Mexico	BioScientific (2008)
Myconate®	Isoflavanone and phormononetin compounds	Corn, soybean	Mexico	Lara (2008)
MYCOgold	Species of 4 genera of AMF	Banana	Malaysia	Malaysian Agri Hi Tech (2008)
Micofert®	*Glomus aggregatum, G. manihotis, G. spurcum*	Sugarcane, tomato, onion, garlic, potato, banana, mandarin, malanga, yam, grains, forest	Cuba	Herrera (1992)
Phosphate solubilizers				
FOSFOSOL®	*Penicillium janthinellum*	Rice	Colombia	Moreno-Sarmiento et al. (2007)
FOSFORINA®	*Pseudomonas fluorescens*	Tomato	Cuba	Ortega (2007)
Biological nitrogen fixers				
PREP PAC	2 kg of rock phosphate, 0.2 kg of urea, 0.13 kg of legume seeds, rhizobial inoculant	Grain legumes	Kenya	Woomer et al. (2003)
BioGro®	*P. fluorescens/P. putida, Klebsiella pneumoniae, Citrobacter freundii*	Rice	Vietnam	Kennedy et al. (2004)
BioPower	Multi-strain	Rice	Pakistan	Kennedy et al. (2004)
BiAgro 10®	*Bradyrhizobium japonicum*	Soybean	Argentina, Brazil, Bolivia	BiAgro (2008)
Dimargon®	*Azotobacter chroococcum*	Rice, cotton	Colombia	Moreno-Sarmiento et al. (2007)
Bio N	Two *Azospirillum* sp.	Rice, maize	Philippines	FNCA (2006)
UPMB 10	*Bacillus sphaericus*	Oil palm	Malaysia	FNCA (2006)
Mixed biofertilizers				
EMAS	Nitrogen fixers: *A. lipoferum, A.beijerinckii*; phosphate solubilizers: *Aeromonas punctata, Aspergillus niger*	Tea, rubber, cocoa, oil palm, sugarcane, rice, maize and potato	Indonesia	FNCA (2006)

Hungria and Campo 2007). The massive use of soybean inoculants in Brazil has stimulated the development of the inoculant industry so as to meet the demand for 30.4 million doses, 98.8% of which are for soybean crops (Hungria and Campo 2007). The financial gain resulting from the use of inoculants represents US$ 3 billion per cropping season which would otherwise be spent on the purchase, transportation, and application of mineral N fertilizer (Menna et al. 2006).

There is some controversy in tropical countries about the utilization of inoculants in crops other than soybean. One reason for this is that some legumes do not have the capacity to form effective nodules. Additionally, it is argued that various legumes are promiscuous hosts and thus do not benefit from inoculation (Vanlauwe and Giller 2006). In spite of these criticisms, it has been established that in any circumstance where rhizobia compatible with a particular host are absent or in small numbers, or when a selected inoculant strain is more effective than indigenous rhizobia, inoculation may result in a crop yield increase (Vanlauwe and Giller 2006; Martinez-Romero 2003). The common bean, *Phaseolus vulgaris*, which is an important source of consumable proteins for humans in tropical countries, has been at the center of this debate for years. However, several studies carried out in Brazil and Mexico have concluded that common bean cultivars with high nitrogen-fixing capacity may be successfully used in combination with *Rhizobium* strains with superior symbiotic and competition potential (Hungria and Campo 2007; Martinez-Romero 2003). On the other hand, the development of promiscuous cultivars of legumes, which would be highly effective in fixing nitrogen, has been proposed as a valuable alternative for some smallholder farmers in tropical Africa.

In addition to soybean and common bean, other legume crops may provide the benefits of nitrogen fixation to tropical farming systems. In the large savanna areas of Latin America, including the Brazilian Cerrado (with an area of more than 200 million ha), and the llanos of Venezuela (28 million ha) and Colombia (20 million ha), beef cattle grazing is the primary agricultural activity (Boddey et al. 1997). Several studies conducted throughout these regions have shown that animal weight gains are usually higher when pastures such as *Brachiaria* sp. are coplanted with forage legumes. In Brazil, weight gains of over 3 kg ha^{-1} d^{-1} at the highest stocking rate have been reported for a mixed *Brachiaria*-legume pasture, compared with maximum gains of 1.0–1.2 kg ha^{-1} d^{-1} for *Brachiaria* sp. pastures at the same site (Boddey et al. 1997). In the eastern llanos of Colombia, the use of legumes such as *Cratylia argentea* has helped growers to maintain cattle weight gains throughout the summer without extra crop inputs (CIAT 2002). In the same area of Colombia, the rice yields following a 10-year period of *Brachiaria* sp. pasture was 1.3 ton ha^{-1}, but this was increased to 3.0 ton ha^{-1} where kudzu (*Pueraria phaseoloides*) was present in the pasture (CIAT 2002; Boddey et al. 1997). These studies suggest that the potential exists in tropical zones for application of inoculant technology to crops such as common bean or forage legumes. Indeed, in some countries such as Argentine and Brazil, nitrogen-fixing rhizobia are now sold for inoculation of forage crops such as alfalfa, red and white clover, *Lotus* and *Melilotus*.

11.2.2 Use and Application of Nonsymbiotic Diazotrophs

Nonsymbiotic bacteria are the other major source of plant growth promotion associated with N uptake by plants. As mentioned previously, Brazil is a world leader in the study and replacement of mineral N fertilizer with BNF in different crops, using not only symbiotic, but also nonsymbiotic bacteria (Hungria and Campo 2007; Kennedy et al. 2004). Brazilian researchers have contributed to this particular field with the description of novel species harboring the BNF function. These species include *Beijerinckia fluminensis*, *Azotobacter paspali*, *Azospirillum lipoferum*, *Azospirillum brasilense*, *Azospirillum amazonense* and some endophitic bacteria such as *Herbaspirillum seropedicae*, *Herbaspirillum rubrisubalbicans*, *Gluconacetobacter diazotrophicus*, *Burkholderia brasilensis*, and *Burkholderia tropica* (Baldani and Baldani 2005; Boddey et al. 2003).

Nonsymbiotic diazotrophic bacteria exert beneficial effects in a number of agricultural crops under a variety of climates and soil conditions (Okon and Vanderleyden 1997). It has been initially believed that the major contribution of such bacteria to plant growth was an increased availability of fixed N. However, further studies indicated that growth promotion was mainly due to physiological and structural changes of the roots which improve plant water and nutrient uptake (Okon and Kapulnik 1986; Dobbelaere et al. 2001).

Several tropical countries have evaluated and developed commercial products based on nonsymbiotic nitrogen fixers. In Mexico, for instance, research along those lines was initiated in the late 1980s and early 1990s, when field experiments were carried out with significant yield increases ranging between 20% and 70% after inoculation of corn seeds with *Azospirillum* sp. in different areas of the country (Caballero-Mellado et al. 1992). Following such success, 450,000 ha of maize and 150,000 ha of other grain crops (wheat, sorghum, and barley) were inoculated, with crop yield increases of 27% (Fuentes-Ramirez and Caballero-Mellado 2005; Dobbelaere et al. 2001). In Colombia, several nonsymbiotic BNF inoculants based mainly on *A. brasilense*, *A amazonense*, and *Azotobacter chroococcum* have been developed and successfully applied to crops such as rice and cotton, with an average yield increase of 5%–10% and a 20%–50% reduction in N fertilizer dose (Moreno-Sarmiento et al. 2007).

Another group of important nonsymbiotic diazothophs are the endophytic bacteria which are able to deliver nitrogen directly into plant tissues, thus ensuring a highly efficient nitrogen uptake by the plant. Some evidence from Brazil suggests that, in rice and sugarcane, fertilizer-N can be reduced by half of the usual 150–250 kg urea-N ha^{-1} doses by exploiting endophytic diazotrophs such as *Acetobacter* spp., *Gluconacetobacter diazotrophicus*, and *Herbaspirillum* spp., alone or in combination (Ladha and Reddy 2003; Oliveira et al. 2003; Oliveira et al. 2002; Elbeltagy et al. 2001; Döbereiner and Baldani 1998). Using the ^{15}N natural abundance technique, Boddey et al. (2001) and Sevilla et al. (2001) have shown that up to 80% of the total nitrogen assimilated by sugarcane may come through BNF by bacterial endophytes. In spite of these promising observations, the development of inoculants for sugarcane in Brazil has been hampered by a lack of demonstrated beneficial effect of endophytic or nonsymbiotic inoculants in field trials. One explanation for this may be the presence of a well-established spontaneous population of N-fixing bacteria within the plant, due to the vegetative mode of propagation of sugarcane (Boddey et al. 2003). Since its initiation in Campos Brazil in the 1930s, the sugarcane breeding program was carried out in soil with low nitrogen contents and no N fertilizer application, suggesting that an unexpected result of the selection process is the production of sugarcane varieties which spontaneously associate with diazotrophic bacteria (Boddey et al. 2003).

Endophytic bacteria have also generated interest in other parts of the world (Muthukumarasamy et al. 2002). In Pakistan for instance, after the evaluation of several strains of *Herbaspirillum* in two rice cultivars (Basmati 385 and Super basmati), BNF levels reached 19%–58% of the N required by rice crops, depending on the strain and rice cultivar tested. Similarly, an increase of wheat straw and grain yield was observed under field conditions, following inoculation with *H. seropedicae* (Kennedy et al. 2004). In India, the combined inoculation of four endophytes, *Burkholderia vietnamiensis*, *G. diazotrophicus*, *H. seropedicae*, and *A. lipoferum*, increased the rice yield between 9.5% and 23.6% under field conditions, whereas the inoculation of *B. vietnamiensis* MGK3 alone produced maximum yield increases of 5.6%–10.7% over the noninoculated control (Govindarajan et al. 2008).

As previously mentioned, the plant effect of non symbiotic BNF bacteria depends on a variety of other plant growth promotion traits besides BNF. Therefore, it has been recognized that the use of multi-strain inoculants may be a suitable strategy for improving crop yields. Individual members of the mixed inoculum may then express complementary functions, such as biological control of phytopathogens, xenobiotic compound degradation, stimulation of non-N nutrient uptake and production of plant growth regulators (Bashan 1998; Bashan and Holguin 1997). For example, in Vietnam a multi-strain inoculant strategy has been developed for rice crops (Table 11.1) that results in a yield increase of 21% over the noninoculated control, with a 50% reduction of applied N doses. Similar strategies have been designed in tropical Australia, Pakistan, and Egypt (Govindarajan et al. 2008; Kennedy et al. 2004; Alam et al. 2001). However, not all combinations of bacterial strains insure large yield increases. This highlights the importance of selecting efficient strains and of carefully examining their interactions.

Despite the impressive amounts of field trial reports and the relatively high number of commercial products available, there is no widespread use of BNF strategies in tropical agriculture. The main reason for the poor integration of nonsymbiotic BNF inoculants to agronomic practice has been an inconsistency in field trial results (Dobbelaere et al. 2001; Bashan and Holguin 1997). In turn, this lack of field reproducibility may be due to non specific host–inoculant relationships, varying physical and chemical soil conditions, poor competitive ability against indigenous soil micro-organisms, and lack of adequate formulation technology (Lucy et al. 2004).

11.3 Phosphate Solubilization

After nitrogen, phosphorus is the major plant growth-limiting nutrient despite its abundance in soils in both inorganic and organic forms (see Chap. 10). Whereas most mineral nutrients in soil solutions are present in millimolar amounts, soluble phosphorus is present only in micromolar or lesser quantities. Phosphate is poorly accessible to plants because of its high reactivity with aluminum, iron, and calcium, which leads to its precipitation (Gyaneshwar et al. 2002). Consequently, in tropical soils, acidity and toxicity of aluminum and iron are typically major constraints to soil fertility (Place et al. 2003). Organic phosphate, which represents 50%–80% of total soil phosphate, is also immobilized by organic soil matter or the formation of complexes with aluminum and iron (Yadav and Tarafdar 2003; Gyaneshwar et al. 2002).

Phosphate-solubilizing micro-organisms (PSM) dissociate the phosphorus from soil complexes through several mechanisms, such as the production of organic acids which dissolve or chelate inorganic phosphate, or the production of phosphatases and phytases which dissociate phosphorus from organic sources (Yadav and Tarafdar 2003; Rodriguez and Fraga 1999). PSM increase the availability of soil phosphate, promoting plant uptake of this element (Gyaneshwar et al. 2002). Other mechanisms besides phosphorus solubilization are involved in plant growth promotion by PSM. These include nitrogen fixation, production of phytohormones and other plant growth-promoting substances, an enhancement of the availability of other trace elements such as iron and zinc, and an enhanced availability of microbially immobilized phosphorus (Lucy et al. 2004; Gyaneshwar et al. 2002; Rodriguez and Fraga 1999).

In India, PSM have been evaluated for their efficiency for hydrolysis of phytin and glycerophosphate, two phosphorus-containing organic compounds that can be mineralized by soil fungi. Three strains of *Aspergillus* sp., two strains of *Emericella* sp., and two strains of *Penicillium* sp. have been selected on the basis of their capacity to produce extracellular phosphatases and phytases with sufficient activity to benefit plant nutrition (Yadav and Tarafdar 2003). In other studies, efforts to identify PSM inoculants resulted in enhanced phosphate uptake by plants and 10%–15% increases in crop yields in 10 out of 37 experiments (Gyaneshwar et al. 2002).

In Brazil, different PSM from maize rhizospheric soil were selected as potential inoculants based on their solubilization efficiency of inorganic and organic sources; however, their usefulness was not evaluated in the field (Oliveira et al. 2008). In Colombia, a huge effort has been conducted over the last 10 years by different NGOs, private and public universities, international research organizations such as the International Center for Tropical Agriculture (CIAT), the government, and the private sector to develop the inoculant industry (Moreno-Sarmiento et al. 2007). Consequently, 15 different microbial inoculant products have been registered by 2006 for use in various crops. While most of these were nonsymbiotic BNF inoculants, one PSM inoculant, FOSFOSOL®, was developed (Table 11.1) and permitted a 5%–38% yield increase in rice crops as compared to the uninoculated control. Both a solid and a liquid formulation of this product are now successfully used by rice growers, with more than 20,000 ha being treated in the major rice production areas of Colombia (Moreno-Sarmiento et al. 2007; Moreno 2007). A similar biofertilizer product, based on the phosphorus-solubilizing fungus *Penicillium bilaii*, has been developed in Canada by Novozymes (formerly Philom Bios) (Rice et al. 1994), and is now being tested in tropical Australia. In Cuba, various microbial inoculants were developed as a consequence of the economic embargo. In particular, the biofertilizer FOSFORINA® (Table 11.1) was developed so as to reduce the need for mineral phosphorus applications (Ortega 2007).

An alternative approach to the use of single PSM as microbial inoculant is the use of mixed cultures or co-inoculation with other micro-organisms which hold different promoting properties. Following inoculation with a composite inoculant comprising *Rhizobium*, the phosphate-solubilizing bacterium *Bacillus megaterium* var. *phosphaticum* and the biocontrol fungus *Trichoderma spp*, chickpea showed increased germination, nutrient uptake, plant height, number of branches, nodulation, seed yield, and total biomass, as compared with either individual inoculation or uninoculated plants (FNCA 2006). Similarly, in a rice–wheat rotation system, the combined inoculation of plant growth-promoting bacteria (PGPB), including phosphate solubilizers, and arbuscular mycorrhizal fungi (AMF) had a significant effect on grain quality due to better phosphorus, protein and iron content (Roesti et al. 2006). Comparable results were reported by Saini et al. (2004), in a sorghum–chickpea rotation system in India, where co-inoculation of AMF with other PGPB including the PSM *Bacillus megaterium* presented an additive effect recorded in terms of grain and stover yield for both crops.

Despite the importance of phosphorus for agriculture, the high cost of this element, and the anticipated depletion of sources of high quality phosphate rocks during this century (Richardson 2001), PSM inoculants are not in widespread use in tropical countries. Again, the lack of consistent and conclusive field results that would demonstrate the benefits of PSM inoculants accounts for this situation. Reasons which have been proposed for the variable effects of PSM on plant growth and crop yields include a lack of knowledge of the plant–micro-organism interactions, the uncontrolled effect of plant varieties and soil types on microbial colonization and functional capacity, and inaccurate strategies for screening of potential PSM strains (Gyaneshwar et al. 2002; Rodriguez and Fraga 1999).

11.4 Arbuscular Mycorrhizal Fungi

The arbuscular mycorrhizal fungi (AMF) are fundamental to the establishment of plants and crops under the limiting conditions prevalent in the tropics (Sieverding et al. 1991). For this reason, AMF inoculants have been used repeatedly, even though limitations exist for their mass production. These limitations may be overcome thanks to recent research developments with respect to multiplication and formulation of the inoculants, and a better understanding of the microbial rhizosphere and micorhizosphere populations (Sánchez et al. 2007). AMF production technologies are mainly adapted to host plants growing in subtropical, temperate, and Mediterranean conditions. In the case of most tropical species, knowledge on mycorrhizal inoculant production remains scarce and such production is largely performed by farmers upon advice from extentionists belonging to academic and research institutions (Corredor 2008).

The response to inoculation with AMF depends on the degree of mycotrophy of the crop, the presence and activity of indigenous mycorrhizal fungi, the nutrient status of the soils (and particularly the available P content), the application of pesticides, the symbiotic effectiveness of the introduced fungal strains, and the ability of these strains to survive in the soil before contact with the roots (Marx et al. 2002).

In Mexico, a sheared-root inoculum of *Glomus intraradices* was tested in wheat field trials, together with different amounts and sources of commercial phosphate. In soil with low phosphate availability, wheat growth capacity was highest in inoculated plants, suggesting that the use of *G. intraradices* may contribute to reducing the need for phosphate fertilizer applications (Mohammad et al. 2004). Similarly, Carrillo et al. (2007) reported a higher production of the exotic vegetable *Abelmoschus esculentus* upon inoculation with AMF in fields with low phosphate and high salt content.

Beginning in the 1980s, CIAT has conducted research in Colombia to evaluate the agronomic importance of native AMF in greenhouses and in the field (Sieverding et al. 1991; Corredor 2008). In Colombia also, Sánchez de Prager (2007) has reported increases in the production of cassava (*Manihot esculenta* Crantz), pepper (*Capsicum annuum* L.), passion fruit (*Passiflora edulis* var. Flavicarpa), banana (*Musa* AAB), bamboo (*Guadua angustifolia* Kunth) at the nursery stage, and sugarcane (*Saccharum* sp.) through AMF inoculation.

In the Democratic Republic of the Congo, Khasa et al. (1992) found that 18 out of 19 tested agricultural plants were successfully colonized by AMF, the exception being amaranth (*Amaranthus*). Eight plant species, African pea (*Sphenostylis stenocarpa*), Acacia (*Acacia* sp.), wild mung (*Vigna* sp.), leucaena (*Leucaena* sp.), onion (*Allium cepa*), sweet potato (*Ipomoea batatas*), tomato (*Lycopersicum esculentum*), and cassava (*Manihot esculenta*), were highly dependent upon AMF colonization for normal growth and development. In Mozambique, a native AMF inoculant reduced the effects produced by drought stress in peanut (*Arachis hypogaea)* (Quilambo et al. 2005).

Research done in India has shown positive results of rice seedling inoculation with *G. intraradices*, *Acaulospora* sp., and *Glomus etunicatum*. These beneficial effects included increased biomass, plant tissue mineral content, and grain yield, both in the Prakash cultivar growing in inundated fields (Secilia and Bagyaraj 1992), and in the Pusa Basmati-1 cultivar sown in soils deficient in phosphorus and zinc (Purakayastha and Chhonkar 2001). Similarly, Krishna et al. (2006) inoculated vine seedlings derived from in vitro culture with a mix of AMF strains of *Acaulospora laevis*, *Acaulospora scrobiculata*, *Entrophospora colombiana*,

Gigaspora gigantea, Glomus manihotis, and *Scutellospora heterogama*, or else with a commercial inoculant containing strains of *Glomus mosseae, G. manihotis*, and *Gigaspora gigantea*. These inoculations resulted in a reduction of the stress generated upon transplantion of the vine seedlings, as evidenced by improved survival percentage, shoot length, fresh and dry weight, leaf area, and photosynthetic rate.

In Brazil, AMF were successfully inoculated to various commercial crops. Upon inoculation with *Gigaspora margarita*, coffee (*Coffea arabica*) seedlings showed increases in growth rate and nutrient concentration (Souza et al. 1991). A total of 24 different AMF species colonized papaya (*Carica papaya*) roots, with *Glomus etunicatum, Paraglomus occultum, Acaulospora scrobiculata*, and *Gigaspora* sp. as the most commonly established species (Trindade et al. 2006).

In view of the positive outcomes of AMF trials performed around the world, the main concern for the incorporation of AMF as a common agricultural practice has become the development of a massive production system (Dalpé and Monreal 2004; Gianinazzi and Vosatka 2004; Marx et al. 2002). Currently, one option for the commercial production of AMF is the multiplication on solid substrate, using trap plants grown on peat, vermiculite, sandy, or clay soils, either in open fields or in greenhouse conditions. This is a simple method suitable for local use at a relatively low cost.

Some commercial products based on AMF prepared on various substrates and using various formulations, or other products containing formononetin and other isoflavanone compounds (which are commonly used to stimulate native AMF) are listed in Table 11.1. For example, MicoFert® is an inoculant developed in Cuba with an organic-mineral substrate that contains AMF (*Glomus aggregatum, G. manihotis*, and *Glomus spurcum*), associated with rootlets of host plants produced in the greenhouse or at a larger scale in a controlled multipliers system. Likewise, researchers in Xalapa (Mexico) developed the product Microfert UV® (formulated in alginate capsules), which improves the productivity of different agricultural crops and ornamentals (Table 11.1).

The aim of on-farm AMF inoculant production is to generate inoculum available to small farmers and to nurserymen at low cost (Sieverding et al. 1991). In Colombia, different institutions and researchers have examined production methods for inoculants based on native AMF. Several multiplication strategies have been tested in greenhouse and field conditions, using various host plants (mainly grasses) grown in a variety of substrates. For example, a model of AMF inoculant on-farm production was developed successfully in cacao (*Theobroma cacao*) by Mejía and Palencia (2005). These workers used compost and sterilized soil as a solid substrate in seedling beds under greenhouse conditions, with field soil containing mycorrhized roots, AMF spores, and mycelium as a fungal source. Kudzu and signal grass (*Brachiaria decumbens*) served as hosts plants. After 4 months, the colonized roots were harvested, and a mixture of soil and roots was used as inoculants in nurseries and fields of cacao plants. In Colombia, AMF inoculants produced in this manner are commercially applied alone or with rock phosphate to cassava and other crops (Sieverding et al. 1991).

In vitro culture of mycorrhizal fungi on root organs involves the inoculation of *Agrobacterium rhizogenes*-transformed roots with AMF propagules to yield a monoxenic cultivation system. This is a promising method for obtaining viable fungal propagules free of contamination (Dalpé and Monreal 2004). Root-organ culture has been used by Jaizme-Vega et al. (2003) for production of *Glomus proliferum*, *Glomus vermiforme*, and *G. intraradices*. The resulting material was encapsulated in calcium alginate spheres for application in the first phase of development of banana seedlings (*Musa* spp. cv. Grande Naine), micropropagated under greenhouse conditions. Under nursery conditions, the inoculated banana plants showed improved AMF root system colonization, increases in the seedling biomass, and better phosphorus uptake as compared to the uninoculated control plants. In India, the TERI (The Energy and Resources Institute) has also promoted the root organ culture technique for large-scale production and commercialization of mycorrhizal inoculants (TERI 2008). This institute has conducted mass in vitro cultivation of a consortium of AMF in a semi-synthetic medium under sterile conditions. The application of the resulting inoculant to vegetable crops and fruits yielded promising field results. Whereas it has been proposed to use bioreactors for production of AMF root organ cultures (Jolicoeur et al. 1999), and it is well established that nutritional parameters determine the productivity in these in vitro systems, more research is needed in order to optimize fungal yields and develop low cost techniques for large-scale production (Sánchez et al. 2007).

11.5 Conclusions

In the tropics, particularly in Asia and Latin America, considerable effort has been made to develop microbial inoculants for important crops such as soybean, rice, sugarcane, corn, and various fruit varieties. Such efforts were driven by the need to combine sustainable management with high productivity in the context of global agriculture. The use of biofertilizers helps to decrease total fertilization costs while improving plant growth and productivity under greenhouse, nursery, and field conditions. As such, biofertilizers are an essential component of an integrated soil fertility management strategy.

Consistent field results are a prerequisite to regular and widespread use of biofertilizer technologies. Therefore, more studies should be done to: (1) understand the fate of the inoculant once it is applied in the field, (2) design quality control procedures and guidelines to monitor microbial activity during the production and formulation of inoculants, (3) improve our knowledge regarding the influence of environmental conditions and plant-associated characteristics on the field performance of a particular inoculant, (4) develop appropriate protocols to precisely select strains and/or combinations of strains able to perform a given desirable function while being associated to a specific crop and exposed to particular growth conditions. This knowledge will be key to the fruitful and widespread integration of biofertilizers in tropical agricultural practices.

Acknowledgments We are grateful to Sandra Espitia and Tiffany Sosa R. for valuable comments on the manuscript. This work was made possible through Ministerio de Agricultura y Desarrollo Rural and Colciencias funds to D. Uribe and J. Vanegas respectively. We do not have any commitment or particular interests to any of the commercial products cited in this chapter.

References

Alam MS, Cui Z, Yamagishi T, Ishii R (2001) Grain yield and related physiological characteristics of rice plants (*Oryza sativa* L.) inoculated with free-living rhizobacteria. Plant Prod Sci 4:126–130

Baldani JI, Baldani VLD (2005) History on the biological nitrogen fixation research in graminaceous plants: special emphasis on the Brazilian experience. An Acad Bras Cienc 77:549–579

Bashan Y (1998) Inoculants of plant growth-promoting bacteria for use in agriculture. Biotechnol Adv 16:729–770

Bashan Y, Holguin G (1997) Azospirillum–plant relationships: environmental and physiological advances (1990–1996). Can J Microbiol 43:103–121

Biagro (2008) Laboratorios Biagro S.A. http://www.biagrosa.com/inocu.htmd. Accessed 20 Nov 2008

BioScientific Inc (2008) BuRIZE — VA-mycorrhizal soil and root inoculant, for use on production agricultural crops. http://www.biosci.com/brochure/BRZBro.pdf. Accessed 20 Nov 2008

Boddey RM, De Moraes Sá JC, Alves BJR, Urquiaga S (1997) The contribution of biological nitrogen fixation for sustainable agricultural systems in the tropics. Soil Biol Biochem 29:787–799

Boddey RM, Polidoro JC, Resende AS, Alves BJR, Urquiaga S (2001) Use of the ^{15}N natural abundance technique for the quantification of the contribution of N_2 fixation to sugar cane and other grasses. Aust J Plant Physiol 28:889–895

Boddey RM, Urquiaga S, Alves BJR, Reis V (2003) Endophytic nitrogen fixation in sugarcane: present knowledge and future applications. Plant Soil 252:139–149

Caballero-Mellado J, Carcano-Montiel MG, Mascarua-Esparza MA (1992) Field inoculation of wheat (*Triticum aestivum*) with *Azospirillum brasilense* under temperate climate. Symbiosis 13:243–253

Carrillo MA, Franco AD, Cano IG (2007) Okra (*Abelmoschus esculentus* L.) mycorrhization under drip irrigation. Revista Fitotecnia Mexicana 30:437–441

Castro-Sowinski S, Herschkovitz Y, Okon Y, Jurkevitch E (2007) Effects of inoculation with plant growth-promoting rhizobacteria on resident rhizosphere microorganisms. FEMS Microbiol Lett 276:1–11

CIAT (2002) Convenio Colombia-CIAT: Un negocio de amplios horizontes para el llano. Feriva, Palmira

Cobos C (2005) Crean con hongos nuevo fertilizante. http://fox.presidencia.gob.mx/buenasnoticias/. Accessed 20 Nov 2008

CORPOICA (2008) Insumos agrícola orgánicos CORPOICA. Presentación productos institucionales. http://www.corpoica.gov.co/SitioWeb/Archivos/Publicaciones/caminosfertiles.pdf. Accessed 20 Nov 2008

Corredor G (2008) Micorrizas arbusculares: aplicación para el manejo de los agroecosistemas. http://www.turipana.org.co/Micorrizas.html. Accessed 20 Nov 2008

Dalpé Y, Monreal M (2004) Arbuscular mycorrhiza inoculum to support sustainable cropping systems. Crop Manag 10:1094–1104

Dobbelaere S et al (2001) Responses of agronomically important crops to inoculation with *Azospirillum*. Aust J Plant Physiol 28:871–879

Döbereiner J (1997) Biological nitrogen fixation in the tropics: social and economic contributions. Soil Biol Biochem 29:771–774

Döbereiner J, Baldani VLD (1998) Biological nitrogen fixation by endophytic diazotrophs in non-leguminous crops in the tropics. In: Malik KA, Mirza MS, Ladha JK (eds) Nitrogen fixation with non-legumes, vol 1. Kluwer Academic Publishers, Dordrecht, pp 3–7

Elbeltagy A et al (2001) Endophytic colonization and in planta nitrogen fixation by a *Herbaspirillum* sp. isolated from wild rice species. Appl Environ Microbiol 67:5285–5293

Esilaba AO, Byalebeka JB, Delve RJ, Okalebo JR, Ssenyange D, Mbalule M, Ssali H (2005) On farm testing of integrated nutrient management strategies in eastern Uganda. Agric Syst 86:144–165

FNCA (2006) Biofertilizer manual. Japan Atomic Industrial Forum, Tokyo

Fuentes-Ramirez LE, Caballero-Mellado J (2005) Bacterial biofertilizers. In: Siddiqui ZA (ed) PGPR: Biocontrol and biofertilization, vol 1. Springer, Dordrecht, pp 143–172

Gianinazzi S, Vosatka M (2004) Inoculum of arbuscular mycorrhizal fungi for production systems: science meets business. Can J Bot 82:1264–1271

Govindarajan M, Balandreau J, Kwon SW, Weon HY, Lakshminarasimhan C (2008) Effects of the inoculation of *Burkholderia vietnamensis* and related endophytic diazotrophic bacteria on grain yield of rice. Microb Ecol 55:21–37

Gyaneshwar P, Naresh Kumar G, Parekh LJ, Poole PS (2002) Role of soil microorganisms in improving P nutrition of plants. Plant Soil 245:83–93

Herrera R (1992) Uso de las micorrizas en las actividades agrícolas y forestales de Cuba. Academia de Ciencias República de Cuba. http://209.85.215.104/search?q=cache:XulgtPBrN0J:www.academiaciencias.cu/paginas/presentacion/reconocimientos/premios.asp%3Fidp%3D321%26nsecc%3DCiencias%2520Agrarias%2520y%2520de%2520la%2520Pesca+MicoFert&hl=es&ct=clnk&cd=24&gl=co. Accessed 18 Nov 2008

Hungria M, Campo RJ (2007) Inoculantes microbianos: Situación en Brasil. In: Izaguirre-Mayoral ML, Labandera-C, Sanjuan J (eds) Biofertilizantes en Iberoamerica: Visión técnica, científica y empresarial, vol 1. Denad Internacional, Montevideo, pp 22–31

Jaizme-Vega MC, Rodriguez-Romero AS, Hermoso CM, Declerck S (2003) Growth of micropropagated bananas colonized by root-organ culture produced arbuscular mycorrhizal fungi entrapped in Ca-alginate beads. Plant Soil 254:329–335

Jolicoeur M, Williams RD, Chavarie C, Fortin JA, Archambault J (1999) Production of *Glomus intraradices* propagules, an arbuscular mycorrhizal fungus, in an airlift bioreactor. Biotechnol Bioeng 63:224–232

Kennedy IR, Choudhury ATMA, Kecskes ML (2004) Non-symbiotic bacterial diazotrophs in crop-farming systems: can their potential for plant growth promotion be better exploited? Soil Biol Biochem 36:1229–1244

Khasa P, Furlan V, Fortin JA (1992) Response of some tropical plant species to endomycorrhizal fungi under field conditions. Trop Agr 69:279–283

Krishna H, Singh SK, Patel VB, Khawale RN, Deshmukh PS, Jindal PC (2006) Arbuscular-mycorrhizal fungi alleviate transplantation shock in micropropagated grapevine (*Vitis vinifera* L.). J Hortic Sci Biotechnol 81:259–263

Ladha JK, Reddy PM (2003) Nitrogen fixation in rice systems: state of knowledge and future prospects. Plant Soil 252:151–167

Lara AM (2008) Nueva herramienta en el tratamiento de semillas de granos y cereales. http://www.phcmexico.com.mx/phcsemillas1.html. Accessed 20 Nov 2008

Lucy M, Reed E, Glick BR (2004) Applications of free living plant growth-promoting rhizobacteria. Antonie Van Leeuwenhoek 86:1–25

Malaysian Agri Hi Tech (2008) Mycogold. http://www.alibaba.com/catalog/11813495/Mycorrhiza_Root_Enhancer.html. Accessed 10 Nov 2008

Martinez-Romero E (2003) Diversity of *Rhizobium–Phaseolus vulgaris* symbiosis: overview and perspectives. Plant Soil 252:11–23

Marx DH, Marrs LF, Cordell CE (2002) Practical use of the mycorrhizal fungal technology in forestry, reclamation, arboriculture, agriculture, and horticulture. Dendrobiology 47:27–40

Matiru VN, Dakora FD (2004) Potential use of rhizobial bacteria as promoters of plant growth for increased yield in landraces of African cereal crops. Afr J Biotechnol 3:1–7

Mejía L, Palencia G (2005) Abono orgánico para el cultivo de cacao. http://www.turipana.org.co/abono_cacao.htm. Accessed 21 Nov 2008

Menna P, Hungria M, Barcellos FG, Bangel EV, Hess PN, Martinez-Romero E (2006) Molecular phylogeny based on the 16S rRNA gene of elite rhizobial strains used in Brazilian commercial inoculants. Syst Appl Microbiol 29:315–332

Mohammad A, Mitra B, Khan AG (2004) Effects of sheared-root inoculum of *Glomus intraradices* on wheat grown at different phosphorus levels in the field. Agric Ecosyst Environ 103:245–249

Moreno N (2007) Producción de biofertilizantes y biocontroladores para una agricultura sostenible. In: Sánchez J (ed) Potencial biotecnológico de microorganismos en ecosistemas y agroecosistemas, vol 1. Departamento de Biología, Facultad de Ciencias, Universidad Nacional de Colombia, Bogotá, pp 261–269

Moreno-Sarmiento N, Moreno-Rodriguez LF, Uribe D (2007) Biofertilizantes para la agricultura en Colombia. In: Izaguirre-Mayoral ML, Labandera-C, Sanjuan J (eds) Biofertilizantes en Iberoamerica: Visión técnica, científica y empresarial, vol 1. Denad Internacional, Montevideo, pp 38–45

Muthukumarasamy R, Revathi G, Seshadri S, Lakshminarasimhan C (2002) *Gluconacetobacter diazotrophicus* (syn. *Acetobacter diazotrophicus*), a promising diazotrophic endophyte in tropics. Curr Sci 83:137–145

OECD – FAO (2008) Agricultural outlook 2008–2017. OECD Publications, Paris

Okon Y, Kapulnik Y (1986) Development and function of *Azospirillum*-inoculated roots. Plant Soil 90:3–16

Okon Y, Vanderleyden J (1997) Root-associated *Azospirillum* species can stimulate plants. ASM News 63:364–370

Oliveira ALM, Urquiaga S, Döbereiner J, Baldani JI (2002) The effect of inoculating endophytic N_2-fixing bacteria on micropropagated sugarcane plants. Plant Soil 242:205–215

Oliveira ALM, Canuto EL, Reis VM, Baldani JI (2003) Response of micropropagated sugarcane varieties to inoculation with endophytic diazotrophic bacteria. Braz J Microbiol 34:59–61

Oliveira CA et al. (2008) Phosphate solubilizing microorganisms isolated from rhizosphere of maize cultivated in an oxisol of the Brazilian Cerrado Biome. Soil Biol Biochem doi:10.1016/j.soilbio.2008.01.012

Omamo SW, Williams JC, Obare GA, Ndiwa NN (2002) Soil fertility management on small farms in Africa: evidence from Nakuru District, Kenya. Food Policy 27:159–170

Ortega E (2007) Una visión del uso de fertilizantes en Cuba. In: Izaguirre-Mayoral ML, Labandera-C, Sanjuan J (eds) Biofertilizantes en Iberoamerica: Visión técnica, científica y empresarial, vol 1. Denad Internacional, Montevideo, pp 46–51

Perry S, Laignelet A, Buitrago G (2004) El programa de biotecnología agrícola para pequeños productores. In: Montoya D (ed) Biotecnología para no biotecnologos. Universidad Nacional de Colombia, Unibiblos, Bogotá, pp 46–63

Place F, Barrett CB, Freeman HA, Ramisch JJ, Vanlauwe B (2003) Prospects for integrated soil fertility management using organic and inorganic inputs: evidence from smallholder African agricultural systems. Food Policy 28:365–378

Purakayastha TJ, Chhonkar PK (2001) Influence of vesicular–arbuscular mycorrhizal fungi (*Glomus etunicatum* L.) on mobilization of zinc in wetland rice (*Oryza sativa* L.). Biol Fertil Soils 33:323–327

Quilambo OA, Weissenhorn I, Doddema H, Kuiper PJC, Stulen I (2005) Arbuscular mycorrhizal inoculation of peanut in low-fertile tropical soil I. Host-fungus compatibility. J Plant Nutr 28:1633–1644

Rice WA, Olsen PE, Leggett ME (1994) Co-culture of *Rhizobium meliloti* and a phosphorus-solubilizing fungus (*Penicillium bilaii*) in sterile peat. Soil Biol Biochem 27:703–705

Richardson AE (2001) Prospects for using soil microorganisms to improve the acquisition of phosphorus by plants. Aust J Plant Physiol 28:897–906

Rodriguez H, Fraga R (1999) Phosphate solubilizing bacteria and their role in plant growth promotion. Biotechnol Adv 17:319–339

Roesti D et al (2006) Plant growth stage, fertiliser management and bio-inoculation of arbuscular mycorrhizal fungi and plant growth promoting rhizobacteria affect the rhizobacterial community structure in rain-fed wheat fields. Soil Biol Biochem 38:1111–1120

Saini VK, Bhandari SC, Tarafdar JC (2004) Comparison of crop yield, soil microbial C, N and P, N-fixation, nodulation and mycorrhizal infection in inoculated and non-inoculated sorghum and chickpea crops. Field Crops Res 89:39–47

Sánchez de Prager M (2007) Las Endomicorrizas: expresión bioedáfica de importancia en el trópico. Universidad Nacional de Colombia, Facultad de Ciencias Agropecuarias, Palmira

Sánchez J, Sosa T, Melgarejo L, Fuentes C (2007) Endomicorrizas: ecología, papel funcional, aplicaciones. In: Sánchez J (ed) Potencial biotecnológico de microorganismos en ecosistemas y agroecosistemas, vol 1. Departamento de Biología, Facultad de Ciencias, Universidad Nacional de Colombia, Bogotá, pp 223–237

Secilia J, Bagyaraj DJ (1992) Selection of efficient vesicular–arbuscular mycorrhizal fungi for wetland rice (*Oryza sativa* L.). Biol Fertil Soils 13:108–111

Sevilla M, Burris RH, Gunapala N, Kennedy C (2001) Comparison of benefit to sugarcane plant growth and $^{15}N_2$ incorporation following inoculation of sterile plants with *Acetobacter diazotrophicus* wild-type and *nif* mutant strains. Mol Plant Microbe Interact 14:358–366

Sieverding E, Friedrichsen J, Suden W (1991) Vesicular-arbuscular mycorrhiza management in tropical agrosystems. Deutsche Gesellschaft fuer Technische Zusammenarbeit (GTZ) Rossdorf

Souza CAS, Siqueira JO, Oliveira E, Carvalho JG (1991) Development and nutrient levels of coffee seedlings inoculated with mycorrhizal fungi: effect of organic matter and simple superphosphate. Pesqui Agropecu Bras 26:1989–2005

Taset M (2004) Alianza entre dos reinos. http://www.elhabanero.cubasi.cu/2004/noviembre/nro1133_04nov/cienc_04nov301.html. Accessed 21 Nov 2008

TERI (2008) The energy and resources institute centre for mycorrhizal research: mycorrhizal technology. http://www.teriin.org/index.php?option=com_content&task=view&id=66. Accessed 21 Nov 2008

Trindade AV, Siqueira JO, Sturmer SL (2006) Arbuscular mycorrhizal fungi in papaya plantations of Espirito Santo and Bahia, Brazil. Braz J Microbiol 37:283–289

Vanlauwe B, Giller KE (2006) Popular myths around soil fertility management in sub-Saharan Africa. Agric Ecosyst Environ 116:34–46

Woomer PL, Okalebo JR, Maritim HK, Obura PA, Mwaura FM, Nekesa P, Mukhwana EJ (2003) PREP-PAC: a nutrient replenishment product designed for smallholders in western Kenya. Agric Ecosyst Environ 100:295–303

Yadav RS, Tarafdar JC (2003) Phytase and phosphatase producing fungi in arid and semi-arid soils and their efficiency in hydrolyzing different organic P compounds. Soil Biol Biochem 35:745–751

Chapter 12
Microbial Degradation of Pesticides in Tropical Soils

Ziv Arbeli and Cilia L. Fuentes

12.1 Introduction

The introduction of pesticides has substantially increased agricultural productivity in industrialized and developing countries over the past 60 years (Peñafiel and Kammerbauer 2001). Increased agricultural production has been necessary to meet the food demands of a growing human population, and crop protection against pests has played a key role in achieving this goal (Oerke and Dehne 2004). This is especially true for tropical zones where high temperatures and humidity favor many pests. Although integrated pest management (IPM) should be encouraged in order to reduce pesticide use, many African farmers view IPM as the preserve of industrialized agriculture in rich countries and do not appreciate its applicability to African situations (Williamson et al. 2008).

Microbial degradation is a principal mechanism for preventing pesticide accumulation in the environment, but prolonged crop protection can be at risk when the degradation rate is too fast. Thus, the ability to predict the rate of degradation and its products, and an understanding of the environmental conditions that affect these factors, are important for environmental risk assessments of pesticide use and for crop protection planning strategies. Since studies on the fate of pesticides in tropical soils are relatively scarce, microbial degradation predictions are probably less accurate in tropical areas than in temperate zones. This is of special concern since tropical regions contain more than one-third of the Earth emerged land and more than half of the world human population. Moreover, the economies of tropical countries are based on agriculture, which is itself heavily dependent on pesticides.

In 1997, the International Union of Pure and Applied Chemistry (IUPAC) commission on agrochemicals and the environment prepared a comprehensive technical

C.L. Fuentes (✉) and Z. Arbeli
Facultad Agronomía, Universidad Nacional de Colombia, Sede Bogotá, A. A. 14490 Bogotá D.C., Colombia
e-mail: clfuentesd@unal.edu.co

P. Dion (ed.), *Soil Biology and Agriculture in the Tropics*, Soil Biology 21,
DOI 10.1007/978-3-642-05076-3_12, © Springer-Verlag Berlin Heidelberg 2010

report on the fate of pesticides in tropical soils (Racke et al. 1997). Since most studies in this field come from temperate zones in Europe and North America, Racke et al. have emphasized the need to evaluate the extent to which studies from temperate zones can be extrapolated to the tropics. The commission has made four recommendations: (1) to encourage research on pesticide fate in tropical soils, (2) to increase studies which compare the fate of pesticides in tropical and temperate soils under the same experimental design, (3) to develop and validate models that predict the fate of pesticides in tropical soils, and (4) to encourage the publication of research results on pesticide fate in tropical soils in international, peer-reviewed journals.

In this review, we will follow the roadmap drawn by Racke et al. (1997). Rather than surveying all the literature on pesticide degradation in the tropics, we will focus on the extent to which it is possible to extrapolate results from microbial degradation research in temperate zones to the tropics. To this end, we will discuss the following sub-questions: (1) What is unique about tropical soils and tropical climates? (2) Do the different agricultural practices used in tropical zones influence pesticide degradation patterns? (3) Are bacterial communities in tropical soils different from those in temperate zones due to geographical barriers or environmental conditions, and, if so, do these differences have an influence at the functional level? (4) To what extent do different soils and environmental conditions in the tropics have an influence on microbial pesticide degradation?

12.2 The Tropical Environment

Any discussion on the uniqueness of tropical soil (micro) biology should begin with an examination of the unique character of tropical soils and tropical conditions (see Chaps. 1 and 2; for other reviews, see Cobley and Steele 1984; Kellman 1997; Racke et al. 1997; Juo and Franzluebbers 2003). Tropical environments have been defined in several ways. The most commonly used definition refers to the land mass between the Tropic of Cancer and the Tropic of Capricorn (23.5°N and S). This area accounts for about 40% of the global surface area (Kellman 1997). The term "tropics" also refers to the continuously warm and frost-free zone of the world (Schade 2005). Geographically, the tropics include the whole region of Southeast Asia, Central America, the islands in the South Pacific Ocean and the Caribbean Basin, a large part of Africa, South America and the Indian subcontinent, and a small part of Northern Australia (Juo and Franzluebbers 2003).

Tropical soils and climate conditions are diverse, depending on location and altitude. As compared to temperate zones, the main distinguishing feature of tropical zones is the low seasonal temperature variation. In addition, temperatures in approximately 87% of the tropical area are usually above 25°C, although this depends on elevation. Several climate types exist within the tropics and they can be roughly categorized according to the amount and distribution of precipitation (Schade 2005): (1) the udic regimes, or the humid tropics, comprising about half of all tropical land closest to the equator, which are characterized by large amounts

Table 12.1 Major tropical crops[a]

Cereals	Tubers	Fruits	Vegetables	Industrial
Rice	Cassava	Avocado	Beans	Coffee
Maize	Yams	Pineapple	Capsicums	Cotton
Sorghum	Potato	Guava	Tomato	Sugarcane
Millets	Sweet potato	Mango	Cucurbits	Soybean
		Passion fruit		Cut flowers
		Papaya		Oil palm
		Banana		Rubber
		Plantain		Tobacco
				Coconut

[a]Adapted from: Racke et al. (1997); Hill (2008)

of rainfall all year round, and annual rainfall rates (1,800–5,000 mm) equating or exceeding those of evapotranspiration (Kellman 1997), (2) the ustic regimes, which receive less precipitation, have one or two dry periods per year and one or two wet seasons with high or low rainfall, and include the monsoon regime, and (3) the torric regimes, which are characterized by low precipitation resulting in dry or desert conditions. Seasonal changes in soil microbial activity in the tropics are therefore mainly influenced by the prevailing moisture conditions (Racke et al. 1997; see Chap. 1). Additionally, these different climatic conditions influence the type of crops and pests which are dominant in each zone and therefore they also correlate with pesticide use and degradation rate. The major crops grown in tropical areas are listed in Table 12.1 (Racke et al. 1997; Hill 2008).

12.3 The Agronomical and Environmental Impact of Pesticides

Oerke and Dehne (2004) have estimated the loss potential due to pests for wheat, rice, maize, barley, potato, soybean, sugar beet, and cotton at the global level (Table 12.2). Overall, weeds have the highest loss potential (23%–50%), with animals, pathogens, and viruses being less important (7%–37%, 11%–22%, and 1%–8% respectively). Crop loss potential from pests varies considerably among regions due to climatic conditions and to geographical distribution of pests. The potential crop damages are usually higher in tropical regions, where climatic conditions promote pest development. A variety of less advanced agronomical practices, such as the adaption of foreign susceptible crop species, insufficiency of crop breeding program, and neglect of crop rotations, also enhance pest action (Oerke and Dehne 2004; Bocquene and Franco 2005).

There is constant controversy over the global dependence on agrochemicals, given their excessive use/misuse, their volatility, their long-distance transport with eventual accumulation in colder climates, and their effect on human health (Sibanda et al. 2000; Ecobichon 2001; Wilson and Tisdell 2001; Abhilash and Singh 2009). Disadvantages of pesticide use include damage to agricultural land,

Table 12.2 Estimated potential crop losses worldwide for major crops due to pathogens, viruses, animal pests, and weeds[a]

Crop	Potential losses, expressed as proportion (%) of the attainable yield in the presence of:				
	Fungal and bacterial pathogens	Viruses	Animals	Weeds	Total losses
Wheat	16	3	9	23	51
Rice	16	2	24	35	77
Maize	11	3	15	37	66
Barley	15	3	7	23	48
Potato	22	8	18	23	71
Soybean	11	1	11	37	60
Sugar beet	14	7	12	50	83
Cotton	11[b]		37	34	82

[a]Summarized from Oerke and Dehne (2004)
[b]Value for combined fungal, bacterial and viral pathogens

fisheries, fauna and flora, unintentional destruction of beneficial predators, and human mortality and morbidity (Wilson and Tisdell 2001). In certain parts of the developing world, pesticide poisoning causes more deaths than infectious diseases (Eddleston et al. 2002). The World Health Organization has estimated that one million people are poisoned annually as a result of pesticide exposure, with 20,000 cases resulting in death. Besides acute effects, long-term effects, for example in the form of cancers, might be even more frequent and should be the focus of future research. The risk is especially high for smallholder farmers who use toxic pesticides without adequate knowledge or appropriate protective clothing (Matthews et al. 2003).

In spite of the negative impacts of pesticides, their benefits to agriculture, human development, and disease control cannot be ignored. It is questionable if organic agriculture is sufficiently productive to supply the world's growing food demands (Connor 2008). Most people eat food grown in a system that uses pesticides and many individuals use pesticides in the house or garden (Cooper and Dobson 2007). The benefits of properly used pesticides include (Whitford et al. 2006):

- Controlling agricultural pests and vectors of plant disease.
- Controlling human and livestock disease vectors and nuisance organisms.
- Preventing or controlling organisms that harm human properties.

Thus, there is an urgent need to develop strategies for facilitating the coexistence of productive, healthy, and economically viable agriculture with the preservation of natural resources (Carvalho et al. 1998; Carvalho 2006). In this context, pesticide safety and regulation, proper application technologies, and IPM are some of the key strategies for minimizing human and environmental pesticide risks (Sibanda et al. 2000; Abhilash and Singh 2009). Assessing the ecological risks of intense pesticide use in tropical countries requires knowledge and evaluation of both pesticide exposure and the biological effects on human and environmental health. Consequently, more knowledge on the fate of pesticides in the tropics is essential.

12.4 Pesticide Usage in the Tropics

Since the 1940s, the use of pesticides has grown steadily at about 11% per year (Eddleston et al. 2002), exceeding 2.2 billion kg worldwide in 2000 and 2001 (Kiely et al. 2004). It has been estimated that in 2008 the global pesticide market surged 29% over the 2007 level of $40.7 billion, to $52.4 billion (Specialists in Business Information 2008). Chemical pesticides have also been crucial to equatorial developing nations in their efforts to control and eradicate insects, diseases, and weeds so as to produce adequate food supplies and to protect forests, plantations, and fiber crops (Ecobichon 2001). In most tropical countries, agrochemicals will remain an essential component of agricultural practices for the foreseeable future (Carvalho et al. 1998).

Pesticide use in the tropics differs from use in temperate zones for several reasons, including differences in the climate, socio-economic conditions, and types of crops (Bloomfield et al. 2006). Tropical conditions allow for the juxtaposition of several crop cycles over a 1-year period. Additionally, constantly high temperatures and moisture levels tend to increase populations of insects, fungi, and other pests. These factors may result in more frequent pesticide applications. Indeed, it has been documented that pesticide use increases with an increase in rainfall and temperature (Chen and McCarl 2001). Moreover, differences in the types of crops grown in the tropics imply a difference in pesticide use, but the evaluation of this effect should be made separately for each crop.

As compared to those that are used in many temperate regions, crop protection methods and agricultural practices prevalent in developing countries are less advanced, have poorer scientific basis, and are not always widespread or strictly followed by smallholder farmers. These factors might either enhance or else reduce pesticide use, and consequently their impact is difficult to predict. For example, poor smallholder growers tend to use substantially fewer pesticides, while cotton farmers in Benin may use pesticides at levels up to twice the recommended dose, with applications being done both with excessive frequencies and excessive dosage (Williamson et al. 2008). Less sophisticated IPM, the introduction of foreign varieties with higher yields but more susceptible to pests, and less effective breeding programs might also result in higher pesticide use. On the other hand, human labor costs are often much lower in tropical countries and thus manual weed control is more common, which ultimately results in a significant reduction of herbicide use. Lastly, it should be noted that in developing countries there is probably a higher tendency to use banned pesticides, such as DDT, in a legal (to control malaria) or illegal manner (Williamson et al. 2008; Abhilash and Singh 2009).

The global pesticide market in 2006 is summarized in Table 12.3 with an emphasis on the tropics (Freedonia Group 2005; FAO 2008). Africa and Oceania only contributed 5% and 2.7% of world pesticide imports respectively, while Europe was the region with the largest pesticide market, representing nearly 50% of the world value (Table 12.4). According to the FAO, the pesticide market in tropical countries, as an average, is relatively small. While the tropics account for

Table 12.3 Summary of the global pesticide market, with an emphasis on tropical regions[a]

Region		Import value			Export value		
		1,000 U$	Total world (%)	Total continent (%)	1,000 U$	Total world (%)	Total continent (%)
Africa	Tropical countries	430,975	2.6	51.7	47,383	0.3	25.8
	Total	834,070	5.1		183,301	1.1	
America	Tropical countries	1,904,073	11.7	48.0	725,800	4.3	23.9
	Total	3,963,911	24.4		3,040,256	18.1	
Asia	Tropical countries	1,028,733	6.3	37.8	1,073,253	6.4	33.9
	Total	2,722,119	16.7		3166332	18.9	
Oceania	Tropical countries	21,719	0.1	5.0	74	0.0004	0.1
	Total	433,403	2.7		122,400	0.7	
Europe	Total	8,320,448	51.1		10,239,308	61.1	
Entire world	Tropical countries	3,385,500	20.8		1,845,610	11.0	
	Total	16,273,951			16,751,597		

[a]Adapted and calculated from http://faostat.fao.org/site/423/default.aspx#ancor

approximately 50% of the world's human population, pesticide imports and exports in tropical countries are only 20.8% and 11% of the global market respectively (Table 12.3). This might result in pesticides having less environmental impact in the tropics, although other factors, such as inadequate pesticide management, may have the opposite effect (Jain 1992). In addition, there are several countries with substantial tropical areas that rank among the leading world agrochemical markets. These include India, Brazil, Mexico, Thailand, Colombia, and Costa Rica (Table 12.4). Studies demonstrate that pesticide use is not uniform among tropical countries. Costa Rica, Colombia, and Ecuador apply an average of 51.2, 16.7, and 6.0 kg, respectively, of active ingredient per ha of arable and permanent cropland, such pesticide application rates representing the first, second and fourth highest worldwide, respectively (Table 12.5). On the contrary, African tropical countries, such as Rwanda, Ethiopia, Cameroon, Senegal and Malawi, use the lowest quantities of pesticides (0.1–0.3 kg ha^{-1}) (Table 12.5).

Another important issue is the type of pesticides used. An analysis of the pesticide market shows a discrepancy in pesticide use patterns between the temperate and tropical climatic zones. In the study of 32 temperate countries, herbicides were the most heavily used pesticides in 21 countries, whereas in nine countries fungicides prevailed. Insecticides were the predominant pesticides in only two countries, Pakistan and Turkey (Table 12.6). By contrast, in many tropical countries the most heavily used pesticides are insecticides. Indeed, out of 18 tropical countries, insecticides represented the pesticide class most commonly used in nine countries, followed by fungicides (five countries) and herbicides (four countries) (Table 12.6). In tropical developing countries, hand weeding is more common and herbicide use is often lower than in developed countries. Interestingly, Pakistan and Turkey, which are the two nontropical countries with predominant insecticide use, have some tropical characteristics.

In summary, the amount and types of pesticides can vary substantially between tropical and nontropical countries, and these variations should be reflected in pesticide risk assessments and pest control strategies. Since the speed of pesticide

Table 12.4 Pesticide market for selected tropical countries in 2006[a]

Country	Import value (1,000 US$)	Export value (1,000 US$)
Tropical African countries		
Angola	2,700	100
Benin	8,613	2
Botswana	3,564	35
Burkina Faso	10,212	217
Burundi	2,134	7
Cameroon	32,957	189
Cape Verde	1,130	8
Central African Republic	352	
Chad	600	
Comoros	110	
Congo	2,500	5
Congo, Democratic Republic of	1,700	
Djibouti	200	
Ethiopia	11,488	1
Gabon	3,764	25
Gambia	960	
Ghana	81,501	77
Guinea	3,359	
Guinea-Bissau	100	
Ivory Coast	34,291	12,541
Kenya	39,250	5,750
Liberia	1,600	
Malawi	11,107	41
Mali	16,744	132
Mauritania	385	
Mozambique	17,749	30
Namibia	5,744	333
Niger	2,622	157
Nigeria	31,536	1,170
Réunion	12,200	220
Rwanda	1734	
Senegal	8,925	11,120
Sierra Leone	670	
Somalia	350	
Sudan	17,610	23
Tanzania, United Republic of	21,003	81
Togo	4,434	30
Uganda	3	14,674
Zambia	17,954	414
Zimbabwe	17,120	1
Sub-total	430,975	47,383
Tropical Asian countries		
Bangladesh	28,243	7
Brunei Darussalam	2,331	3
Cambodia	1,058	3
India	6,440	635,131
Indonesia	95,136	89,889
Lao People's Democratic Republic	118	
Malaysia	75,695	128,408

(*continued*)

Table 12.4 (continued)

Country	Import value (1,000 US$)	Export value (1,000 US$)
Maldives	863	
Myanmar	11,500	
Oman	10,291	2,502
Philippines	119033	1649
Singapore	104,218	136,397
Sri Lanka	22,200	885
Thailand	344,522	66,270
Timor-Leste	59	3
Vietnam	199,359	12,066
Yemen	7,667	40
Sub-total	1,028,733	1,073,253
Tropical countries in the Americas		
Antigua and Barbuda	2,196	17
Bahamas	2,969	43
Barbados	5,466	11,899
Belize	7,465	
Bermuda	1,600	
Bolivia	79,714	5
Brazil	516,941	242,235
British Virgin islands	450	
Cayman islands	800	
Colombia	146,338	178,312
Costa Rica	144,341	65,109
Cuba	42,915	820
Dominican Republic	967	1,817
Ecuador	129,101	9,971
El Salvador	35,784	832
French Guiana	3,000	15
Grenada	1,009	4
Guadeloupe	11,000	120
Guatemala	84,096	69,781
Guyana	5,020	220
Haiti	1,500	
Honduras	52,917	815
Jamaica	14,982	909
Martinique	19,000	100
Mexico	365,890	107,843
Montserrat	147	
Netherlands Antilles	2,000	
Nicaragua	30,991	602
Panama	41,681	132
Peru	66,542	11,700
Saint Kitts and Nevis	1,003	2
Saint Lucia	2,899	100
Saint Vincent and Grenadines	2,410	
Suriname	4,203	
Trinidad and Tobago	10,463	4,057
US Virgin islands	250	
Venezuela	65,056	16,523
Sub-total	1,904,073	725,800

(*continued*)

Table 12.4 (continued)

Country	Import value (1,000 US$)	Export value (1,000 US$)
Tropical Oceanian countries		
Cook Islands	288	
Fiji	3,927	7
French Polynesia	6,238	6
Guam	320	20
Kiribati	148	
New Caledonia	5,250	22
Papua New Guinea	2,700	13
Samoa	1,301	6
Solomon Islands	290	
Tonga	577	
Vanuatu	600	
Wallis and Futuna Islands	80	
Sub-total	21,719	74
Total tropical world	3,385,500	1,846,510

[a]Adapted from: http://www.faostat.fao.org/site/423/default.aspxancor

degradation depends on the type, rate, and frequency of pesticide application (Arbeli and Fuentes 2007), it is recommended that studies on pesticide degradation differences between tropical and temperate zones take pesticide use patterns into account.

12.5 Biogeography of Bacteria

Bacteria are often considered as the major catalysts of pesticide degradation. In the following sections we will compare microbial degradation processes as they occur in the tropics and in temperate zones. A central question will be whether unique microbial populations exist in tropical soils. Ramette and Tiedje (2007) have defined prokaryotic biogeography as "the science that documents the spatial distribution of prokaryotic taxa in the environment at local, regional, and continental scales." In the current discussion on pesticide degradation in the tropics, this definition will be expanded, as proposed by Green et al. (2008), to include the spatial distribution of functional traits. Our discussion of pesticide degradation will be limited to large-scale geographic distribution, as it will deal with tropical vs temperate zones. Its object will be to examine to what extent geographical barriers or distance (and therefore life history) and environmental conditions affect the structure and function of microbial communities involved in pesticide degradation.

12.5.1 Does Everything Have the Potential to Be Everywhere?

Until recently, prokaryotic biogeography research concentrated on the basic question: "does every micro-organism have the potential to be everywhere?" (Whitfield 2005). On the one hand, it is evident that global transport of micro-organisms is

Table 12.5 Average pesticide use in selected countries, calculated in kg of active ingredient per total ha of arable and permanent cropland, 2000[a]

Position	Country[b]	Kg active ingredient ha^{-1}
1	Costa Rica (T)	51.2
2	Colombia (T)	16.7
3	Netherlands	9.4
4	Ecuador (T)	6.0
5	Portugal	5.3
6	France	4.6
7	Greece	2.8
8	Uruguay	2.7
9	Suriname (T)	2.6
10	Germany	2.5
11	Honduras (T)	2.5
12	Austria	2.4
13	Dominican Republic (T)	2.1
14	Ireland	1.8
15	Slovakia	1.8
16	Paraguay	1.5
17	Jordan	1.4
18	Denmark	1.4
19	Pakistan	1.3
20	Czech Republic	1.3
21	Turkey	1.3
22	Peru (T)	1.2
23	Oman (T)	1.2
24	Thailand (T)	1.1
25	Romania	1
26	Yemen (T)	0.8
27	Sweden	0.6
28	Poland	0.5
29	Finland	0.5
30	Bangladesh (T)	0.4
31	Norway	0.4
32	Malawi (T)	0.3
33	Lithuania	0.2
34	Senegal (T)	0.2
35	Cameroon (T)	0.1
36	Bhutan	0.1
37	Estonia	0.1
38	Iraq	0.1
39	Ethiopia (T)	0.1
40	Rwanda (T)	0.1
General average		3.29

[a]Summarized from: World Resources Institute. http://www.nationmaster.com/red/graph/agr_pes_use-agriculture-pesticide-use&int=-1
[b]*T*: Tropical country

much easier than that of most macro-organisms. Due to their small size, their high numbers, and their ability to form stable dormant forms, micro-organisms can be easily transported across the planet (Finlay 2002; Kellogg and Griffin

Table 12.6 Pesticide consumption in selected tropical and nontropical countries, 2001. Data refer to the quantity of pesticides used by, or sold to the agricultural sector, in metric tons of active ingredients[a]

Country	Pesticide consumption (metric tons of active ingredients)			
	Fungicides and bactericides	Herbicides	Insecticides	Total
Tropical countries				
Belize	155	365	405	925
Brazil	9,527	44,620	21,544	75,691
Burundi	0	13	22	35
Cameroon	451	117	113	681
Cape Verde	1		4	5
Costa Rica	5,237	2,980	1,468	9,685
Dominican Republic	3,069	2,904	900	6,873
Ecuador	1,702	3,068	1,034	5,804
Eritrea			28	28
Ethiopia	15	454	153	622
Kenya	711	562	303	1,576
Madagascar	41	22	22	85
Mauritius	177	1,042	605	1,824
Nicaragua	600	1,452	1,904	3,956
Niger	0		31	31
Rwanda	1	1	69	71
Senegal	7	76	198	281
Vietnam	5,391	3,982	9,661	19,034
Nontropical countries				
Austria	1,336	1,436	99	2871
Belgium	2,302	4,908	892	8,102
Czech Republic	1,050	2,590	157	3,797
Denmark	538	2,313	71	2,922
Estonia	12	298	3	313
Finland	192	1,120	35	1,347
France	54,130	32,122	2,487	88,739
Germany	7,912	14,942	1,255	24,109
Greece	4,860	2,650	2,638	10,148
Hungary	1,637	3,149	298	5,084
Iceland	0	5	1	6
Ireland	429	1,405	65	1,899
Italy	48,221	9,982	8,874	67,077
Lithuania	102	533	7	642
Malta	68	18	23	109
Netherlands	3,951	3,093	275	7,319
New Zealand	616	2,269	249	3,134
Norway	119	378	11	508
Pakistan	229	1,030	10,611	11,870
Paraguay	643	7,389	1,706	9,738
Poland	2,815	4,748	549	8,112
Portugal	11,562	2,235	415	14,212
Romania	2,802	3,960	1,110	7,872
Slovakia	537	2,136	175	2,848
Slovenia	921	362	81	1,364
Sweden	209	1,435	21	1,665
Switzerland	706	656	93	1,455

(continued)

Table 12.6 (continued)

Country	Pesticide consumption (metric tons of active ingredients)			
	Fungicides and bactericides	Herbicides	Insecticides	Total
Syrian Arab Republic	1,248	705	994	2,947
Turkey	4,046	5,964	13,169	23,179
United Kingdom	4,962	22,841	1,596	29,399
Uruguay	942	2,812	187	3,941

[a]Adapted and calculated from: http://faostat.fao.org/site/423/default.aspxancor

2006; Griffin 2007). A good example that supports the famous saying of Baas Becking — "Everything is everywhere, but nature selects" — is that of thermophiles, which are found easily in cold soils, where they are inactive unless temperature increases above 40°C. These bacteria have been found in both air and rain samples, suggesting a mechanism for their transport to cold soils (Marchant et al. 2008). In accordance with this finding, there are numerous examples of cosmopolitan prokaryotes (Ramette and Tiedje 2007) and recent studies suggest that environmental conditions, especially pH (Fierer and Jackson 2006; Fierer et al. 2007) and salinity (Lozupone and Knight 2007), are far more important than geographical distance in controlling bacterial diversity and community structure. However, microbial endemism was observed in hot springs (Papke et al. 2003; Whitaker et al. 2003) and pristine soils (Cho and Tiedje 2000; Fulthorpe et al. 2003). Again in apparent contradiction with the microbial ubiquity concept, it was noted that distance between habitats, and therefore limitations in microbial dispersal, is an important factor in shaping microbial communities, at least at the sub-species level. Yet another departure from the ubiquity notion is that specialized microbial populations are found in air environments (Fierer et al. 2008; Tringe et al. 2008), and it is likely that microbial taxa vary in their dispersal abilities according to their population size and their ability to form stable dormant forms, colonize new habitats, and resist extinction (Martiny et al. 2006). Thus, forces may well be at play to shape microbial populations, but the relative importance of geographical distance, dispersal barriers, and environmental heterogeneity in delineating microbial biogeographic patterns remains difficult to assess (Foissner 2006; Fierer 2008).

12.5.2 *The Phenotypic Dimension*

Fenchel (2003) has argued that the studies suggesting that microbial endemism exists (Cho and Tiedje 2000; Fulthorpe et al. 2003; Papke et al. 2003; Whitaker et al. 2003) were limited to genetic analysis and that they may only reflect neutral or nearly neutral mutations. Nevertheless, the above-cited studies do show that the diversification rate of some bacteria is greater than the rate of worldwide microbial mixing and that these changes may well have biological meaning in terms of local adaptation. In fact, it has been shown that local adaptation exists even on a much smaller scale, such as 1 ha of old-growth forest (Belotte et al. 2003). By impeding generalized colonization, local adaptation would establish an ecological barrier to bacterial mixing.

12.5.3 Microbial Diversity

While discussing microbial biogeography in relation to tropical soils, one cannot avoid wondering whether microbial diversity is higher in the tropics, as is the case for macro-organisms. Since many bacteria are specifically associated with other organisms such as plant and animals (Andrews and Harris 2000; Graystona et al. 2001; Marschner et al. 2001; Ehrenfeld et al. 2005; Lambais et al. 2006; Godoy-Vitorino et al. 2008; Ohkuma 2008), a higher diversity of macro-organisms may impact on microbial diversity. Nevertheless, it is yet to be demonstrated that soil micro-organisms are more diverse in the tropics, especially as regards agricultural soils devoid of native vegetation.

In an analysis of 98 soil samples taken across North and South America, Fierer and Jackson (2006) used a ribosomal DNA-fingerprinting method to compare bacterial community composition and diversity. Bacterial diversity was unrelated to site temperature and latitude, and community composition was largely independent of geographic distance. The diversity, richness, and microbial community differed according to ecosystem type, and these differences could largely be explained by soil pH. Bacterial diversity was highest in neutral soils and lowest in acidic soils, with soils from the Peruvian Amazon being the most acidic and least diverse in the study. In keeping with this, a meta-analysis of richness gradients across latitudes for 581 eukaryotes (including unicellular organisms) revealed that the strength and slope of the latitude-dependent diversity gradient is inversely related to body mass (Hillebrand 2004).

Conversely, other studies have found a significant increase in the richness of oceanic bacterial plankton from the poles towards the tropics (Pommier et al. 2007; Fuhrman et al. 2008), indicating that marine bacteria behave similarly to macro-organisms in that they demonstrate higher diversity in tropical zones. Thus, it appears that latitude exerts more influence on microbial communities colonizing the relatively homogenous marine environment than on those that inhabit the highly heterogeneous soil environments.

It should be stressed that the link between microbial diversity and function is not yet well established and thus it is not clear whether a putative higher microbial diversity in the tropics could account for possible differences in pesticide degradation rates or pathways. Studies with macro-organisms have evidenced a positive relationship between biodiversity and ecosystem functionality (Hector et al. 1999; Cardinale et al. 2006), up to a saturation level beyond which further increases in biodiversity do not impact functionality, presumably because of functional redundancy (Srivastava and Velland 2005). The apparent value of this saturation level rises as the analysis includes an increasing number of functions (Hector and

Bagchi 2007). In the case of micro-organisms, it is assumed that both species and functional diversities are very high, and recent studies which attempt to elucidate the functional relevance of microbial diversity are sometimes contradictory (van der Heijden et al. 1998; Naeem et al. 2000; Bell et al. 2005; Girvan et al. 2005; Reed and Martiny 2007; Wertz et al. 2007; Kemmitta et al. 2008). Hence, it appears difficult to correlate variations in microbial diversity with the capacity of an agricultural ecosystem to perform ecological services, let alone pesticide degradation.

Biodegrading bacterial strains or biodegradation genes with very high similarity (98%–100%) have often been isolated from distant geographical zones, including temperate and tropical soils. Among these are: 2,4-D degradation genes (*tfd*), which were isolated from Australia, the US, Canada, Denmark, India, and China (Bhat et al. 1994; McGowan et al. 1998; Jacob Baelum personal communication); hexachlorocyclohexane degradation genes (*lin*), which were isolated from Japan, India, France, Spain and Germany (Dogra et al. 2004; Boltner et al. 2005); and parathion degradation genes (*opd*), which were isolated from the Philippines, the US, and India (Mulbry and Karns 1989; Serdar et al. 1989; Somara et al. 2002). Biodegradation of atrazine is one of the best documented cases of recent evolution and dispersion of new catabolic genes (Seffernick and Wackett 2001; Shapir et al. 2007). These genes have been found in Europe (Rousseaux et al. 2001; Piutti et al. 2003), North America (de Souza et al. 1998; Topp et al. 2000a,b), New Zealand (Aislabie et al. 2005), China (Cai et al. 2003), Japan (Satsuma 2006), and Colombia (Arbeli and Fuentes, in preparation).

The factors that are most likely to reduce differences in pesticide-degrading bacteria and pesticide degradation genes across distinct geographical regions include: (1) agricultural practices such as vegetation removal, soil tillage and commercial seed distribution, which probably enhance aerial microbial transport (Lighthart 1984 and references therein), (2) the integration of many degradation genes in transposable elements; this may enhance the mobility of degradation genes through their chance association with bacterial hosts exhibiting good transport abilities and/or good local adaptation, and (3) convergent evolution occurring under strong selective pressure, coincident with the independent appearances, at different times and places, of similar mutations leading to improved degradation ability (Senior et al. 1976).

12.6 The Effects of Tropical Conditions

12.6.1 The Effects of Tropical Conditions on Microbial Populations

From the discussion above it appears that any difference between microbial communities in tropical and temperate agricultural soils would be the result of fine-tuning processes. Bacterial communities from these two types of environments may well comprise bacteria of the same species, with very similar genes, with only subtle differences. The abundance of each species or subspecies in the community would be different from soil to soil, but no common denominator among microbial communities from tropical soils would differentiate these from microbial communities in temperate soils. On the contrary, it seems that under similar soil conditions, microbial communities from the tropics are similar to those in temperate zones (Fierer and Jackson 2006; Fierer et al. 2007; Lozupone and Knight 2007).

12.6.2 The Effects of Tropical Conditions on Microbial Activity

The two principal parameters defining tropical conditions, temperature and soil moisture (see above), exert a strong influence on microbial activity (Lavelle et al. 1993; Linshi et al. 2003; Pietikäinen et al. 2005; Cook and Orchard 2008), and particularly on pesticide degradation (Walker 1978; Ghadiri et al. 1995; Awasthi et al. 2000; Alletto et al. 2006; Rodríguez Cruz et al. 2008). In fact, evapotranspiration, which depends on both temperature and humidity, is often considered as the best predicting parameter for litter decomposition in soils (Meentemeyer 1978; Berg et al. 1993; Aerts 1997). According to the model of Walker and Eagle (1983) (as described in Bailey 2004), an increase in temperature of 10°C, or an increase in soil moisture content (as a percentage of field capacity) by a factor of 2, decreases pesticide half-life in the soil by a factor of 2. Obviously, this relationship remains only a rough estimate, and a more accurate measure of temperature influence on pesticide degradation in soil may be found elsewhere (EFSA 2007). Since temperature is relatively constant in the tropics, the moisture regime should be the main controlling factor for microbial activity (Cook and Orchard 2008), including pesticide degradation activity. In addition, it should be kept in mind that other factors such as soil pH (Suett et al. 1996; Houot et al. 2000; Bending et al. 2003) may also influence the pesticide degradation rate.

Leaf litter decomposition in the humid tropics is considered to be faster and more complete than in temperate zones (Lavelle et al. 1993) and pesticide degradation probably follows a similar pattern (Racke et al. 1997). In other tropical regimes, microbial activity is expected to be controlled by soil humidity (Davidson et al. 2000), so that in aridic tropical regimes microbial activity might be lower than in humid temperate zones. Thus, it is difficult to generally predict pesticide degradation rates and these should be modeled or tested independently in any particular region of interest. From the considerations above, it may be concluded that, under the same environmental conditions, there is no significant difference in either kinetics or pathways of pesticide degradation in soils from tropical and temperate zones (Racke et al. 1997). Generally, however, the acceleration of microbial degradation, volatilization and chemical degradation by high temperatures contributes to make dissipation of pesticides significantly faster under tropical conditions (Simon et al. 1992; Racke et al. 1997). The combined effect of these stimulations of pesticide degradation may be viewed positively from an environmental, but not necessarily a crop protection, perspective (Arbeli and Fuentes 2007).

When discussing pesticide degradation, it is essential also to consider microbial community turnover (Arbeli and Fuentes 2007). This factor counteracts the degrader population buildup that would occur following regular applications of a particular pesticide: insofar as microbial succession would occur faster under tropical conditions, populations of pesticide degraders would be less stable in the tropics, impeding population buildup and the ensuing accelerated degradation of the regularly used pesticide. Conversely, in the case where microbial survival would be enhanced under the relatively constant tropical conditions, increased persistence of the degrading microbial populations may allow for accelerated degradation upon repeated pesticide use. Obviously, the rate of microbial turnover might differ in udic, ustic, and torric regimes.

Genetic evolution may bear special relevance to the degradation of pesticides, which, as xenobiotics, often have unusual structures. The degradation rate of recently commercialized pesticides may be rather slow, and in some cases there is evidence of the evolution of new degradation genes (Senior et al. 1976; Poelarends et al. 2000; Seffernick and Wackett 2001). Evolution and speciation rates have been predicted to be higher in the tropics (Rohde 1992; Cardillo 1999). This hypothesis is based on the assumption that at higher temperatures the generation times are reduced, mutation rates are higher, and selection is faster (Rohde 1992; Allen et al. 2006). Thus, although pesticide degradation genes are usually discovered in temperate zones, they may well have originated in the tropics.

12.7 Pesticide Dissipation Under Tropical and Temperate Conditions

Pesticide dissipation from soil depends on various mechanisms, such as volatilization, leaching, surface runoff, photo- or chemical reaction, and microbial degradation. Microbial degradation is often considered to be the predominant mechanism, but abiotic processes may be comparatively more important under certain circumstances. The particular rate of each mechanism depends on environmental conditions (e.g., temperature, moisture), soil properties (e.g., pH, organic matter content, clay content, biomass, redox potential), pesticide type (e.g., chemical structure, water solubility, sorption to soil), and pesticide application strategy (e.g., application rate and frequency, formulation).

12.7.1 Influence of Soil Properties on Pesticide Dissipation

Pesticide dissipation rates are best studied in the laboratory under controlled conditions. Soils are not uniform, and therefore it is difficult to generalize particular results on pesticide dissipation. For example, half-lives of chloropyrifos and flumetsulam in North American soils range from 10 to 325 days ($n = 24$ soils) (Lehman et al. 1992) and 13 to 130 days ($n = 21$ soils) (Laskowski 1995) respectively. Similarly, mineralization studies in our laboratory have demonstrated high variability in glyphosate half-lives in Colombian tropical soils, ranging from 40 days to several years (Cuervo 2007; Figueroa del Castillo 2008; Melgarejo Prieto 2008), and Smith et al. (2006) reported that the half-lives of carbofuran and diazinon in Sri Lankan tropical soil ranged from 87 to 622 and from 276 to 2,302 days respectively. This high natural variance demonstrates the difficulty of comparing pesticide dissipation results between tropical and temperate soils (Racke et al. 1997).

Racke et al. (1997) reviewed cases in which pesticide degradation was compared between tropical and temperate soils under the same experimental conditions in the same laboratory. Under these conditions, no clear differences were observed in the degradation rates of fenamiphos (in seven tropical and nine temperate soils from America, Asia and Europe (Simon et al. 1992)), atrazine (in soils from Thailand and Japan (Korpraditskul et al. 1992, 1993)), and simazine (in three tropical and 13 temperate soils from different countries (Walker et al. 1983)). Similarly, soil properties and environmental conditions were much more important than geographical location (Australia, India, and Ecuador) in determining the fenamiphos dissipation rate (Cáceres et al. 2008) and the sorption of diuron, imidacloprid, and thiacloprid by soil (Oliver et al. 2005). All these studies, however, have clearly showed that these pesticides were dissipated more rapidly in all soils (temperate and tropical) with an increase in the incubation temperature.

12.7.2 Influence of the Climate on Pesticide Dissipation

As discussed earlier, temperature and moisture are primordial factors that influence microbial activity and pesticide degradation. In addition, higher temperatures enhance pesticide volatilization and the rates of chemical reactions (hydrolysis, oxidation, and reduction), while the extent of leaching varies with precipitation, and the extent of photochemical reactions with sun radiation (reviewed by Racke et al. 1997). These relationships suggest that under field conditions the pesticide dissipation rate should be faster in the tropics, and field studies on pesticide dissipation rates support these predictions. In a set of coordinated studies financed by the International Atomic Energy Agency that included 17 soils from 14 tropical and subtropical countries, half-lives of DDX (DDT + DDE + DDD) ranged from 22 to 365 days with the single exception of an acidic (pH 4.5) Brazilian soil where the half-life was above 672 days (Hassan 1994). In comparison, half-lives for DDT in temperate regions ranged from 837 to 6,087 days (Racke et al. 1997). Another comprehensive study concluded that simazine half-lives in all four tropical soils investigated (from Indonesia, The Philippines, and Taiwan) were shorter than 20 days, while the half-lives in 17 temperate soils generally ranged from 20 to more than 120 days, although some exceptionally short half-lives of less than 20 days were also observed (Walker et al. 1983). Laabs et al. (2002) reported that the dissipation rate of ten pesticides in two tropical Brazilian soils were typically five to ten times faster than in temperate climates, and Karlsson et al. (2000) observed a relatively fast degradation rate of chlorinated pesticides in Lake Malawi. More generally, Karlsson et al. (2000) suggested that the tropics can contribute to global pesticide elimination because of the rapid removal process, while Sarkar et al. (2008) suggested that pesticide bioaccumulation in tropical India is slower in comparison to temperate zones due to higher dissipation rates and a higher elimination rate by the biota.

12.7.3 Influence of Agricultural Practice on Pesticide Dissipation

Agricultural practice is not uniform across the tropics. In general, different crops and regulations may imply a use of distinct pesticides. Pesticide structure determines volatility, solubility, tendency to adsorb to soil, and persistence to degradation. For example, to the extent that greater amounts of persistent chlorinated pesticides are used in the tropics, (Williamson et al. 2008; Abhilash and Singh 2009), the overall pesticide dissipation rate would be slower in tropical regions. On the other hand, higher frequency and quantity of pesticide applications in the tropics may cause higher bioavailability and faster adaptation of microbial communities to the pesticides in use. This can accelerate dissipation rate, but it could also imply a higher accumulation of pesticides in the environment.

12.8 Conclusions

It appears that pesticide dissipation rates are faster under tropical humid conditions than under temperate conditions, but that the mechanisms involved are similar. Although pesticides might behave differently under different environmental conditions, their behavior in tropical regions can be roughly predicted by using models based on temperate climate studies. However, many uncertainties remain with regard to the consequences of higher pesticides application rates and other differentiating factors, making it essential to continue studying pesticide behavior under tropical conditions, to document the concentrations of pesticides and their degradation products in the environment, including ground water and agricultural products, and finally to examine pesticide exposure pathways, particularly through drinking water and food. The objective of these studies would be to adapt environmental risk assessments and pest control programs to tropical conditions.

Recommendations on pesticide application in the tropics are often adapted from studies in temperate countries. However, these recommendations may not be optimal if dissipation rates are faster. Thus, in order to insure adequate crop protection while minimizing pesticide use and risks, it is recommended that integrated pest control studies under tropical conditions include knowledge of local pesticide dissipation rates.

Pesticide degradation in the tropics is of global concern. Prediction models of global pesticide accumulation are often based on parameters obtained from temperate zone studies. For example, Woodwell et al. (1971) assumed in their global model that the mean residence time of DDT in soils was 4.5 years, although this residence time could actually be much less in 40% of the total land area, corresponding to the tropical regions where, according to the same authors, DDT remains most heavily used.

Acknowledgments We are thankful to Sebastian Reinhold Sørensen, Matthew Robert Alexander and Patrice Dion for their helpful comments on the manuscript. This work was supported by the Dirección de Investigación — Sede Bogotá (DIB) and the Escuela de Posgrados, Facultad de Agronomía, Universidad Nacional de Colombia.

References

Abhilash PC, Singh N (2009) Pesticide use and application: an Indian scenario. J Hazard Mater 165:1–12

Aerts R (1997) Climate, leaf litter chemistry and leaf litter decomposition in terrestrial ecosystems: a triangular relationship. Oikos 79:439–449

Aislabie J, Bej AK, Ryburn J, Lloyd N, Wilkins A (2005) Characterization of *Arthrobacter nicotinovorans* HIM, an atrazine-degrading bacterium, from agricultural soil New Zealand. FEMS Microbiol Ecol 52:279–286

Allen AP, Gillooly JF, Savage VM, Brown JH (2006) Kinetic effects of temperature on rates of genetic divergence and speciation. Proc Natl Acad Sci USA 103:9130–9135

Alletto L, Coquet Y, Benoit P, Bergheaud V (2006) Effects of temperature and water content on degradation of isoproturon in three soil profiles. Chemosphere 64:1053–1061

Andrews JH, Harris RF (2000) The ecology and biogeography of microorganisms on plant surfaces. Annu Rev Phytopathol 38:145–80

Arbeli Z, Fuentes CL (2007) Accelerated biodegradation of pesticides: an overview of the phenomenon, its basis and possible solutions; and a discussion on the tropical dimension. Crop Prot 26:1733–1746

Awasthi N, Ahuja R, Kumar A (2000) Factors influencing the degradation of soil-applied endosulfan isomers. Soil Biol Biochem 32:1697–1705

Bailey SW (2004) Climate change and decreasing herbicide persistence. Pest Manag Sci 60: 158–162

Bell T, Newman JA, Silverman BW, Turner SL, Lilley AK (2005) The contribution of species richness and composition to bacterial services. Nature 436:1157–1160

Belotte D, Curien JB, Maclean RC, Bell G (2003) An experimental test of local adaptation in soil bacteria. Evolution 57:27–36

Bending GD, Lincoln SD, Sørensen SR, Morgan JAW, Aamand J, Walker A (2003) In-field spatial variability in the degradation of the phenyl–urea herbicide isoproturon is the result of interactions between degradative *Sphingomonas* spp. and soil pH. Appl Environ Microbiol 69:827–834

Berg B et al (1993) Litter mass loss rates in pine forests of Europe and Eastern United States: some relationship with climate and litter quality. Biogeochemistry 20:127–159

Bhat MA, Tsuda M, Horiike K, Nozaki M, Vaidyanathan CS, Nakazawa T (1994) Identification and characterization of a new plasmid carrying genes for degradation of 2, 4-dichlorophenoxyacetate from *Pseudomonas cepacia* CSV90. Appl Environ Microbiol 60:307–312

Bloomfield JP, Williams RJ, Gooddy DC, Cape JN, Guha P (2006) Impacts of climate change on the fate and behaviour of pesticides in surface and groundwater — a UK perspective. Sci Total Environ 369:163–177

Bocquene G, Franco A (2005) Pesticide contamination of the coastline of Martinique. Mar Pollut Bull 51:612–619

Boltner D, Moreno-Morillas S, Ramos JL (2005) 16S rDNA phylogeny and distribution of *lin* genes in novel hexachlorocyclohexane-degrading *Sphingomonas* strains. Environ Microbiol 7:1329–1338

Cáceres T, Megharaj M, Naidu R (2008) Degradation of fenamiphos in soils collected from different geographical regions: the influence of soil properties and climatic conditions. J Environ Sci Health B43:314–322

Cai B, Han Y, Liu B, Ren Y, Jiang S (2003) Isolation and characterization of an atrazine-degrading bacterium from industrial wastewater in China. Lett Appl Microbiol 36:272–276

Cardillo M (1999) Latitude and rates of diversification in birds and butterflies. Proc R Soc Lond B 266:1221–1225

Cardinale BJ, Srivastava DS, Duffy JE, Wright JP, Downing AL, Sankaran M, Jouseau C (2006) Effects of biodiversity on the functioning of trophic groups and ecosystems. Nature 443: 989–992

Carvalho FP (2006) Agriculture, pesticides, food security and food safety. Environ Sci Policy 9:685–692

Carvalho FD, Nhan D, Zhong C, Tavares T, Klaine S (1998) Tracking pesticides in the tropics. IAEA bulletin 40:24–30. http://www.iaea.org/Publications/Magazines/Bulletin/Bull403/40305692430.pdf. Accessed 6 Oct 2008

Chen CC, McCarl BA (2001) An investigation of the relationship between pesticide usage and climate change. Clim Change 50:475–487

Cho JC, Tiedje JM (2000) Biogeography and degree of endemicity of fluorescent *Pseudomonas* strains in soil. Appl Environ Microbiol 66:5448–5465

Cobley LS, Steele WM (1984) An introduction to the botany of tropical crops. Longman, New York

Connor DJ (2008) Organic agriculture cannot feed the world. Field Crops Res 106:187–190

Cook FJ, Orchard VA (2008) Relationships between soil respiration and soil moisture. Soil Biol Biochem 40:1013–1018

Cooper J, Dobson H (2007) Pesticides and humanity: the benefits of using pesticides. Natural Resources Institute, University of Greenwich, UK

Cuervo JL (2007) Interacción del glifosato (Roundup®) con la biota del suelo y comportamiento de este herbicida en tres suelos del Tolima-Colombia, bajo condiciones controladas. Doctoral Thesis, Facultad Agronomía, Universidad Nacional de Colombia Sede Bogotá

Davidson EA, Verchot LV, Cattânio JH, Ackerman IL, Carvalho JEM (2000) Effects of soil water content on soil respiration in forests and cattle pastures of eastern Amazonia. Biogeochemistry 48:53–69

de Souza ML, Seffernick J, Martinez B, Sadowsky MJ, Wackett LP (1998) The atrazine catabolism genes *atzABC* are widespread and highly conserved. J Bacteriol 180:1951–1954

Dogra C et al (2004) Organization of *lin* genes and IS6100 among different strains of hexachlorcyclohexane degrading *Sphingomonas paucimobilis* strains: evidence of natural horizontal transfer. J Bacteriol 186:2225–2235

Ecobichon DJ (2001) Pesticide use in developing countries. Toxicology 160:27–33

Eddleston M et al (2002) Pesticide poisoning in the developing world — a minimum pesticides list. Lancet 360:1163–1167

EFSA (European Food Safety Authority) (2007) Opinion on a request from EFSA related to the default *Q*10 value used to describe the temperature effect on transformation rates of pesticides in soil. EFSA J 622:1–32
Ehrenfeld JG, Ravit B, Elgersma K (2005) Feedback in the plant–soil system. Annu Rev Env Resour 30:75–115
FAO (2008) — FAOSTAT, ResourceSTAT-Pesticides Trade. http://www.faostat.fao.org/site/423/default.aspx#ancor. Accessed 13 Aug 2008
Fenchel T (2003) Biogeography for bacteria. Science 301:295–296
Fierer N (2008) Microbial biogeography: patterns in microbial diversity across space and time. In: Zengler K (ed) Accessing uncultivated microorganisms: From the environment to organisms and genomes and back. ASM Press, Washington, DC, pp 95–115
Fierer N, Jackson R (2006) The diversity and biogeography of soil bacterial communities. Proc Natl Acad Sci USA 103:626–631
Fierer N, Morse J, Berthrong S, Bernhardt ES, Jackson RB (2007) Environmental controls on the landscape-scale biogeography of stream bacterial communities. Ecology 88(9):2162–2173
Fierer N, Liu Z, Rodríguez-Hernández M, Knight R, Henn M, Hernandez MT (2008) Short-term temporal variability in airborne bacterial and fungal populations. Appl Environ Microbiol 74:200–207
Figueroa del Castillo L (2008) Potencial de degradación bacteriano de 14C glifosato en tres suelos del Tolima, Colombia sometidos a diferentes uso. M.Sc. Thesis, Facultad Agronomía, Universidad Nacional de Colombia Sede Bogotá
Finlay BJ (2002) Global dispersal of free-living microbial eukaryote species. Science 296: 1061–1063
Foissner W (2006) Biogeography and dispersal of micro-organisms: a review emphasizing protists. Acta Protozool 45:111–136
Freedonia Group (2005) World Pesticides to 2009 — demand and sales forecast, market share, market size, market leaders. Study no 1927
Fuhrman JA, Steele JA, Hewson I, Schwalbach MS, Brown MV, Green JL, Brown JH (2008) A latitudinal diversity gradient in planktonic marine bacteria. PNAS 105:7774–7778
Fulthorpe RR, Rhodes AN, Tiedje JM (2003) High levels of endemicity of 3-chlorobenzoate-degrading soil bacteria. Appl Environ Microbiol 64:1620–1627
Ghadiri H, Rose CW, Connell DW (1995) Degradation of organochlorine pesticides in soils under controlled environment and outdoor conditions. J Environ Manage 43:141–151
Godoy-Vitorino F et al (2008) Bacterial community in the crop of the Hoatzin, a neotropical folivorous flying bird. Appl Environ Microbiol 74:5905–5912
Girvan MS, Campbell CD, Killham K, Prosser JI, Glover LA (2005) Bacterial diversity promotes community stability and functional resilience after perturbation. Environ Microbiol 7:301–313
Graystona SJ, Griffith GS, Mawdsleyb JL, Campbella CD, Bardgettc RD (2001) Accounting for variability in soil microbial communities of temperate upland grassland ecosystems. Soil Biol Biochem 33:533–551
Green JL, Bohannan BJM, Whitaker RJ (2008) Microbial biogeography: from taxonomy to traits. Science 320(5879):1039–1043
Griffin DW (2007) Atmospheric movement of microorganisms in clouds of desert dust and implications for human health. Clin Microbiol Rev 20:459–477
Hassan A (1994) Appraisal of overall program accomplishments. J Environ Sci Health B29: 205–226
Hector A, Bagchi R (2007) Biodiversity and ecosystem multifunctionality. Nature 448:188–191
Hector A et al (1999) Plant diversity and productivity experiments in European grasslands. Science 286:1123–1127
Hill DS (2008) Pests of crops in warmer climates and their control. Springer, Netherlands, p 750
Hillebrand H (2004) On the generality of the latitudinal diversity gradient. Am Nat 163:192–211
Houot S, Topp E, Yassir A, Soulas G (2000) Dependence of accelerated degradation of atrazine on soil pH in French and Canadian soils. Soil Biol Biochem 32:615–625

Jain V (1992) Disposing of pesticides in the third world. Environ Sci Technol 26:226–228
Juo ASR, Franzluebbers K (2003) Tropical soils: Properties and management for sustainable agriculture. Oxford University Press, New York
Karlsson H et al (2000) Persistent chlorinated pesticides in air, water, and precipitation from the Lake Malawi area, southern Africa. Environ Sci Technol 34:4490–4495
Kellman MC (1997) Tropical environments: The functioning and management of tropical ecosystems. Routledge, Florence, KY
Kellogg CA, Griffin DW (2006) Aerobiology and the global transport of desert dust. Trends Ecol Evol 21:638–644
Kemmitta SJ et al (2008) Mineralization of native soil organic matter is not regulated by the size, activity or composition of the soil microbial biomass — a new perspective. Soil Biol Biochem 40:61–73
Kiely T, Donaldson D, Grube A (2004) Pesticides industry sales and usage: 2000 and 2001 market estimates. U.S. Environmental Protection Agency, Washington, DC
Korpraditskul R, Korpraditskal V, Kuwatsuka S (1992) Degradation of the herbicide atrazine in five different Thai soils. J Pestic Sci 17:287–289
Korpraditskul R, Katayama A, Kuwatsuka S (1993) Chemical and microbiological degradation of atrazine in Japanese and Thai soils. J Pestic Sci 18:77–83
Laabs V, Amelung W, Pinto A, Zech W (2002) Fate of pesticides in tropical soils of Brazil under field conditions. J Environ Qual 31:256–268
Lambais MR, Crowley DE, Cury JC, Büll RC, Rodrigues RR (2006) Bacterial diversity in tree canopies of the Atlantic Forest. Science 312:1917
Laskowski DA (1995) EPA guidelines for environmental fate studies: meaningful data for assessing exposure to pesticides. In: Leng ML, Leovey EMK, Zubkoff PL (eds) Agrochemical environmental fate: State of the art. Lewis Publishers, Boca Raton, FL, pp 117–128
Lavelle P, Blanchart E, Martin A, Martin S, Spain A (1993) A hierarchical model for decomposition in terrestrial ecosystems: application to soils of the humid tropics. Biotropica 25:130–150
Lehman RG, Miller JR, Fontaine DD, Laskowski DD, Hunter JH, Cordes RC (1992) Degradation of a sulfonamide herbicide as a function of soil sorption. Weed Res 32:197–205
Lighthart B (1984) Microbial aerosols: estimated contribution of combine harvesting to an airshed. Appl Environ Microbiol 47:430–432
Linshi J, Nissinen A, Erhard M, Taskinen O (2003) Climatic effects on litter decomposition from arctic tundra to tropical rainforest. Global Change Biol 9:575–584
Lozupone CA, Knight R (2007) Global patterns in bacterial diversity. Proc Natl Acad Sci USA 104:11436–11440
Marchant R et al (2008) Thermophilic bacteria in cool temperate soils: are they metabolically active or continually added by global atmospheric transport? Appl Microbiol Biotechnol 78:841–852
Marschner P, Yang CH, Lieberei R, Crowley DE (2001) Soil and plant specific effects on bacterial community composition in the rhizosphere. Soil Biol Biochem 33:1437–1445
Martiny JBH et al (2006) Microbial biogeography: putting microorganisms on the map. Nat Rev Microbiol 4:102–112
Matthews G, Wiles T, Balegue P (2003) A survey of pesticide application in Cameroon. Crop Prot 22:707–714
McGowan C, Fulthorpe R, Wright A, Tiedje JM (1998) Evidence for interspecies gene transfer in the evolution of 2, 4-dichlorophenoxyacetic acid degraders. Appl Environ Microbiol 64: 4089–4092
Meentemeyer V (1978) Macroclimate and lignin control of litter decomposition rate. Ecology 59:465–472
Melgarejo Prieto MR (2008) Determinación de residuos de 14C glifosato y de AMPA en tres suelos del Tolima sometidos a diferentes uso. M.Sc. Thesis, Facultad Agronomía, Universidad Nacional de Colombia Sede Bogotá

Mulbry WW, Karns JS (1989) Parathion hydrolase specified by the *Flavobacterium opd* gene – relationship between the gene and protein. J Bacteriol 171:6740–6746
Naeem S, Hahn DR, Schuurman G (2000) Producer–decomposer co-dependency influences biodiversity effects. Nature 403:762–764
Oerke EC, Dehne HW (2004) Safeguarding production — losses in major crops and the role of crop protection. Crop Prot 23:275–285
Ohkuma M (2008) Symbioses of flagellates and prokaryotes in the gut of lower termites. Trends Microbiol 16:345–352
Oliver DP, Kookana RS, Quintana B (2005) Sorption of pesticides in tropical and temperate soils from Australia and the Philippines. J Agric Food Chem 53:6420–6425
Papke RT, Ramsing NB, Bateson MM, Ward DM (2003) Geographical isolation in hot spring cyanobacteria. Environ Microbiol 5:650–659
Peñafiel W, Kammerbauer H (2001) Evaluación del uso y manejo de pesticidas en una zona subtropical del Alto Beni de Bolivia. Ecología en Bolivia 36:55–63
Pietikäinen J, Pettersson M, Bååth E (2005) Comparison of temperature effects on soil respiration and bacterial and fungal growth rates. FEMS Microbiol Ecol 52:49–58
Piutti S et al (2003) Isolation and characterisation of *Nocardioides* sp. SP12, an atrazine-degrading bacterial strain possessing the gene trzN from bulk and maize rhizosphere soil. FEMS Microbiol Lett 221:111–117
Poelarends GJ, Kulakov LA, Larkin MJ, Vlieg JETV, Janssen DB (2000) Roles of horizontal gene transfer and gene integration in evolution of 1, 3 dichloropropene and 1, 2-dibromoethane degradative pathways. J Bacteriol 182:2191–2199
Pommier T et al (2007) Global patterns of diversity and community structure in marine bacterioplankton. Mol Ecol 16:867–880
Racke KD, Skidmore MW, Hamilton DJ, Unsworth JB, Miyamoto J, Cohen SZ (1997) Pesticide fate in tropical soils. Pure Appl Chem 69:1349–1371
Ramette A, Tiedje J (2007) Biogeography: an emerging cornerstone for understanding prokaryotic diversity, ecology, and evolution. Microb Ecol 53:197–207
Reed HE, Martiny JBH (2007) Testing the functional significance of microbial composition in natural communities. FEMS Microbiol Ecol 62:161–170
Rodríguez Cruz MS, Jones JE, Bending GD (2008) Study of the spatial variation of the biodegradation rate of the herbicide bentazone with soil depth using contrasting incubation methods. Chemosphere 73:1211–1215
Rohde K (1992) Latitudinal gradients in species diversity: the search for the primary cause. Oikos 65:514–527
Rousseaux S, Hartmann A, Soulas G (2001) Isolation and characterization of new Gram-negative and Gram-positive atrazine degrading bacteria from different French soils. FEMS Microbiol Ecol 36:211–222
Sarkar SK et al (2008) Occurrence, distribution and possible sources of organochlorine pesticide residues in tropical coastal environment of India: an overview. Environ Int 34:1062–1071
Satsuma K (2006) Characterisation of new strains of atrazinedegrading *Nocardioides* sp. isolated from Japanese riverbed sediment using naturally derived river ecosystem. Pest Manag Sci 62:340–349
Schade G (2005) Tropical soils and agriculture: nutrients, soil organic matter, and sustainable management practices. http://www.met.tamu.edu/class/atmo613/Tropical%20Soils%20and%20Agriculture.doc. Accessed 10 Nov 2008
Seffernick JL, Wackett LP (2001) Rapid evolution of bacterial catabolic enzymes: a case study with atrazine chlorohydrolase. Biochemistry 40:12747–12753
Senior E, Bull T, Slater JH (1976) Enzyme evolution in a microbial community growing on herbicide Dalapon. Nature 263:476–479
Serdar CM, Murdock DC, Rohde MF (1989) Parathion hydrolase gene from *Pseudomonas diminuta* MG: subcloning, complete nucleotide sequence, and expression of the mature portion of the enzyme in *Escherichia coli*. Biotechnology 7:1151–1155

Shapir N, Mongodin EF, Sadowsky MJ, Daugherty SC, Nelson KE, Wackett LP (2007) Evolution of catabolic pathways: genomic insights into microbial *s*-triazine metabolism. J Bacteriol 189:674–682

Sibanda T, Dobson HM, Cooper JF, Manyangarirwa W, Chiimba W (2000) Pest management challenges for smallholder vegetable farmers in Zimbabwe. Crop Protect 19:807–815

Simon L, Spiteller M, Wallnofer PR (1992) Metabolism of fenamiphos in 16 soils originating from different geographic areas. J Agric Food Chem 40:312–317

Smith LH, Liyanage JA, Watawala RC, Aravinna AGP, Kookana RS (2006) Degradation of the pesticides carbofuran and diazinon in tropical soils from Sri Lanka. CSIRO Land and Water Science Report 67/06, CSIRO, Australia

Somara S, Manavathi B, Tebbe C, Siddavattam D (2002) Localisation of identical organophosphorus pesticide degrading (*opd*) genes on genetically dissimilar indigenous plasmids of soil bacteria: PCR amplification, cloning and sequencing of the *god* gene from *Flavobacterium balustinum*. Indian J Exp Biol 40:774–779

Specialists in Business Information (2008) Emerging trends and opportunities in the world pesticides market. Publication SB1928574

Srivastava DS, Velland M (2005) Biodiversity–ecosystem function: is it relevant to conservation? Annu Rev Ecol Evol Syst 36:267–294

Suett DL, Jukes AA, Parekh NR (1996) Non-specific influence of pH on microbial adaptation and insecticide efficacy in previously-treated field soils. Soil Biol Biochem 28:1783–1790

Topp E, Zhu H, Nour SM, Houot S, Lewis M, Cuppels D (2000a) Characterization of an atrazine-degrading *Pseudaminobacter* sp. isolated from Canadian and French agricultural soils. Appl Environ Microbiol 66:2773–2782

Topp E, Mulbry WM, Zhu H, Nour SM, Cuppels D (2000b) Characterization of s-triazine herbicide metabolism by a *Nocardioides* sp. isolated from agriculture soils. Appl Environ Microbiol 66:3134–3141

Tringe SG et al (2008) The airborne metagenome in an indoor urban environment. PLOS One 3:e1862

van der Heijden MGA et al (1998) Mycorrhizal fungal diversity determines plant biodiversity, ecosystem variability and productivity. Nature 396:69–72

Walker A (1978) Simulation of the persistence of eight soil-applied herbicides. Weed Res 18: 305–313

Walker A, Eagle DJ (1983) Prediction of herbicide residues in soil for advisory purposes. Asp Appl Biol 4:503–509

Walker A et al (1983) EWRS herbicide-soil working group: collaborative experiment on simazine persistence in soil. Weed Res 23:373–383

Wertz S, Degrange V, Prosser JI, Poly F, Commeaux C, Guillaumaud N, Le Roux X (2007) Decline of soil microbial diversity does not influence the resistance and resilience of key soil microbial functional groups following a model disturbance. Environ Microbiol 9:2211–2219

Whitaker RJ, Grogan DW, Taylor JW (2003) Geographic barriers isolate endemic populations of hyperthermophilic Archaea. Science 301:976–978

Whitfield J (2005) Biogeography: is everything everywhere? Science 310:960–961

Whitford F, Pike D, Hanger G, Burroughs F, Johnson B, Blessing A (2006) The benefits of pesticides: a story worth telling. Purdue University, Purdue Extension, West Lafayette, IN

Williamson S, Ball A, Pretty J (2008) Trends in pesticide use and drivers for safer pest management in four African countries. Crop Prot 27:1327–1334

Wilson C, Tisdell C (2001) Why farmers continue to use pesticides despite environmental, health and sustainability costs. Ecol Econ 39:449–462

Woodwell GM, Craig PP, Johnson HA (1971) DDT in the biosphere: where does it go? Science 174:1101–1107

World Resources Institute (2008) Agriculture statistics pesticide use. http://www.nationmaster.com/red/graph/agr_pes_use-agriculture-pesticide-use&int=-1. Accessed 23 Nov 2008

Chapter 13
Soil and Crop Contamination Through Wastewater Irrigation and Options for Risk Reduction in Developing Countries

Robert C. Abaidoo, Bernard Keraita, Pay Drechsel, Priyanka Dissanayake, and Akple S. Maxwell

13.1 Introduction

Wastewater irrigation is widely practiced globally. Estimates show that at least 20 million ha are irrigated with wastewater, and about 200 million farmers are involved (Raschid-Sally and Jayakody 2008; Jimenez and Asano 2008). Wastewater irrigation creates both opportunities and problems. The opportunities of wastewater irrigation are that it facilitates convenient disposal of waste products and has the beneficial aspects of adding valuable plant nutrients and organic matter to soil (van der Hoek et al. 2002). Wastewater also provides reliable irrigation water supplies and, as a farming practice, contributes to the improvement of urban food supply and the livelihood of many farmers and produce traders.

However, wastewater, especially when used in irrigation without adequate treatment, has constituents that cause soil and crop contamination, hence posing public health and environmental risks. This is commonly seen in most developing countries which lack resources for effective wastewater treatment facilities; hence, large volumes of wastewater generated, especially in urban areas, remain untreated. Estimates show median levels of treated wastewater to be about 35% in Asia, 14% in Latin America and the Caribbean and 0% in sub-Saharan Africa (WHO 2000). Therefore, in these countries partially treated wastewater and large amounts of untreated wastewater are discharged into urban drainage systems and other natural

R.C. Abaidoo (✉)
International Institute of Tropical Agriculture, Ibadan, Nigeria; College of Agriculture and Natural Resources, Kwame Nkrumah University of Science and Technology, University Post Office, Kumasi, Ghana
e-mail: rcabaidoo@yahoo.com

B. Keraita and A.S. Maxwell
International Water Management Institute (IWMI), West Africa Office, Ghana; Department of International Health, University of Copenhagen, Denmark

P. Drechsel and P. Dissanayake
International Water Management Institute, Colombo, Sri Lanka

P. Dion (ed.), *Soil Biology and Agriculture in the Tropics*, Soil Biology 21,
DOI 10.1007/978-3-642-05076-3_13, © Springer-Verlag Berlin Heidelberg 2010

waterways and used by farmers. A recent survey suggests that in and around three out of four cities in the developing world, wastewater without any significant treatment is used for irrigation purposes (Raschid-Sally and Jayakody 2008).

With such extensive use of wastewater in irrigated agriculture, contamination of soils and crops is expected. In this chapter, the occurrence and extent of crop and soil contamination as well as the related negative impacts are discussed. In addition, we highlight some of the measures that could help reduce the health risks posed by the practice.

13.2 Soil and Crop Contaminants in Wastewater

13.2.1 Pathogenic Micro-organisms

13.2.1.1 Categories of Pathogens and their Associated Risks

The common human microbial pathogens most frequently detected in wastewater are enteric in origin. Enteric pathogens enter the environment through the feces of infected hosts and can enter water through defecation into water, contamination with sewage effluent, or from run-off from soil and other land surfaces (Feachem et al. 1983). The numbers and types of pathogens found in wastewater vary both spatially and temporally depending on season, water use, economic status of the population, disease incidence in the population producing the wastewater, awareness of personal hygiene, and quality of water or food consumed. Table 13.1 shows the main categories of pathogens found in wastewater, including viruses, bacteria, protozoa, and helminths, their excreted levels, and associated risks.

Bacteria are the most common of the microbial pathogens found in wastewaters. Whereas most of these are innocuous and pose no threat to humans, a wide range of bacterial pathogens and opportunistic pathogens can also be detected in wastewaters. Among the enteric strains, *Escherichia coli* O157:H7, which causes severe haemorrhagic colitis and haemolytic uraemia in humans (Ratnam et al. 1988), has generated tremendous concern but has not been widely studied in underdeveloped environments where the potential of infection is high. Of equally great concern in regard to soil contamination are helminths, especially ascaris and hookworms. Studies from wastewater-irrigated fields in Mexico, India and Pakistan clearly associate helminth-contaminated soils and wastewater to increased human infections (Ensink 2006; Peasey 2000). Some of the helminths require an intermediate host for development prior to becoming infectious for humans. Of lesser relevance in wastewater irrigation are protozoans and viruses. Most enteric viruses have a narrow host range, meaning that most viruses of interest in wastewater only infect humans (Haas et al. 1999). Being obligate intracellular parasites, they are present in water as inactive particles.

Table 13.1 Pathogen levels and diseases associated with raw domestic wastewater[a]

Taxon and pathogen	Numbers of pathogen in raw wastewater	Disease caused by pathogen	Symptoms of disease
Bacteria			
Escherichia coli	10^0–10^{10}	Gastroenteritis	Diarrhea, nausea, dehydration
Salmonella typhi	1–10^5	Typhoid fever	Headache, fever, malaise, anorexia, cough
Salmonella spp.	1–10^5	Salmonellosis	Diarrhea, fever, abdominal stomach cramps
Shigella spp.	10–10^4	Shigellosis	Bloody diarrhea, vomiting, cramps, fever
Vibro cholerae	10^2–10^5	Cholera	Watery diarrhea, severe dehydration, vomiting
Campylobacter	10–10^4	Campylobacteriosis	Diarrhea, cramping, abdominal pain, fever, arthritis
Viruses			
Adenovirus	10^2–10^6	Respiratory disease	Respiratory symptoms, coughing
Poliovirus	10^2–10^6	Poliomyelitis	Often asymptomatic; fever, nausea, vomiting, paralysis
Rotavirus	10^2–10^5	Gastroenteritis	Diarrhea, nausea, dehydration
Hepatitis A	10^2–10^6	Infectious hepatitis	Fever, malaise, jaundice, abdominal discomfort
Protozoans			
Cryptosporidium parvum	10^2–10^5	Cryptosporisiosis	Watery diarrhea, abdominal cramps, pain
Giardia intestinalis	10^2–10^5	Giardiasis	Diarrhea, abdominal cramps, malaise, weight loss
Entamoeba histolytica	10^2–10^5	Amoebiasis (amoebic dysentery)	Bloody diarrhea, abdominal discomfort, fever, chills
Helminths			
Ascaris lumbricoides	1–10^3	Ascariasis	Often asymptomatic; wheezing, coughing, fever
Enterobius vericularis	1–10^3	Enterobiasis (Pinworm)	Perianal pruritis, vulvovaginitis, urinary tract infection
Taenia saginata	1–10^3	Taeniasis (Beef tapeworm)	Abdominal discomfort and weight loss
Trichuris trichiura	10^0–10^2	Trichuriasis (Whipworm)	Abdominal cramping, nausea, vomiting, flatulence, diarrhea, tenesmus, and weight loss

[a]Adapted from: Feachem et al. (1983); Yates and Gerba (1998); Ottoson (2005)

13.2.1.2 Persistence of Pathogens in Soils and Crops

Most gastro-intestinal pathogens hardly reproduce outside the human or animal digestive tract, and outside of their host they usually die off or lose their infectivity in an exponential manner. Depending on the environment in which they are found, they resist die-off to a varying extent and this affects causation and transmission of

Table 13.2 Survival time of selected pathogens in soils and on crops[a]

Taxon and pathogen	Survival time[b]:	
	In soils	On crops
Viruses		
Enteroviruses	<100 but usually <20	<60 but usually <15
Bacteria		
Fecal coliform	<70 but usually <20	<30 but usually <15
Salmonella spp	<70 but usually <20	<30 but usually <15
Vibrio cholerae	<20 but usually <10	<5 but usually <2
Protozoa		
E. histolytica	<20 but usually <10	<10 but usually <2
Helminths		
A. lumbricoides	Many months	<60 but usually <30
Hookworm larvae	<90 but usually <30	<30 but usually <10
T. saginata (eggs)	Many months	<60 but usually <30
Trichuris triciura (eggs)	Many months	<60 but usually <30

[a]*Source*: Feachem et al. (1983)
[b]Survival time is in days unless otherwise stated

Table 13.3 Factors affecting pathogen survival in the environment[a]

Factor	Comment
Humidity	Humid environments favor pathogen survival Dry environments facilitate pathogen die-off
Soil content	Clay soils and soils with high organic content favor pathogen survival
Temperature	The most important factor in pathogen die-off. High temperatures lead to rapid die-off; low temperatures lead to prolonged survival. Freezing temperatures can also cause pathogen die-off
pH	Some viruses survive longer in lower pH soils, while alkaline soils are associated with more rapid die-off of viruses; neutral to slightly alkaline soils favor bacterial survival
Sunlight (UV radiation)	Direct sunlight leads to rapid pathogen inactivation through desiccation and exposure to UV radiation
Foliage/plant type	Certain plants have sticky surfaces (e.g., zucchini) or can absorb pathogens from the environment (e.g., lettuce, sprouts), leading to prolonged survival of some pathogens; root crops such as carrots are more prone to contamination, and facilitate pathogen survival
Competition with native flora and fauna	Antagonistic effects from bacteria or algae may enhance die-off; bacteria may be preyed upon by protozoa

[a]*Source*: Strauss (1985)

diseases (Bastos 1992). The most environmentally resistant pathogens are helminth eggs, which in some cases can survive for several years in the soil. Table 13.2 shows the survival times of pathogens in soils and on plants.

Pathogen survival depends upon a number of factors, some of which are outlined in Table 13.3. The effect of temperature on pathogen survival has been discussed widely. For example, in experiments with lettuce spiked with *Cryptosporidium* oocysts and incubated for 3 days at 20°C and 4°C, Warnes and Keevil (2003) detected zero viable oocysts on the lettuce incubated at 20°C, but found 10% of the oocysts still viable from the lettuce incubated at 4°C. Gagliard and Karns (2000) reported the positive effect of low temperatures on *E. coli* survival in soils. Biological factors, such as competition and antagonism by indigenous bacteria, might be involved in this temperature effect, but this remains poorly studied.

13.2.1.3 Transfer of Pathogens to Crops

Pathogen transfer to crops depends on a number of factors. Salad crops (e.g., tomatoes, lettuce, pepper, cabbage, and garden eggs), root crops (e.g., radishes, onions, carrots) or those crops that grow close to the soil (e.g., zucchinis) present high potential for wastewater-related pathogen contamination. In addition, crops that have certain surface properties (hairy, sticky, rough, or with crevices) protect pathogens from exposure to radiation and from mechanical removal by rain or during post-harvesting washing. The amount of water each crop holds is also an important factor in exposure to pathogens. A study conducted by Shuval et al. (1997) revealed that a mature lettuce plant could retain 10.8 ml of irrigation water compared to the 0.36 ml estimated for a cucumber plant. Stine et al. (2005) also showed that lettuce and cantaloupe surfaces retained pathogens from irrigation water spiked with *E. coli* and a bacteriophage (PRD1), whereas bell peppers, which are smooth, did not. Fortunately, transfer of helminths from soils to crops is very limited. Levels of contamination of 100 g lettuce with helminth eggs were 0.6 egg at harvest following irrigation with wastewater containing >10 eggs per liter, while no nematode eggs were detected on lettuces irrigated with effluent with <0.5 egg per liter despite growing in heavily contaminated soil containing >1,200 ascaris eggs per 100 g (WHO 2006).

Pathogen detection on crop surfaces (rather than within internal tissue) has been widely used as an indication of potential health risks to populations consuming crops. Though some studies have pointed out pathogen uptake by roots (Solomon et al. 2002), others have indicated that pathogen uptake by roots can be considered as negligible compared to surface contamination (Van Ginneken and Oron 2000). Several studies have shown that human pathogens can survive and grow in tomatoes and tomato products (Zhuang and Beuchat 1996; Tsai and Ingham 1997). Coliforms were found in *Amaranthus* (up to 35 colony-forming units (cfu) g^{-1}) and eggplant (35 cfu g^{-1}) grown with sewage water in the peri-urban zone of Hubli-Dharwad, India (Brook et al. 2001). Investigations done by Jablasone et al. (2004) revealed that salmonella can survive inside tomato tissue and reach the fruit. In a study done in Kampala, Uganda, *E. coli* was isolated from both the surface and inside of plant tissues (Serani et al. 2008). Higher contamination levels were recorded on the surface of plants studied (yam, dodo, and tomato) than inside those crops. Wounds/cuts on plant leaves, stems, and roots will increase the likelihood of pathogen presence in the plants. Further investigations are needed on this subject so that the relevance of pathogens in crops can be assessed and factored in risk modeling.

13.2.2 Heavy Metals

Heavy metals are of particular concern due to their adverse effects on agricultural productivity and human health. In a review of the use of reclaimed water in Australian horticultural production industry, Hamilton et al. (2005) classified potentially phytotoxic metals in wastewater (reclaimed water) into four groups based on their retention in soil, translocation in plants, phytoxicity, and potential risk to the food chain (Table 13.4). In this classification, Cd, Co, Mo, and Se (Group 4) were cited as posing the greatest risk to human and animal health, even though they may appear in wastewater-irrigated crops at concentrations that are not generally phytotoxic. Plants that are grown in metal-polluted environments often do not show visible symptoms of intoxication even if they contain elevated concentrations of toxic metals (Clemens 2001). This is supported by the WHO, which lists B and Cd as heavy metals of particular concern because of their high level of toxicity and bioaccumulation in crops (WHO 2006). The heavy metals mercury (Hg), cadmium (Cd), and lead (Pb) do not have any essential function, but they are detrimental, even in small quantities, to plants, animals, and humans and accumulate because of a long biological half-life (Gothberg et al 2002). In plants, Pb inhibits growth; Hg stunts seedling growth and root development and inhibits photosynthesis; and Cd interferes with photosynthesis and mineral assimilation with leaf chlorosis, necrosis, and abscission as toxicity symptoms (Paris and Jones 1997). Human fetuses and small children are especially susceptible to Pd and Hg, which adversely affect the central nervous system, impacting the neurologic, psychomotor, and intellectual development of the child (Richardson and

Table 13.4 Group classification of metals according to their bioavailability[a]

Group	Metal	Soil adsorption	Phytotoxicity	Food chain risk
1.	Ag, Cr, Sn, Ti, Y, Zr	Low solubility and strong retention in soil	Low	Little risk because they are not taken up to any extent by plants
2.	As, Hg, Pb	Strongly sorbed by soil colloids	May be absorbed by plant roots but not translocated to shoots; generally not phytotoxic, except at very high concentrations	Pose minimal risks to the human food chain
3.	B, Cu, Mn, Mo, Ni, Zn	Less strongly sorbed by soil than metals of groups 1 and 2	Readily taken up by plants and phytotoxic at concentrations that pose little risk to human health	Soils and plants form a barrier that protects the food chain from these elements
4.	Cd, Co, Mo, Se	Least abundant of all metals	Pose human or animal health risks at plant tissue concentrations that are generally phytotoxic	Bioaccumulation through the soil–plant–animal food chain

[a]*Source*: Modified from Hamilton et al. (2005)

Gangoli 1995; Gothberg et al 2002). Cadmium accumulates in the liver and, above all, in the kidneys, where injuries first appear at a more or less old age, depending on the amount that has been accumulated (Richardson and Gangoli 1995; Gothberg et al. 2002)

Various studies have dealt with the impacts of wastewater irrigation on heavy metal contamination of soils and crops. Sharma et al. (2007) concluded that the use of wastewater for irrigation increased the contamination of Cd, Pb, and Ni in the edible portion of vegetables, potentially causing health risk in the long term. Similar findings have been documented from a study conducted in Harare, Zimbabwe, where farmers use wastewater for irrigating leafy vegetables (Mapanda et al. 2005; Mutchweti et al 2006). Gothberg et al. (2002) found that mercury concentrations were very high in aquatic vegetables in the Bangkok region. Cadmium and Pb concentrations in crops harvested from Central Mexico had concentrations higher than the maximum values allowed in Dutch and German standards for cereals, forage, and fresh vegetables (Lucho-Constantino et al. 2005). However, in many locations, especially in developing countries, sources of heavy metal contamination are scanty. A comprehensive study conducted in Pakistan at locations that have been subjected to wastewater irrigation for over 30 years revealed no significant increases in levels of Cd, Pb, and Zn on soils and wheat crops, and concentrations were even below the European Union (EU) maximum permissible levels (Simmons et al. 2009). Karnchanawong et al. (2006) also concluded that heavy metal (Cd, Pb, Cu, Zn) contamination was very low and within safety limits for consumption in a study done in Chiang Mai Province, Thailand. Other studies have shown that, even when there are elevated levels in soils, relatively low concentrations appear in plant tissues. This could be due to the cation adsorption to surfaces of soil particles, which would limit plant uptake. Hence, while some studies clearly show that wastewater irrigation can lead to elevated levels of heavy metals in soils and crops, others find no evidence for this, probably because of a lack of sources or due to other factors that influence heavy metal accumulation in soils and plant uptake. Where potential sources have been identified, care should be taken to assess heavy metal accumulation in plant and soil under wastewater irrigation and to determine their biotoxic potential, since bioavailability of the different fractions cannot be taken for granted.

13.2.3 Other Organic and Inorganic Compounds

Continuous agricultural use of wastewater can lead to accumulation in soils, not only of heavy metals, but also of organic and inorganic compounds such as various kinds of salts, pesticides, pharmaceutically active compounds, and endocrine-disrupting chemicals (Hamilton et al. 2007). Most of these chemical compounds concentrate in biosolids, and pose more risks if biosolids from untreated wastewater are used. They accumulate in soils over time (Xiong et al. 2001). The accumulation rate varies largely, but it mainly depends on the source of irrigation water, with industrial wastewater being a source of most of these contaminants. When their concentration is elevated, they can lead to reduced agricultural productivity, and they can cause negative human and animal health effects upon their absorption by edible plants.

Some of the adverse plant effects of contaminants found in wastewater come from salinization and toxicity. Salinization of agricultural lands is perhaps one of the most critical negative environmental impacts of wastewater reuse. Salinization and sodicity of soils are caused by inorganic salts and affect soil productivity by changing the osmotic pressure of the root zone or directly affecting soil properties through swelling and dispersion. This may then cause specific ion toxicity and interference with uptake of essential nutrients such as nitrate and potassium, because of the antagonism with sodium, chloride, and sulfate (Rengasamy 2006). Salinization and sodification processes have destroyed millions of ha of irrigated land over the world. Currently, at least 20% of the world's irrigated land is salt-affected and/or irrigated with waters containing elevated levels of salts (Ghassemi et al. 1995; Qadir et al. 2008). As the agricultural use of salt-affected land and saline water resources increases, their sustainable use for food and feed production will become a more serious issue (Suarez 2001; Wichelns and Oster 2006; Qadir et al. 2008). Whereas salinity acts at the soil–plant interface, toxicity results from reactions in the plant itself. It occurs when certain ions are taken up by plants with the soil water and accumulate in the leaves during water transpiration to such an extent that the plant is damaged. The degree of damage depends upon time, concentration of toxic material, crop sensitivity and crop water use: if damage is severe enough, crop yield is reduced. Common toxic ions in irrigation water are chloride, sodium, and boron, all of which will be contained in sewage. Damage can be caused by each of these ions, either individually or else in combination. Not all crops are equally sensitive to these toxic ions.

While nutrients and biosolids increase agricultural productivity, excessive amounts as evident in untreated wastewater can lead to negative environmental impacts. Generally, nitrogen and sodium levels in untreated wastewater often exceed plant requirements and change soil physical and chemical properties. For example, in a study conducted in Faisalabad, Pakistan, soil sodium adsorption ratio (SAR), exchangeable sodium percentage (ESP), residual sodium carbonate (RSC), electrical conductivity (EC), and the electrical conductivity of the saturation extract (ECe) were 63.3%, 37.0%, 31.1%, 50.6% and 50.6% higher respectively under wastewater irrigation than in canal water-irrigated plots (Simmons et al. 2009). In addition, soil alkalinity increased marginally under wastewater irrigation, with mean pH values for wastewater and canal water of 8.92 and 8.75 respectively. Proper management is required for reducing negative effects such as undesirable plant vegetative growth, delayed and uneven maturity, poor crop quality, and pollution of water resources. For instance, applying high concentrations of nitrogen on farms can lead to reduced fruit size and quality, while making plants more prone to pest attacks and less resistant to diseases (Hamilton et al. 2007). Being very soluble, nitrates are easily transported through soils and may contaminate aquifers, therefore affecting the quality of groundwater. Excessive biodegradable organic matter in wastewater has also been associated with clogging of soil pores, which disrupts soil structure and causes offensive odors.

13.3 Human Health Risks

Human health risks posed by wastewater irrigation have been extensively studied. Most of these studies form the basis of the Guidelines for safe Use of Wastewater in Agriculture, published by the WHO (WHO 2006). Table 13.5 is a simplified presentation of wastewater-related human health risks and the affected groups.

13.3.1 Occupation-Related Risks

Helminth infections, especially ascaris and hookworms, have more importance in relation to occupation-related risks than do bacterial, viral and protozoan infections (Blumenthal et al. 2000). The most affected groups are farm workers, owing to the long duration of their contact with wastewater and contaminated soils. For instance, in Haroonabad, Pakistan, prevalence rates for hookworm infections as high as 80% have been reported for farmers using untreated wastewater (Van der Hoek

Table 13.5 Simplified presentation of the main human health risks from wastewater irrigation

Kind of risk	Health risk	Who is at risk	How contamination occurs
Occupational risks (contact)	• Mostly parasitic worms such as ascaris and hookworm infections • Diarrhoeal diseases especially in children • Skin infections – itching and blister on the hands and feet • Nail problems such as koilonychias (spoon-formed nails)	• Farmers/field workers • Children playing on the farm • Market women	• Contact with irrigation water and contaminated soils • Contact with irrigation water and contaminated soils • Contact with irrigation water and contaminated soils • While harvesting, being exposed to contaminated soils • Washing vegetables, especially in wastewater
Consumption-related risks	• Mainly bacterial and viral infections such as cholera, typhoid, ETEC[a], hepatitis A, viral enteritis which mainly cause diarrhoeas • Parasitic worms such as ascaris	• Vegetable consumers • Children playing on the farm	• Eating contaminated vegetables, especially raw produce • Licking soil

[a]ETEC: enterotoxigenic *Escherichia coli*

et al. 2002). In general, there is overwhelming evidence from epidemiological studies of the high risk presented by helminths, hence the strict WHO guideline of <1 egg l^{-1} of irrigation water (WHO 2006). Nevertheless, recent epidemiological studies conducted in Vietnam among rice farmers using wastewater found evidence for increased diarrhea and skin problems, but not for enhanced risk of helminth infections (Trang et al. 2007a). Wastewater farmers seldom associate infections and diseases with their irrigation practices (Gerstl 2001). This has implications for awareness campaigns and calls for more comprehensive assessments considering compounding factors as well as the financial implications of the health impact.

Recent studies from Vietnam and Cambodia have strongly associated skin diseases such as dermatitis (eczema) to contact with untreated wastewater (Van der Hoek et al. 2005; Trang et al. 2007b). In a study conducted in the Katmandu Valley, more than half of about 110 farmers interviewed who were using wastewater had experienced skin problems (Rutkowski et al. 2007). The reported skin problems included itching and blisters on the hands and feet. Similar problems were reported by rice farmers along the Musi River in Hyderabad, Pakistan and urban vegetable farmers using wastewater in Ghana (Obuobie et al. 2006; Buechler et al. 2002). Nail problems such as koilonychias (spoon-formed nails) have also been reported and have been associated with hookworm infections, which cause iron deficiency (anemia), damaging the formation of nails (van der Hoek et al. 2002). Studies done in Vietnam did not find any association between the risk of eye ailments (conjunctivitis or trachoma) and wastewater-related exposure but recommended more investigations for linking skin infections to particular water pollutants (Trang et al. 2007b).

13.3.2 Consumption-Related Risks

The primary concern relates to vegetables eaten uncooked, such as salads. Several studies including a prospective cohort study conducted by Peasey (2000), an analytical descriptive study by Cifuentes (1998), and several descriptive studies including one done in Jerusalem (Shuval et al. 1984) have shown higher ascaris infections for both adults and children consuming uncooked vegetables irrigated with wastewater. Several studies on diarrhoeal diseases resulting from consumption of contaminated vegetables have been published and extensively reviewed (Harris et al. 2003; Shuval et al. 1986). Enterotoxigenic *E. coli* (ETEC) is mostly associated with diarrhea (travelers' diarrhea) in developing countries. In addition, viral enteritis (especially that caused by rotaviruses) and hepatitis A are the most frequently reported viral infections resulting from vegetable consumption.

Not much has been documented in relation to risks that would be specifically associated with organic and inorganic compounds in wastewater. This could be attributed to the fact that most of these risks are chronic. They usually occur only after many years of exposure and may also result from a variety of several other exposures not related to wastewater irrigation (WHO 2006). Nevertheless, health risks associated with irrigation using industrial discharges have been reported. In Japan, Itai–itai disease, a bone and kidney disorder associated with chronic cadmium poisoning, occurred in areas where paddy rice was irrigated with the contaminated Jinzu River (WHO 2006). In China, industrial wastewater irrigation was associated with a 36% increase in hepatomegaly (enlarged liver) and a 100% increase in both cancer and congenital malformation rates (Yuan 1993).

13.4 Measures to Reduce Risks

Wastewater treatment is the best choice. However, the costs involved in engineering-based technologies at any significant scale are in general unbearably high for developing countries in sub-Saharan Africa (where most threats derive from pathogenic contaminations) or South Asia (where threats associated with heavy metal and organic contaminants are increasing). Even where treatment plants are externally funded, they usually only treat a small fraction of the produced wastewater and face significant maintenance problems. However, simple farm-based measures, low-cost treatment options, and so-called "nontreatment" options are supported by the WHO (2006), and can reduce the risk significantly. Where this is not possible, the affected areas might have to be taken out of production (Simmons et al. 2005, 2009).

13.4.1 Soil Phytoremediation

Because the costs of growing a crop are minimal compared to those of soil removal and replacement, the use of plants to remediate hazardous soils is seen as having great promise. Several recent reviews on many aspects of soil metal phytoremediation are available (Moffat 1995; Cunningham et al. 1995; Salt et al. 1996). Phytoremediation, also often referred to as bioremediation, botanical bioremediation, and green remediation, is the use of plants to detoxify soil contaminants. The idea of using rare plants which hyperaccumulate metals to selectively remove and recycle excessive soil metals was introduced in 1983 (Chaney 1983) and has increasingly been examined as a practical and more cost-effective technology than the standard soil replacement, solidification, or washing strategies (Cunningham et al. 1995; Salt et al. 1996). Currently, there are about 420 species belonging to about 45 plant families recorded as hyperaccumulators of heavy metals (Cobbett 2003). Vetiver is known to absorb much higher volumes of nutrient and heavy metals from contaminated lands than most hyperaccumulators, due to its very fast growth and high yield (Truong et al. 2008). The bioremediation potential of plants might be improved by using plants together with associated rhizosphere organisms or through bioengineering (Cunningham et al. 1996). Categories of phytoremediation include:

1. Phytoextraction – plants remove contaminants from soils;
2. Phytovolatilization – plants produce volatile chemical species of soil elements;
3. Rhizofiltration – plant roots remove contaminants from flowing water;
4. Phytostabilization – plants transform soil metals to less toxic forms, without removing the metal from the soil.

13.4.2 Land Grading and Cultivation

Land grading is important for achieving good uniformity of application from surface irrigation methods and acceptable irrigation efficiencies in general (Pescod 1992). If the wastewater is saline, it is very important that the irrigated land be appropriately graded. Salts accumulate in the high spots which have too little water infiltration and leaching, while in the low spots water accumulates, causing water-logging and soil crusting. Stratified soils are difficult to irrigate, as layers of clay, sand, or hardpan frequently impede or prevent free movement of water through and beyond the root zone. This will lead to water saturation and salt accumulation in the root zone. Irrigation efficiency as well as water movement in the soil can be greatly enhanced by sub-soiling and chiseling of the land. The effects of sub-soiling and chiseling remain for about 1–5 years, but, if long term effects are required, the land should be deep- and slip-ploughed.

13.4.3 Field Water Management

Proper water management during application can minimize soil and crop contamination. Control of timing and frequency of irrigation is not only important for pathogen reduction but also for reducing salinity and other toxicities. One of the most widely documented water management measures used to reduce pathogen contamination is cessation of irrigation, where irrigation is stopped a few days before crops are harvested to allow for pathogen die-off due to exposure to unfavorable environmental factors such as sunlight (Shuval et al. 1986). As much as 99% elimination of detectable viruses has been reported after 2 days of exposure to sunlight, supporting regulations that a suitable time interval should be maintained between irrigation and crop handling or grazing (Feigin et al. 1991). In another study, it was revealed that when trickling filter effluent with 10^6 thermotolerant coliform per 100 ml was used to spray-irrigate lettuce, initial concentrations exceeded 10^5 thermotolerant coliform per 100 g fresh weight. Once irrigation ceased, no salmonella could be detected after 5 days, and, after 7–12 days, the levels of thermotolerant coliform were comparable to those detected on lettuce irrigated with fresh water (Vaz da Costa-Vargas et al. 1996). In Ghana, studies from field trials show an average daily reduction of 0.65 log units of thermotolerant coliforms on lettuce (Keraita et al. 2007a). However, the studies from Ghana show that cessation of irrigation is accompanied by high yield losses (1.4 tons ha^{-1} of fresh weight), which may dissuade farmers from adopting this practice (Keraita et al. 2007a).

Good field water management can also reduce salinity and other toxicities. Salts accumulated in the soils can be displaced from the root zone by leaching, i.e., by using extra amount of irrigation water (leaching fraction) to enable deep percolation of salts. In other cases and especially in arid and semi-arid areas, salinity and other toxicities are often associated with shallow water tables (Pescod 1992). So drainage in this case is important to lower water tables to desirable levels, which will prevent transport of salts to the root zone and soil surface by capillary movement.

13.4.4 Field-Based Measures to Improve Irrigation Water Quality

The many technologies that exist for treating wastewater are comprehensively reviewed by the WHO (2006), together with their effectiveness to reduce pathogens. However, most of these technologies are applicable for source treatment and need to be modified if they are to be used on the farms. This section highlights three treatment technologies which have been tested and have shown some potential, although further modifications remain necessary to make them more effective.

13.4.4.1 Waste Storage and Treatment Reservoirs

Waste storage and treatment reservoirs (WSTRs), also called effluent storage reservoirs, have been used in several countries. They offer the advantage of storing wastewater until it can be used in the growing season (or until the dry season) and they offer an additional treatment stage. Pond systems are very effective in removing pathogens and can achieve 100% removal especially for helminths and protozoa. Due to their larger sizes, helminth eggs and protozoa in WSTRs are mostly removed by sedimentation while bacteria and viruses are removed mostly from die-off in the ponds (Mara 2004). Sedimentation may also contribute to coliform bacteria and virus removal (Karim et al. 2004). Water quality can be improved by using three WSTRs in parallel, and using batch-fed processes to ensure that all of the wastewater is retained for an adequate time, so as to reduce pathogen concentrations to the appropriate level prior to irrigation (Mara and Pearson 1992). Research by Mara et al. (1996) in northeast Brazil showed that treating wastewater in an anaerobic pond followed by a fill, rest, and use sequential cycle of treatment in WSTR was very efficient at reducing fecal coliform concentrations.

WSTRs may provide a low-cost option for treating wastewater which would otherwise be used without treatment. For example, in Mexico, irrigation with untreated or partially treated wastewater was estimated to be directly responsible for 80% of all ascaris infections and 30% of diarrhoeal disease in farm workers and their families. However, when wastewater was retained in a series of reservoirs, there was minimal risk of either ascaris infection or diarrhoeal disease (Cifuentes et al. 2000). Procedures for designing WSTR are outlined in Mara (2004). Compared to WSTR, on-farm sedimentation ponds are much smaller with surface areas of 3–4 m^2 and depths of about 1 m. These ponds have been tested in wastewater irrigated farms in Ghana, and proved to significantly reduce pathogen levels, especially helminths (Keraita et al. 2008a). In a study done in southern Italy (Lopez et al. 2005) it was demonstrated that after appropriate storage periods in reservoirs, the number of pathogen indicators decreased by 2–3 log units; salmonella populations decreased from a mean value of about 28 cfu 100 ml^{-1} to 4 cfu 100 ml^{-1}; and helminth eggs present in inflow wastewater were no longer detected in the effluent.

13.4.4.2 Filtration Techniques

The use of slow sand filters is probably the most effective, simplest, and least expensive water treatment process, especially for developing countries (Muhammad et al. 1996). Sand filters remove pathogenic micro-organisms from polluted water, first by retaining them in the filtration media and then by promoting their elimination (Stevik et al. 2004). Retention is achieved mainly through straining, whereby micro-organisms are physically blocked as they move through the filter media, and through adsorption when they get attached to the filtration media. Generally, parasites and other large-sized pathogens are retained in filters mainly by straining, while smaller-sized agents such as bacteria and viruses are retained by adsorption. Pathogen elimination is then achieved by exposure to unfavorable environmental conditions, such as high temperature, and through predation. The typical pathogen removal range reported by the WHO based on a review on several studies for slow sand filters is 0–3 log units and 1–3 log units for bacteria and helminths, respectively (WHO 2006). In Ghana, column slow sand filters removed more than 98% and 71%–96% of bacteria and helminths respectively (Keraita et al. 2008b). This removal was significant but not sufficient as irrigation water had very high initial levels of indicator organisms. It is worth mentioning that soils also can act as biofilters.

13.4.5 Use of Appropriate Irrigation Methods

Under normal conditions, the type of irrigation method selected will depend on water supply, climate, soil, crops to be grown, cost of irrigation method, and the ability of the farmer to manage the system. However, when using wastewater as the source of irrigation, other factors, such as pathogen contamination of plants, harvested product, farm workers, and the environment, as well as salinity and toxicity hazards, need to be considered. There is considerable scope for reducing the undesirable effects of wastewater use in irrigation through selection of appropriate irrigation methods. Table 13.6 presents an analysis of the parameters that are useful for evaluation of the risk-reduction potential of four widely practiced irrigation methods.

The use of appropriate irrigation methods has also been recommended under the so-called "non-(conventional) treatment" options for health protection in the WHO guidelines for safe wastewater irrigation (WHO 2006). Based on health impacts from wastewater, the WHO has classified irrigation methods in three distinct categories: flood and furrow, spray and sprinkler, and localized irrigation methods (WHO 2006). Flood and furrow irrigation methods apply water on the surface and pose the highest risks to field workers, especially when protective clothing is not used (Blumenthal et al. 2000). Spray and sprinkler are overhead irrigation methods and have the highest potential for pathogen transfer to crop surfaces, as water is applied on edible parts of most crops. They also promote wide movement of pathogens through aerosols. Localized techniques such as drip and trickle irrigation offer farm workers the best possible health protection and also ensure minimal pathogen transfer to crop surfaces, since water is directly applied to the root (Pescod 1992). However, localized techniques are comparatively the most expensive and are also prone to clogging as polluted water has high particulate levels. They can reduce contamination on crops by 2–4 log units (WHO 2006). In this respect, recently introduced low-cost drip irrigation techniques, such as bucket drip kits from Chapin Watermatics (USA) and International Development Enterprises (India), have more potential for use and adoption in low-income countries (Kay 2001). Studies done in Ghana using bucket drip kits show massive reduction in contamination (up to 6 log units), especially during the dry season (Keraita et al. 2007b). These studies from Ghana also demonstrated that the traditional watering can system could be modified so as to diminish splashing of contaminated soils to the crops, which in turn would reduce crop contamination.

13.4.6 Crop Selection

Some crops are more prone than others to contamination with pathogens, or more sensitive to salinity and toxicity. Thus, proper crop selection will help decrease human health risks. For example, crops with their edible parts more exposed to contaminated soils and irrigation water, such as low-growing crops and root tubers, will be more prone to pathogen contamination. The World Health Organization, in its guidelines for safe use of wastewater in agriculture, advises for crop restrictions, especially targeting vegetables and other crops eaten raw (WHO 2006). In Mexico, farmers could shift from leafy vegetables to vegetables such as onions and garlic, which provide a less favorable environment for bacteria survival. However, a shift in crop choice is only feasible if the market value of the alternative crop is similar to

Table 13.6 Parameters for evaluation of commonly used irrigation methods in relation to risk reduction[a]

Parameter of evaluation	Irrigation method			
	Furrow irrigation	Border irrigation	Sprinkler irrigation	Drip irrigation
Foliar wetting and consequent leaf damage resulting in poor yield	No foliar injury as the crop is planted on the ridge	Some bottom leaves may be affected, but the damage is not so serious as to reduce yield	Severe leaf damage can occur, resulting in significant yield loss	No foliar injury occurs under this method of irrigation
Salt accumulation in the root zone with repeated applications	Salts tend to accumulate in the ridge, which could harm the crop	Salts move vertically downwards, and are not likely to accumulate in the root zone	Salt movement is downwards, and root zone is not likely to accumulate salts	Salt movement is radial along the direction of water movement. A salt wedge is formed between drip points
Ability to maintain high soil water potential	Plants may be subject to water stress between irrigations	Plants may be subject to water stress between irrigations	Not possible to maintain high soil water potential throughout the growing season	Possible to maintain high soil water potential throughout the growing season and minimize the effect of salinity
Suitability to handle brackish wastewater without significant yield loss	Fair to medium. With good management and drainage, acceptable yields are possible	Fair to medium. Good irrigation and drainage practices can produce acceptable yields	Poor to fair. Most crops suffer from leaf damage, and yield is low	Excellent to good. Almost all crops can be grown with very little reduction in yield

[a]*Source*: Pescod (1992)

that of the original crop. Crop restrictions can be hard to implement if necessary conditions such as law enforcements, market pressure, and demand for cleaner vegetables are not in place. Hence, while there have been successful crop restriction schemes in India, Mexico, Peru, and Chile (Buechler and Devi 2003; Blumenthal et al. 2000), this has not been possible in sub-Saharan Africa and other countries where wastewater irrigation is informal.

The salt tolerance of agricultural crops varies in an eight to tenfold range. This wide tolerance range allows for greater use of moderately saline water, much of which was previously thought to be unusable. It also greatly expands the acceptable range of water salinity (EC_w) considered suitable for irrigation. Vegetables such as carrots, onions, and okra are known to be very sensitive to salinity, while others such as asparagus are very tolerant (Ayres and Westcot 1985). Based on salt tolerance of plant species, there are emerging examples of crop diversification and management for optimal utilization of salt-affected soils and saline-sodic waters (Qadir et al. 2008). For example, forage grass in general is more salt-tolerant as compared to most field crops and can provide additional income sources for farmers in marginal lands. Crop diversification and production systems based on salt-tolerant plant species are likely to be the key to future agricultural and economic growth in regions where salt-affected soils exist, saline drainage waters are generated, and/or saline aquifers are pumped for irrigation (Qadir et al. 2008). Similarly, plants have different tolerances for toxicity. A review on a number of studies on toxicity tolerance by various crops has been prepared by Pescod (1992).

13.4.7 Postharvest Measures

It is generally recognized that the above farm-based interventions can only reduce and not eliminate crop contamination at farm level. Furthermore, several studies show that further contamination (mainly microbial) can occur during postharvest handling at markets and consumption points (Amoah et al. 2007; Ensink et al. 2007). Therefore, post-harvest interventions are equally important to ensure food safety. Improving stakeholders' knowledge and awareness of personal hygiene in relation to food safety has been advocated. It is customary that vegetables be washed, but this only removes much of the adhering soil and some dirt. Washing has little effect in reducing the number of bacteria and achieves reductions of only 0.1 to 1 $\log_{10}$ unit (Beuchat et al. 1998). Approximately 1 and 2 log reductions (depending on the nature of the surface) can be achieved when leaves are washed vigorously in tap water in the markets. A review done on the effectiveness of postharvest risk reduction options is presented in Table 13.7 (WHO 2006).

In Ghana, perception studies show that food vendors are usually confident that their cleaning and treatment process is sufficient to eliminate contaminants. However, assessments done in West African cities show that only a few of the methods used in cleaning vegetables achieved some reductions (Amoah et al. 2007). For many of them, some adjustments were needed to achieve pathogen reductions of

Table 13.7 Postharvest measures for pathogen reduction[a]

Control measure	Extent of pathogen reduction (log units)	Comments
Produce washing	1	Washing salad crops, vegetables and fruit with clean water
Produce disinfectant	2–3	Washing salad crops, vegetables and fruit with a disinfectant solution and rinsing with clean water
Produce peeling	2	Fruits, root crops
Produce cooking	6–7	Immersion in boiling or close-to-boiling water until the food is cooked ensures pathogen destruction

[a]*Source*: Adapted and modified from WHO (2006)

2–3 log units. Salt is considered the cheapest disinfectant and the most likely to be adopted, but it causes deterioration of lettuce at the high concentrations of 23 and 35 ppm. A weak 7 ppm NaCl solution is recommended to achieve some pathogen reduction while preserving freshness of lettuce (Amoah et al. 2007). At an acetic acid concentration of 12,500 ppm (approximately one part vinegar in five parts water), a significant reduction in pathogen levels can be achieved with a contact time of at least 5 min. However, vinegar is considered expensive by most street food vendors, and is mainly used by restaurants and middle/upper class households. Locally-promoted chlorine tablets are cheaper than vinegar and very efficient, allowing for 2–3 log unit reductions (Amoah et al. 2007).

Various studies have shown that none of the available washing and sanitizing methods, including some of the newest sanitizing agents such as chlorine dioxide and ozone, can guarantee the microbiological quality of minimally processed vegetables without compromising their organoleptic quality (Beuchat et al. 1998). This is probably due to insufficient knowledge of the proper application conditions of the sanitizing agent. Indeed, whereas the efficacy of a particular disinfection method is dependent on the type and physiology of the target micro-organisms, other parameters such as characteristics of produce surfaces, exposure time, concentration of cleaners and sanitizers, pH and temperature conditions also influence the effectiveness of washing. However, only a few studies have been conducted to optimize these parameters, hence the varied and often ineffective methods for washing of vegetables.

13.5 Conclusions

The use of (untreated) wastewater in irrigated agriculture is a source of soil and crop contamination. Contamination levels vary widely depending on local sources and tend to be higher where untreated wastewater is used for irrigation. In many developing countries, peri-urban farmers have little choice apart from using surface water sources. Soil salinization poses one of the greatest challenges in terms of the extent and complexity of the corrective measures to be adopted. Crop diversification may represent a means of mitigating this situation, and may provide economic and on-farm benefits to farming communities. A different type of risk comes from heavy metal pollution and specific ion toxicities, especially where industrial wastewater is used. However, industrial wastewater is often localized and hence much easier to clean or exclude from farming.

Most studies conducted on wastewater irrigation have focused on microbiological contaminants and assessment of health risks. Typically, such studies address pathogen characterization, pathogen survival in the environment, exposure and transmission routes, infection risks for exposed groups, and disease outbreaks. Diarrheal diseases and helminth infections (especially hookworm and ascaris) have been found to pose the greatest human health risks. These diseases are transmitted mainly by consumption of wastewater-irrigated produce or contact with contaminated soils and irrigation water. Though a number of risk reduction measures have been suggested, examples of actual field testing and implementation are scarce. Studies from Ghana highlight the great potential of low-cost farm-based risk reduction measures for alleviation of the negative human health impacts of untreated wastewater use. By adopting the WHO multiple barrier approach, suitable complementary measures can be implemented along the food chain to achieve maximum aggregate contamination reduction. Equally important is to encourage local risk reduction innovations and involve end-users in developing and testing risk reduction options.

References

Amoah P, Drechsel P, Abaidoo RC, Klutse A (2007) Effectiveness of common and improved sanitary methods in selected cities of West Africa for the reduction of coliform bacteria and helminth eggs on vegetables. Trop Med Int Health 12:40–50

Ayres RS, Westcot DW (1985) Water quality for agriculture. Irrigation and drainage paper 29 Rev 1. FAO, Rome

Bastos RKX (1992) The bacteriological quality of drip and furrow irrigation with treated wastewater. PhD Thesis, Department of Civil Engineering, The University of Leeds, England

Beuchat LR, Nail BV, Adler BB, Clavero MRS (1998) Efficacy of spray application of chlorine in killing pathogenic bacteria on raw apples, tomatoes, and lettuce. J Food Protect 61:1305–1311

Blumenthal UJ, Peasey A, Ruiz-Palacios G, Mara DD (2000) Guidelines for wastewater reuse in agriculture and aquaculture: recommended revisions based on new research evidence. WELL study, Task No. 68, Part 1, London and Loughborough

Buechler S, Devi G (2003) Household food security and wastewater dependent livelihood activities in Andhra Pradesh, India. Unpublished background document prepared for the WHO in preparation of Guidelines for safe Use of Wastewater in Agriculture

Buechler S, Devi G, Raschid-Sally L (2002) Livelihoods and wastewater irrigated agriculture along the Musi River in Hyderabad City, Andhra Pradesh, India. Urban Agric Magazine 8:14–17

Brook RM, Purushothaman S, Hunsal C (2001) Changing frontiers: the peri-urban interface, Hubli-Dharwad, India. Books for Change, Bangalore

Chaney RL (1983) Plant uptake of inorganic waste constituents. In: Parr JF, Marsh BP, Kla JM (eds) Land treatment of hazardous wastes. Noyes Data Corp, Park Ridge, NJ, pp 50–76

Cifuentes E (1998) The epidemiology of enteric infections in agricultural communities exposed to wastewater irrigation: perspectives for risk control. Int J Environ Health Res 8:203–213

Cifuentes E, Blumenthal U, Ruiz-Palacios G, Bennett S Quigley M (2000) Health risks in agricultural villages practicing wastewater irrigation in Central Mexico: perspectives for protection. In: Chorus I, Ringelband U, Schlag G, Schmoll O (eds) Water Sanitation & Health, IWA Publishing, London, pp. 249–256

Clemens S (2001) Molecular mechanisms of plant metal tolerance and homeostatis. Planta 212:475–486

Cobbett C (2003) Heavy metals and plants — model systems and hyperaccumulators. New Phytol 159:289–293

Cunningham SD, Berti WR, Huang JW (1995) Phytoremediation of contaminated soils. Trends Biotechnol 13:393–397

Cunningham SD, Anderson TA, Schwab AP, Hsu FC (1996) Phytoremediation of soils contaminated with organic pollutants. Adv Agron 56:55–114

Ensink JHJ (2006) Water quality and the risk of hookworm infection in Pakistani and Indian sewage farmers. PhD thesis, London School of Hygiene and Tropical Medicine, University of London, UK

Ensink JHJ, Mahamood T, Dalsagaard A (2007) Wastewater irrigated vegetables: market handling versus irrigation water quality. Trop Med Int Health 12:2–7

Feachem DG, Bradley DJ, Garelick H, Mara DD (1983) Sanitation and disease: Health aspects of excreta and wastewater management. Wiley, Bath

Feigin A, Ravina I, Shalhevet J (1991) Irrigation with treated sewage effluent: management for environmental protection. Springer, Berlin

Gagliard JV, Karns JS (2000) Leaching of *Escherichia coli* O157:H7 in diverse soils under various agricultural management practices. Appl Environ Microbiol 66:877–883

Gerstl S (2001) The economic costs and impact of home gardening in Ouagadougou, Burkina Faso. PhD Dissertation, University of Basel, Switzerland

Ghassemi F, Jakeman AJ, Nix HA (1995) Salinisation of land and water resources: Human causes, extent, management and case studies. CABI Publishing, Wallingford

Gothberg A, Greger M, Bengtsson BE (2002) Accumulation of heavy metals in water spinach (*Ipomoea aquatica*) cultivated in the Bangkok region, Thailand. Environ Toxicol Chem 21:1934–1939

Haas CN, Rose JB, Gerba CP (1999) Quantitative microbial risk assessment. Wiley, New York

Hamilton AJ et al (2005) Position of the Australian horticultural industry with respect to the use of reclaimed water. Agric Water Manage 71:181–209

Hamilton AJ, Stagnitti F, Xiong X, Kreidl SL, Benke KK, Maher P (2007) Wastewater irrigation: the state of play. Vadose Zone J 6:823–840

Harris LJ, Beuchat FJM, LR PME, Suslow TV, Garrett EH, Busta FF (2003) Outbreaks associated with fresh produce: incidence, growth, and survival of pathogens in fresh and fresh-cut produce. Compr Rev Food Sci Food Safety (Suppl.) 2:78–89

Jablasone J, Brovko LY, Griffiths MW (2004) A research note: the potential for transfer of *Salmonella* from irrigation water to tomatoes. J Sci Food Agric 84:287–289

Jimenez B, Asano T (2008) Water reclamation and reuse around the world. In: Jimenez B, Asano T (eds) Water reuse: An international survey of current practice, issues and needs. IWA Publishing, London, pp 1–26

Karim MR, Manshadi FD, Karpiscak MM, Gerba CP (2004) The persistence and removal of enteric pathogens in constructed wetlands. Water Res 38:1831–1837

Karnchanawong S, Silprasert A, Keawvichit R, Prapamontol T, Wongwarapat K (2006) Contamination study on domestic wastewater reuse for vegetable cultivation. Conference proceedings: An International Perspective on Environmental and Water Resources, Bangkok, Thailand
Kay M (2001) Smallholder irrigation technology: Prospects for Sub Saharan Africa. IPRTRID, FAO, Rome
Keraita B, Konradsen F, Drechsel P, Abaidoo RC (2007a) Reducing microbial contamination on lettuce by cessation of irrigation before harvesting. Trop Med Int Health 12:8–14
Keraita B, Konradsen F, Drechsel P, Abaidoo RC (2007b) Effect of low-cost irrigation methods on microbial contamination of lettuce. Trop Med Int Health 12:15–22
Keraita B, Drechsel P, Konradsen F (2008a) Using on-farm sedimentation ponds to reduce health risks in wastewater irrigated urban vegetable farming in Ghana. Wat Sci Technol 57:519–525
Keraita B, Drechsel P, Konradsen F (2008b) Potential of simple filters to improve microbial quality of irrigation water used in urban vegetable farming in Ghana. J Environ Sci Health A 43:1–7
Lopez A, et al. (2005) Agricultural wastewater reuse in southern Italy. Proceedings from International Conference on Integrated Concepts in Water Recycling, Wollongong
Lucho-Constantino CA, Prieto-Garcia F, Razo LMD, Rodríguez-Vázquez R, Poggi-Varaldo HM (2005) Chemical fractionation of boron and heavy metals in soils irrigated with wastewater in central Mexico. Agric Ecosyst Environ 108:57–71
Mapanda F, Mangwayana EN, Nyamangara J, Giller KE (2005) The effect of long-term irrigation using wastewater on heavy metal contents of soils under vegetables in Harare, Zimbabwe. Agric Ecosyst Environ 107:151–165
Mara DD (2004) Domestic wastewater treatment in developing countries. Earthscan Publications, London
Mara DD, Pearson HW (1992) Sequential batch-fed effluent storage reservoirs: a new concept of wastewater treatment prior to unrestricted crop irrigation. Water Sci Technol 26:1459–1464
Mara DD, Pearson HW, Oragui JI, Crawley LR, de Oliveira R, Silva SA (1996) Wastewater storage and treatment reservoirs in northeast Brazil. TPHE Research Monograph No. 12. Department of Civil Engineering, University of Leeds, England
Moffat A (1995) Plants proving their worth in toxic metal cleanup. Science 269:302–303
Muhammad N, Ellis K, Smith MD (1996) Optimization of slow sand filtration. In: Reaching the unreached: Challenges for the 21st century. Proceedings of the 22nd WEDC Conference, New Delhi, pp 283–285
Mutchweti M, Birkett JW, Chinyanga E, Zvauya R, Scrimshaw MD, Lester JN (2006) Heavy metal content of vegetables irrigated with mixtures of wastewater and sewage sludge in Zimbabwe: implications for human health. Agric Ecosyst Environ 112:41–48
Obuobie E, Keraita B, Danso G, Amoah P, Cofie O, Raschid-Sally DP (2006) Irrigated urban vegetable production in Ghana: characteristics, benefits and risks. IWMI-RUAF-CPWF, IWMI, Accra, Ghana
Ottoson J (2005) Comparative analysis of pathogen occurrence in wastewater: management strategies for barrier function and microbial control. PhD thesis, Royal Technical University, Stockholm
Paris J, Jones JB (1997) The handbook of trace elements. CRC Press, Boca Raton, FL
Peasey A (2000) Human exposure to *Ascaris* infection through wastewater reuse in irrigation and its public health significance. PhD thesis, University of London, England
Pescod (1992) Wastewater treatment and use in agriculture. Irrigation and drainage, paper 47, FAO, Rome
Qadir M, Tubeileh A, Akhtar J, Labri A, Minhas PS, Khan MA (2008) Productivity enhancement of salt-affected environments through crop diversification. Land Degrad Dev 19:429–453
Raschid-Sally L, Jayakody P (2008) Drivers and characteristics of wastewater agriculture in developing countries: Results from a global assessment. IWMI Research Report 127, International Water Management Institute Colombo, Sri Lanka

Ratnam S, March SB, Ahmed R, Bezanson GS, Kasatyiya S (1988) Characterization of *Eschrichia coli* serotype O157:H7. J Clin Microbiol 26:2006–2012

Rengasamy P (2006) Soil salinity and sodicity. In: Stevens D (ed) Growing crops with reclaimed wastewater. CSIRO Publishing, Collingwood, pp 125–138

Richardson ML, Gangoli S (1995) The dictionary of substance and their effects. Royal Society of Chemistry, Cambridge, UK

Rutkowski T, Raschid-Sally L, Buechler S (2007) Wastewater irrigation in the developing world — two case studies from Katmandu Valley in Nepal. Agric Water Manage 88:83–91

Salt DE, Blaylock M, Kumar PBAN, Dushenkov S, Ensley BD, Chet I, Raskin I (1996) Phytoremediation: a novel strategy for the removal of toxic metals from the environment using plants. Biotechnology 13:468–474

Serani S, Nasinyama GW, Nabulo G, Lubowa A, Makoha M (2008) Biological hazards associated with vegetables grown on untreated sewage-watered soils in Kampala. In: Cole DC, Lee-Smith D, Nasinyama GW (eds) Healthy city harvests: generating evidence to guide policy on urban agriculture. CIP/Urban Harvest and Makerere University Press, Lima, pp 151–169

Sharma RK, Agrawal M, Marshall F (2007) Heavy metal contamination of soil and vegetables in suburban areas of Varanasi, India. Ecotox Environ Safe 66:258–266

Shuval HI, Lampert Y, Fattal B (1997) Development of a risk assessment approach for evaluating wastewater reuse standards for agriculture. Water Sci Technol 35:15–20

Shuval HI, Adin A, Fattal B, Rawitz E, Yekutiel P (1986) Wastewater irrigation in developing countries: health effects and technical solutions. World Bank Technical Paper No. 51, Washington, DC

Shuval HI, Yekutiel P, Fattal B (1984) Epidemiological evidence for helminth and cholera transmission by vegetables irrigated with wastewater: Jerusalem — a case study. Water Sci Technol 17:433–442

Simmons RW, Pongsakul P, Saiyasitpanich D, Klinphoklap S (2005) Elevated levels of cadmium and zinc in paddy soils and elevated levels of cadmium in rice grain downstream of a zinc mineralized area in Thailand: implications for public health. Environ Geochem Health 27:501–511

Simmons RW, Noble AD, Pongsakul P, Sukreeyapongse O, Chinabut N (2009) Cadmium-hazard mapping using a general linear regression model (Irr-Cad) for rapid risk assessment. Environ Geochem Health 31:71–79

Solomon EY, Yaron S, Mathews KR (2002) Transmission of Escherichia coli O157:H7 from contaminated manure and irrigation water to lettuce plant tissue and its subsequent internalization. Appl Environ Microbiol 68:397–400

Stevik TK, Aa K, Ausland G, Hanssen JF (2004) Retention and removal of pathogenic bacteria in wastewater percolating through porous media. Water Res 38:1355–1367

Stine SW, Song I, Choi CY, Gerba CP (2005) Application of microbial risk assessment to development of standards for enteric pathogens in water used to irrigate fresh produce. J Food Protect 68:913–918

Strauss M (1985) Health aspect of nightsoil and sludge use in agriculture and aquaculture – Part II: survival of excreted pathogens in excreta and faecal sludges. IRCWD News 23:4–9

Suarez DL (2001) Sodic soil reclamation: modelling and field study. Aust J Soil Res 39: 1225–1246

Trang DT, Hien BTT, Mølbak K, Cam PD, Dalsgaard A (2007a) Epidemiology and aetiology of diarrhoeal diseases in adults engaged in wastewater-fed agriculture and aquaculture in Hanoi, Vietnam. Trop Med Int Health 12:23–33

Trang DT et al (2007b) Skin disease among farmers using wastewater in rice cultivation in Nam Dinh, Vietnam. Trop Med Int Health 12:51–58

Truong P, Van TT, Pinners E (2008) The vetiver system for improving water quality: the prevention and treatment of contaminated water and land. The Vetiver Network International, Bellingham, WA

Tsai Y, Ingham SC (1997) Survival of *Escherichia coli* O157:H7 and *Salmonella* spp. in acidic condiments. J Food Protect 60:751–755

Van der Hoek W, Tuan Anh V, Dac Cam P, Vicheth C, Dalsgaard A (2005) Skin diseases among people using urban wastewater in Phnom Penh. Urban Agric Magazine 14:30–31

Van der Hoek W, Ul-Hassan M, Ensink JHJ, Feenstra S, Raschid-Sally L, Munir S, Aslam MR (2002) Urban wastewater: a valuable resource for agriculture. International Water Management Institute Research Report 63, Colombo

Van Ginneken M, Oron G (2000) Risk assessment of consuming agricultural products irrigated with reclaimed wastewater: An exposure model. Wat Resources Res 36(9):2691–2699

Vaz da Costa-Vargas S, Bastos RKX, Mara DD (1996) Bacteriological aspects of wastewater irrigation. TPHE Research Monogram. Department of Civil Engineering, University of Leeds, England

Warnes S, Keevil CW (2003) Survival of *Cryptosporidium parvum* in faecal wastes and salad crops. *Cryptosporidium parvum* in food and water. Teagasc Irish Agriculture and Food Development Authority, Carlow, pp 13–22

Wichelns D, Oster JD (2006) Sustainable irrigation is necessary and achievable, but direct costs and environmental impacts can be substantial. Agric Water Manage 86:114–127

WHO (2000) Global water supply and sanitation assessment 2000 report. World Health Organization (WHO) and United Nations Children's Fund (UNICEF), Geneva and New York

WHO (2006) Guidelines for the safe use of wastewater, excreta and grey water: Wastewater use in agriculture, vol 2. WHO Technical Report Series 778. World Health Organization, Geneva

Xiong X, Stagnitti F, Peterson J, Allinson G, Turoczy N (2001) Heavy metal contamination of pasture soils by irrigated municipal sewage. B Environ Contam Toxicol 67:535–540

Yates MV, Garba CP (1998) Microbial considerations in wastewater reclamation and reuse. In: Asano T (ed.) Wastewater reclamation and reuse. Water Quality Management Library, Vol 10, Technomic Publishing, Lancaster, PA, pp 437–488

Yuan Y (1993) Aetiological study of high stomach cancer incidents among residents in wastewater irrigated areas. Environ Protect Sci 19:70–73

Zhuang RY, Beuchat LR (1996) Effectiveness of trisodium phosphate for killing *Salmonella montevideo* on tomatoes. Lett Appl Microbiol 22:97–100

Chapter 14
Towards a New Purpose for Traditional and Other Forms of Soil Knowledge

Patrice Dion

14.1 Introduction: The Genesis of Soil Knowledge

Being an agricultural microbiologist, the author of the current chapter will not pretend to offer more than a candid overview of the subject of traditional knowledge as it applies to tropical soils. The chapter draws on an enormous body of literature that could not be cited in its entirety. Although not referred to systematically, Brookfield (2001) has been a constant inspiration.

The invention of agriculture during the Neolithic revolution depended on profound ecological, economical and cultural changes in human societies (Mazoyer and Roudart 2002). At the root of this revolution may lie a "revolution of the symbols," consisting in a shift of the perception humans had of the arrangement of the world. Space was endowed with a vertical symmetry, being given a "top," to which humans remaining at the "bottom" could aspire through prayer. Hence, dissatisfaction with everyday life was generated, and eventually resolved by performing a transition to subsistence agriculture (Cauvin 1998). Although it is likely not to have been wholly linear nor exclusive, this causal relationship between a postulated shift in the human relationship to the imaginary and the invention of agriculture remains crucial to an appreciation of the nature of traditional soil knowledge.

The cult of the Earth spirit has been hypothesized as the oldest and most common element of all religions, and, as such, to predate agriculture (Hillel 1991). Along with inventing agriculture, humans of the Neolithic developed an even closer relationship with the soil, which became an instrument sustaining both material life and the spiritual. In ancient Greece, dedication of local cults to particular deities was inspired by economic activities, which, in these early societies, were rather strictly dictated by soil types (Retallack 2008). Thus, the soil was

P. Dion
Département de phytologie, Université Laval, Pavillon Charles-Eugène Marchand, 1030, avenue de la Médecine, Québec G1V 0A6, Canada
e-mail: Patrice.Dion@fsaa.ulaval.ca

P. Dion (ed.), *Soil Biology and Agriculture in the Tropics*, Soil Biology 21,
DOI 10.1007/978-3-642-05076-3_14,

participant to the harsh reality of agricultural labor, but also to an imaginary world populated by aspirations of humans and their intuitions of other states or conditions. Of the various biases that may distort our assessments of traditional soil knowledge, some may well be elicited by the ambivalent nature of our ongoing relationship with the soil itself. Depending on the object of our attention, we may perceive the spiritual dimension of the soil as exaggeratedly expanded or contracted: the novelty of its manifestations among foreign groups makes this spiritual perception of the soil appear to us as strikingly prevalent over practical knowledge; on the contrary, as we consider our own dwellings, obviousness reduces the apparent dimension of our spiritual relationship with the soil to an insignificant scale.

With a precise knowledge neither of plant nutrition principles nor of the determinants of soil fertility, ancient peoples had concepts of soil fertility, which they characterized by means of specific words or by means of deities (Krogh and Paarup-Laursen 1997). The first farmers may well have been more concerned with managing plants and animals than with natural resources. Relying on recently domesticated rice, wheat, chicken and cattle, they maintained soil fertility through a combination of means that included fallowing, organic manuring, flooding and mixed cultivation. Thus, early farmers strengthened and dynamized their relationship with the soil, to which they attributed a prominent place in their conception of nature and humanity. Indeed, the words "human" and "humus" share the same Latin root, and Jewish tradition named the first man after the Hebrew word for soil, "adama." Early soil knowledge was recorded in oral tradition and texts. Such bodies of traditional wisdom have contributed little to the construction of modern soil science, whose birth in the seventeenth century was triggered by the European transition to naturalism.

14.2 Soil Science: Current Views and Some Interpretations

Current science envisions the soil as a self-organizing structure with interlocked chemical, physical, and biological frameworks. In the present section, the two principles driving self-organization of the soil, superposition and fractal order, will be briefly presented. It will then be remarked that these same principles may also structure the most fundamental elements of human identity, particularly the human brain. Modern methods for detection and characterization of micro-organisms have highlighted the complexity of soil biology, by revealing, in a single gram of soil, almost immeasurably diverse populations of organisms establishing intricate interactions (Quince et al. 2008). A chronological approach, whereby the successive processes of soil colonization are reconstructed through an evolutionary time course, may contribute to the grasping of this complexity (Dion 2008). This analysis makes it apparent that land was sequentially colonized by a series of organisms initially characterized by adaptations to ancient, anoxic environments. As the ambient conditions changed, new organisms joined the initial soil colonists, but without replacing them. Thus, a series of complementary phenomena, corresponding to the persistence of biological mechanisms, organisms, and environments, became assembled in a process of evolutionary layer accretion. In this process, the current biological diversity of soil developed through the superposed evolution of organisms which appeared at different times. The evolutionary layers composing the extant soil biota are linked to each other by symbiogenesis, and to their physicochemical support through feedback mechanisms.

Feedback between the biological and physicochemical components of the soil is expressed across a wide scale range (Young and Crawford 2004), corresponding to the fractal scaling of the soil itself (Crawford et al. 1997). In other words, because of structural self-similarity, auto-regulatory properties are expressed at different scales within the highly heterogeneous soil system, although the relative importance of various auto-regulatory mechanisms may be scale-dependent.

Parallels may be drawn between the accretive and fractal structure of the soil and the accretive and fractal structure of the human brain. The triune brain hypothesis of Paul D. MacLean (1990); see also Cory (2000) presents the human brain as an evolved three-level modular structure. This postulated structure comprises three interconnected components, the protoreptilian complex, the paleomammalian complex and the neocortex. The accretion of brain components results from evolution taking place in the vertebrate line, during which the simple protoreptilian complex, and later the paleomammalian complex were not replaced, but rather provided the substructures and homologues for subsequent brain development. As such, an understanding of the soil, with its own accretive structure, appears as a reflexive experiment espousing the very nature of human thinking. In so far as such fundamental correspondences exist between soil knowledge and the underlying mental processes, constants between soil knowledge belonging to different peoples should be identified, as they have been for example in the case of yet another fundamental cerebral function, language (Whitfield 2008). One such universal trait in our understanding of the soil is the notion that "land is part of existence" (Östberg 1995). The antiquity and ubiquity of burial rituals is a testimony to this anastomosis between human existence and that of the soil. Yet another commonly encountered principle is that of verticality and of vertical movements of soil horizons, which accompanies both scientific and traditional conceptualizations (Östberg 1995; Legros 2007). To recognize such constants requires that the nature and general characters of traditional soil knowledge be better examined. Thus, in addition to its practical value as a promoter of technological improvements for resource-poor farmers (Altieri 2002), traditional soil knowledge may hold a profound conceptual significance as an indicator of mental traits shared by all humans.

14.3 A Brief Presentation of Traditional Soil Knowledge

Traditional soil knowledge arises from the repeated observations and experimentations of persons cultivating or otherwise using the particular soil which is the object of this knowledge. Approximate synonyms commonly found in the literature

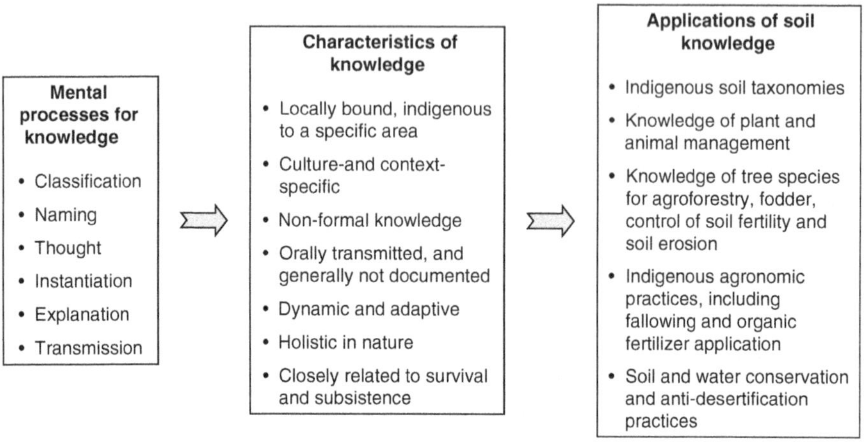

Fig. 14.1 Progression towards construction and instantiation of traditional knowledge. Through a series of mental processes (Sanga and Ortalli 2003) a traditional knowledge endowed with certain characteristics is constructed, used and conserved (Boven and Morohashi 2002). Some of this knowledge is directly relevant to soil management (Lalonde 1993)

include "indigenous soil knowledge," "local soil knowledge," and "folk soil knowledge" (Winklerprins 1999). Admittedly, these various terms are not strictly equivalent, and the expression "traditional knowledge" was chosen here because it appears neutral and unrestrictive.

Whereas applying a unique definition to the multiple forms of independently developed traditional knowledge appears difficult, it is generally agreed that such knowledge, as it occurs in its various forms delimited in time and space, is endowed with a series of diagnostic characteristics. It arises from a complementary series of mental actions, and is variously used to act on its object (Fig. 14.1).

It may be tempting to distinguish scientific from traditional knowledge on the basis of their relative insistence on the questions of "how things happen" and "why things happen," the implication being that traditional knowledge is more focused on the "why" than is science: because it implicitly recognizes a mythological component to traditional knowledge, such a distinction will not be made here. Indeed, it appears essential not to summon mythology to our current discussion of knowledge. Rather than representing a component of a certain form of knowledge, mythology relates to the mind as does knowledge itself. Mythology follows a course additive and parallel to that of knowledge, allowing both systems of thought to become disconnected. Indeed, Descola (2005) provides telling examples of myths contradicting extant and well-integrated knowledge.

On the other hand, magic, which is the art of producing new causes for natural phenomena, truly relates to cognition. In fact, a functional relationship exists between magic and science, since a body of knowledge may pass from the magical to the scientific domain once it is provided with a rational explanation. Contrasting the views of Durkheim and those of Malinowski, Vyse (1997) discusses magic either as being "prescientific" or else as accompanying science and being alternative to it: although somewhat contradictory, both views make it apparent that magic should be considered on a par with science, that is, as a genuine component of knowledge. Both magic and science create mental order, which then represents a sizable entity to be shared within a society (Gellner 1992).

The attempt has often been made to move beyond utilitarian accounts of traditional knowledge and to provide a conceptualization. Sillitoe (1998) has made a particularly eloquent contribution along those lines:

> [Traditional knowledge] is local in extent and embedded in parochial cultural traditions. Its distribution is fragmentary. Although on the whole more widely shared locally than specialized scientific knowledge, no one person, authority or social group knows it all. There may be a certain asymmetry here, some clustering of certain knowledge within populations (e.g., by gender, age etc., or according to specialist status, maybe reflecting political or ritual power). It exists nowhere as a totality, there is no grand repository, and hence no coherent overall theoretical model (. . .). It is as much skill as knowledge, and its learning across generations is characterized by oral transmission and learning through experience and repetitive practice. It is the heritage of practical everyday life, with its functional demands, and is fluid and constantly changing, being subject to ongoing negotiation between people and their environments.

Similarities between traditional knowledge and science have been stressed before (Tsuji and Ho 2002), and accordingly various aspects of Sillitoe's statement apply to both scientific and traditional knowledge. For example, the mention of traditional knowledge arising and being maintained as a result of interactions between people and their environments, is also valid for scientific knowledge. Indeed, the main argument of the current chapter will stem from this central concept of all forms of knowledge being generated at the interface of humans and their environment. This shared origin of all identities of knowledge provides a framework for integration, not strictly of these different identities themselves, but more generally of humanity with its multiple, proteiform knowledge. The constancy of its origin reflects accurately the status of knowledge as an intrinsic component of the human essence.

14.4 The Relevance of Traditional Soil Knowledge

Numerous studies have demonstrated the relevance of traditional knowledge to various aspects of natural resource management in the tropics, including the maintenance of biological diversity, biological crop pest control strategies, soil fertility management and soil and water conservation (Lalonde 1993). Most conspicuous among the various expression forms of traditional knowledge are the vernacular soil classification systems. The Hanunóo of the Philippines distinguished general and specific soil types, using eight criteria: moisture content, sand content, rock content, general texture, firmness, structure in dry season, stickiness in wet season, and color. The relative fertility ratings that the Hanunóo associated with the resulting categories coincided rather well with the results from scientific soil analysis (Conklin 1957). Following this early account of vernacular soil classification given by Conklin, the correspondence between the respective evaluations of soils by farmers and scientists has been confirmed (Mairura et al. 2007), and the systematic study of traditional soil classifications gave birth to the field of ethnopedology. However, the very use of the term “ethnopedology” and of related expressions, all of which imply a discriminating process that has no equivalent in traditional cultures, has been criticized (Sillitoe 1998).

There are many examples of long-term uses of traditional knowledge to ensure sustainable agricultural production in a seemingly challenging terrain. Examples include anthrosol-producing techniques of terracing and soil carbon management. Terrace agriculture, as practiced for more than fifteen centuries in the Colca Valley, Peru, has greatly altered properties of natural Mollisols in the area. These changes are inferred to result primarily from management practices such as bench terracing, tillage, fertilization, and irrigation. Crop management practices such as intercropping, fallowing, and use of legumes in crop rotations may also have had long-term influences on soil properties. Uninterrupted agricultural use over fifteen centuries has resulted in increased fertility and tilth, and traditional agriculture management practices have conserved soils (Sandor and Eash 1995). With respect to soil carbon management, black-earth-like anthropogenic soils with enhanced fertility, known as “Terra Preta do Indio,” have been described in the Amazonian Basin. These are human-made soils derived from the surrounding poor soils, and were created alongside permanent or semi-permanent human occupation and agriculture. Their mode of creation has been partly reconstituted, and involved accumulation of incompletely burnt organic material (or black carbon) and of fresh organic material such as human excrement or biomass taken from the surrounding land (Glaser et al. 2001).

The zai system of land rehabilitation (Mando et al. 2006) is a method for water harvesting practiced by Mossi farmers in the Yatenga province in the northern part of Burkina Faso. It involves restoring desirable physical and chemical qualities to the soil by mixing small quantities of organic material in small holes that have been dug into the crusted soil, while piling the removed earth along the lowest side of the hole. Water is retained by the resulting half-moon mound and flows into the hole, while termites attracted by the organic matter dig deep tunnels where water becomes trapped. An effective but labor-intensive technique, zai is also found in other regions of the Sahel. Like zai, the ngoro or matengo pit system, found on steep slopes of the Mbinga District of Southwest Tanzania, consists of a series of regular holes. The ridges are built on top of lines of cut grass which decomposes to release nutrients. Crops are grown on the ridges, and not in the holes (Kayombo et al. 1999).

Various soil conservation measures are used by farmers in the Dogon plateau of Mali. These include the use of stone bunds, trash lines of millet stalks, micro-basins and mounds, as well as the creation of artificial fields on barren rock by means of soil transportation (Kassogué et al. 1996).

Largely conducted by women, the vinyungu valley bottom cultivation plays a crucial role in providing food security in South-Western Tanzania. It consists of raised-bed cultivation in wetland valley bottoms, with burning of initial spontaneous vegetation and construction of variously oriented ditches to provide drainage. Organic matter and nutrients accumulate in vinyungu, which also has higher moisture levels during the dry season (Lema 1996).

These traditional techniques are used in various combinations according to the type of soil. Other factors orienting the choice of soil conservation methods are related to economic returns from land, capital and labor investments (Kayombo et al. 1999). Within this co-evolutionary interplay of social and ecological forces shaping agricultural practices, changes in international prices exert a critical influence (Koning and Smaling 2005). These observations remind us that traditional soil knowledge, at least as it is made apparent through its instantiation, should be interpreted against an interfering body of economic incentives. Taking a further step, knowledge could also be viewed through the whole fabric of life, as is suggested by a revision of traditional knowledge used by women from Ogun State, Nigeria, to enhance food security. Practices relating specifically to soil management, such as slash and burn cultivation, are accompanied by methods for farm pest control, seed selection, food preservation, nutrition, and various actions of personal life (Olatokun and Ayanbode 2008).

Many other traditional soil conservation measures have been documented from different parts of the world, as efforts are made to preserve and promote their use. Numerous such case stories bring an eloquent testimony to the capacity of smallholder farmers for combining various elements of traditional knowledge and enriching these through observation and innovations. For example, a farmer in Touroum, Burkina Faso, established a highly integrated system of agro-sylvo-pastoralism, that included various soil protection or rehabilitation measures such as the construction of stone contour bunds, composting, adoption of the zai system, and manure management (Boven and Morohashi 2002). Elsewhere, farmers of the Jos plateau, Nigeria, made an innovative use of mixed organic fertilizers, legume crops and soil preparation techniques to achieve agricultural intensification without soil fertility loss (Phillips-Howard and Lyon 1994).

In interpreting these practices of smallholder farmers, efforts have been made to distinguish between knowledge as specifically applied to soil and that related to management of plants or water. Depending on conditions, the same technique may be used for different reasons; for example, stone bunds may be built to clear stony fields, to prevent erosion or to retain run-off water (Asrat et al. 1996). Furthermore, various soil conservation measures may be used primarily because they also serve other purposes (Okoba and De Graaff 2005). Discussing Conklin's findings on the Hanunóo, Brookfield (2001) remarked that farmers are not interested in the soil per se, but more on the use of soil to grow plants and produce income. This has two implications, the first being that traditional soil knowledge, which may take the form of fairly accurate classification systems and fertility ratings, may not lead to direct soil management measures. Secondly, various operations intended to prevent erosion or maintain soil fertility in fact involve plant manipulations, such as spreading of slashed underbrush or plant material burning as a prelude to swidden cultivation (Conklin 1957).

Traditional farmers often have a distinct perception of soil degradation symptoms, and to them, for example, gullies may represent new opportunities for irrigation. In this sense, it has been proposed that land productivity may be a better measure of soil degradation than soil loss per se (Phillips-Howard and Lyon 1994). However, to unequivocally sanction an apparent indifference to obvious land degradation symptoms may represent a somewhat extreme attitude. Recent research has demonstrated that gullies, in particular, are the source of social disruptions and health problems, in addition to presaging an unsustainable agricultural system (Moges and Holden 2008). This discrepancy between the views of agronomists and soil scientists, on the one hand, and those of participatory research proponents, on the other, has been carefully examined by Koning and Smaling (2005), who testified to the reality and importance of soil degradation processes in sub-Saharan Africa.

14.5 Traditional Versus Scientific Soil Knowledge, and Other Systems of Opposition

Whereas some authors clearly see a dichotomy between science and traditional knowledge (Brodnig and Mayer-Schönberger 2000), others have questioned the very existence of traditional knowledge, qualifying the perception of this knowledge as a culturally biased interpretation of performance skills (Richards 1993). Taking yet another stance, some have doubted whether science and traditional knowledge should in fact be distinguished from each other (Agrawal 1995). A fourth opinion has it that science and traditional knowledge are indeed fundamentally distinguishable, but also complementary. Having recognized that there are "different ways of knowing about the soil," researchers and local people will proceed together to reconstruct knowledge and make it an integrated ensemble (Winklerprins 1999).

In a sense, all these views are culturally oriented, even if some are in fact defined against culture itself. It is an inescapable conclusion that any judgment on knowledge whether it concerns its mere existence or the apparent relationships between its manifestations, is the result of a culturally biased process.

One possible culturally neutral stance may be to simply state that there are different ways of generating valuable soil knowledge: this premise leads us to the alternative view that in fact there are not two distinguishable types of knowledge, but rather a continuum lying between two opposing poles in a "system of reality." While this defining polarity may be variously described, for example as religiosity versus secularity, magic versus science, or culture versus reason, it appears most appropriate here to propose the association of enchantment and disenchantment. Enchantment refers to "cultural practices that mark the marvelous erupting amid the everyday" (Bennett 2001). One extreme of this proposed spectrum is defined exclusively by enchantment: there lie superstition and irrationalism. At the other extreme, that would be the preserve of disenchantment, reside reductionism and a regressive form of systematized knowledge. In between these extremes, the enchanted and disenchanted components of the system of reality contribute in various proportions to the generation and the instantiation of differentiated forms of knowledge.

The system of reality is set to a particular value between exclusive enchantment and exclusive disenchantment as the result of a calibration process driven by ontology. The term "ontology" is used here following Descola (2005), to designate a coherent body of mental processes conferring a series of corresponding identities to existing things and to humanity within it. Considered in this sense, ontology prolongs a perspective along which are organized various aspects of traditional knowledge. Forcing our attention to the foreground, practices appear as the immediate and most obvious elements of knowledge. Behind practices exist various concepts as the hidden constituents of knowledge, to be acted upon for adaptation and conservation purposes (Brodt 2001). It is now proposed that ontology resides farther still than concepts and fundamental to them. It intervenes both in the upstream and downstream directions of a knowledge valorization process (Fig. 14.2a).

In the downstream direction, ontology presides over the calibration of a system of reality between exclusive enchantment and exclusive disenchantment, on the basis of which certain knowledge becomes valued. The emerging valuable knowledge is situated at some degree between the magic and the reductionism extremes.

In the upstream direction, ontology acts as a relay between the mind and the outside world, this world presenting itself under the form of a threat. Ontology executes a transition between the objective intensity of the threat and a particular subjective state of certainty or uncertainty. The system of reality alluded to before is then calibrated according to subjectivity (or a balanced certainty–uncertainty state), and not objectivity (or the real intensity of the threat).

In proposing this set of relationships, we are approaching the provocative "anything goes" of Paul Fereyabend, insofar as any currently available knowledge may be instrumented by a particular ontology, providing that this knowledge corresponds to an established balance within the disenchantment–enchantment continuum. The function of traditional soil knowledge becomes to provide risk adjustment options. It is also the function of all forms of knowledge.

Having said that knowledge is calibrated along a continuum from disenchantment to enchantment, the notion forcefully tends to emerge that traditional knowledge is more "enchanted" than scientific knowledge. However, the commonly recognized specificity of indigenous knowledge, conceived as being "tacit, intuitive, experiential, informal, uncodified" (Ellen and Harris 2000) would apply, not to the "substantive domain knowledge," that is to the knowledge immediately generated in response to a particular threat, but rather to the "external knowledge," or to the knowledge which is required to instantiate a particular substantive knowledge. Since this instantiation knowledge as it is expressed in Western societies is itself different from science (Fischer 2005), it appears that the perceived differences between science and indigenous knowledge may be related to the fact that we

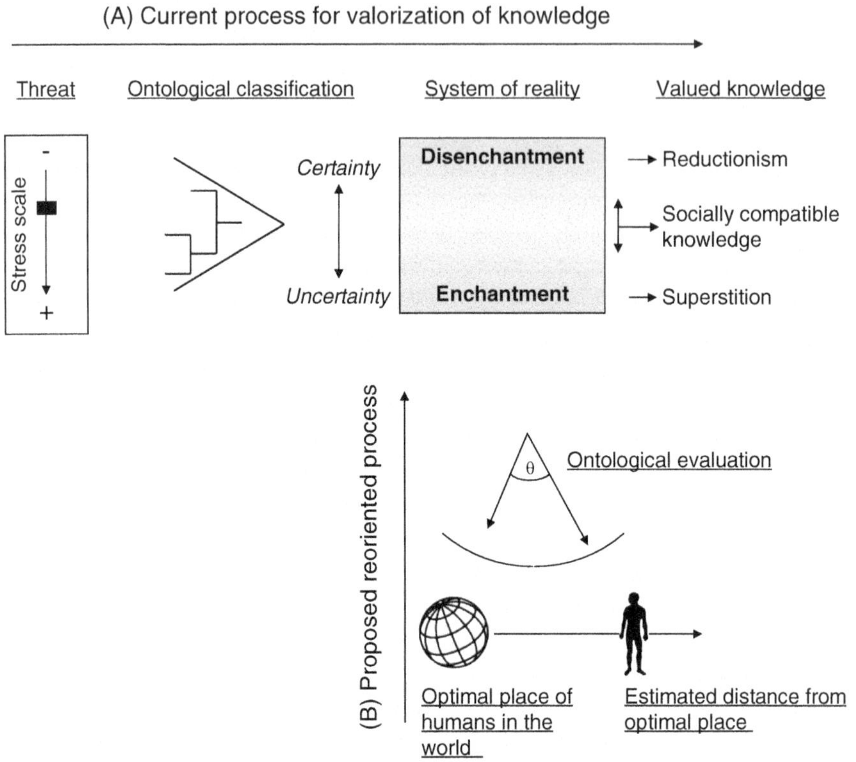

Fig. 14.2 Unified processes for a nature-induced valorization of knowledge. (**a**) From left to right: inferred extant process. Threats are objectively organized along a stress scale, according to the risk they present to human survival. These threats are then evaluated according to an ontological system, which subjectively places a natural situation along a scale of certainty and uncertainty, and sets the stage for further mental processes. Certainty favors activation of processes based on disenchantment, whereas uncertainty stimulates processes that rely on enchantment. The balance between disenchanted and enchanted mental processes finally leads to a valorization of knowledge, the most valorized knowledge being that which is deemed to provide the best possible response to the inducing threat. This valued knowledge is positioned along a scale ranging from regressive reductionism to superstition. Any mechanism for valorization of knowledge may disintegrate when a particular threat imposes excessive stress, which overrides the interpretative capacity of the ontological matrix. (**b**) From bottom to top: proposed reoriented process for valorization of knowledge. According to this proposed scheme, the ontological discriminating power is applied, not to the negative values of threat evaluation and fear of death, but to the positive determination of an optimal place for humans in the world (represented here as the angle θ between two vectors, one pointing towards the estimated optimal place for humans and the second towards their perceived current position). The disenchanted and enchanted systems of reality will then be balanced so as to value knowledge that will best contribute to guide humans closer to the perceived optimum

culturally confuse instantiation of traditional knowledge with traditional knowledge itself, whereas we distinguish between our own knowledge and its instantiation. Thus, culture may make us overly sensitive to the enchanted component of indigenous knowledge, while minimizing the perceived contribution of enchantment to instantiation of our own, scientific, knowledge.

14.6 Threats, Knowledge and Persistence of Knowledge

We have seen that the relative proportion of enchantment and disenchantment that at a specific moment corresponds to equilibrium within the system of reality is set according to intensity of a perceived threat. The balanced association between threat and magic was well observed by Malinowski (2004):

> We find magic wherever the elements of chance and accident, and the emotional play between hope and fear have a wide and extensive range. We do not find magic wherever the pursuit is certain, reliable and well under the control of rational methods and technological processes. Further, we find magic where the element of danger is conspicuous.

Following Malinowski's pioneering work, this balancing tendency (and consequently the system of reality within which it operates) has been confirmed as deeply entrenched in human behavior and has been observed across cultures (Subbotsky and Quinteros 2002). As we shall see, it will fade out only when the threat is so intense and generates so much uncertainty that it falls outside the realms of the ontological matrix with which a particular human group is functioning.

Within this process of threat evaluation and ontological valorization of knowledge, space is provided for innovation. The need for innovation may stem from a newly perceived threat, to which no pre-existing form of knowledge would seem optimally adapted. Alternatively, it may arise from a persisting threat that would not dissolve through the instantiation of current knowledge. Thus, an interaction exists between perceived threats and currently existing knowledge, establishing between the two a transient state of balance or imbalance. Innovation occurs within a particular window of disequilibrium between threat and knowledge. Outside of this window, innovation is discouraged if current knowledge is deemed to respond perfectly to the threat or, at the other extreme, in cases where knowledge under all its current or foreseeable forms appears as totally unrelated either to a new or to a persisting threat. Such a circumstance of catastrophic disequilibrium would signal the demise of all forms of knowledge, and thus of the very system used for valorization of knowledge: this would amount to ontological collapse.

Traditional ontological matrices may show significant resilience in the face of threats. For example, through a blend of old practices and innovations, the Kofyar of central Nigeria succeeded in maintaining a diversified and intensive form of agriculture following their migration to a new agricultural frontier (Netting and Stone 1996).

Various propositions have been made as to the nature of the agents leading to disintegration of a particular ontological matrix. These agents may be institutional, demographic, ecological, social or economic (Thrupp 1989; Gyasi et al. 1995;

Brookfield 2001). In particular, population threshold effects clearly operate. Increases in rural population numbers may result in land extensification, corresponding to occupation of forested or otherwise marginal land for low-output agricultural activities, or alternatively to agricultural intensification, this taking the form of diminished fallows and increased use of chemical fertilizers. Both extensification and intensification will lead to soil degradation (Bilsborrow 1992). However, in some cases where environmental degradation was concomitant with rapid population growth, alternative explanations, rather relating to institutions, changing values or ecological dynamics, may better account for the observed changes (Leach and Fairhead 2000). In particular, tenurial constraints and property rights are often at the origin of land degradation. Tenants are often compelled to overexploit the soil, whereas land inheritance systems may result in the generation of small and overcropped plots. Uncertainties regarding ownership of trees, especially timber species with commercial value, also reduce conservation incentives (Gyasi et al. 1995).

Population movement is yet another frequent phenomenon leading to the destruction of an ontological matrix. Massive immigration may result in occupation of the land by unskilled farmers, initially at least with fairly negative consequences (Brookfield 2001; Lopez-Gonzaga 2002). While bringing poorly adapted knowledge to their new region, settlers may then engage in an adaptation process of learning by doing: this process has occurred, for example, in the Brazilian Amazon (Pichon 1997). It is likely that, as the Amazonian agricultural frontier becomes socially and politically integrated, institutions and market forces will cooperate in absorbing these "performance skills" and link them as components of knowledge related to a dominant ontology.

Emigration, in causing population reduction, can also threaten ecological equilibria by weakening indigenous institutions, such as collective labor arrangements (García-Barrios and García-Barrios 1990). The adoption of a new cash crop may also create the conditions for resource degradation, as farmers may initially experience difficulties in applying new methods of production (Brookfield 2001).

Ontological disintegration may take various forms, from a brutal and complete collapse (Brookfield 2001; Lopez-Gonzaga 2002) to partial maintenance within a fraction or all of a population (Zimmerer 1996a). It is not knowledge that may disappear initially, but the need or the means for instantiation of this knowledge (Jacob 1998). Unused knowledge will then be expected to gradually fade out. It should be remarked, however, that perception of threats itself depends to some extent on the existence of an ontological matrix, thus generating a self-sustained loop, whereby perceived threats and an ontology justify each other. This phenomenon will lead to the remanence of an ontology, which will persist for some time or even indefinitely, at least in some of its elements, after the circumstances under which it was functional have disappeared. More generally, it can be stated that the life cycle of ontologies shows hysteresis, this delay in birth and death of a mental scheme with respect to the causes that elicit or destroy it, having both biological and social roots. Hysteresis itself will lead to a bistable state, whereby two ontologies coexist in space and time (Hountondji 2002).

Among Cochabamba (Bolivia) peasants, a detailed knowledge of soil erosion coexists with supernatural explanations of this phenomenon, invoking religious beliefs and customs. This dual system tends to disappear among younger peasants, who cast soil erosion less in terms of religion and more in terms of human-induced cause. However, the younger peasants retain a broader view of the erosion problem than the strictly technical stance of governmental institutions or NGOs. Indeed, the young peasants link their search of a solution to the soil erosion problem to cultural and political demands. Hence, even among young peasants with extensive exposure to modernity, the original ontological matrix, with its generalizing character, persists to some degree in its opposition with a technical and disenchanted ontological matrix (Zimmerer 1996b).

14.7 Conclusion: Proposal for a Reorientation of Ontological Processes

The difficulties in applying "top-down" scientific assessment and intervention procedures to smallholder agriculture are amply recognized. For example, formal scientific experiments involving controlled trials and a series of repetitions in space and time are impractical for the design of agroforestry systems (Rocheleau et al. 1989). More generally, agricultural science attempts to transcend time and space in drawing up its principles, whereas traditional knowledge is deeply rooted in both factors (Richards 1993). Scientific assessments and interventions are often large-scale, and either derived from or else intended for state planning and policy (Sillitoe 1998). Attempts to encourage permanent adoption of soil conservation measures often lead to the establishment of a bistable state in a particular zone, where well-preserved and irreversibly damaged soils coexist (Antle et al. 2006). A particularly telling case study of top-down intervention in the Burunge Hills in the Kondoa district, Tanzania, shows how the implementation of a ready-made soil conservation technique led to social conflict (Östberg 1995). Whereas adapted procedures might be designed for integrating local knowledge and management systems (Braimoh 2002; Gowing et al. 2004; Mairura et al. 2007), practical difficulties arise, such as the need for site-specific management strategies and for changes in policies and scientific practices (Warren et al. 2003). It has been proposed that these difficulties be resolved through the framing of experimental situations where the user conducts exploratory research on the basis both of new elements and of his own experience, and without adopting a proven package (Rocheleau et al. 1989).

The legitimization of traditional knowledge, conceived as a means of empowerment of the people who possess it (Thrupp 1989), will require a reorientation of the ontological classification system, away from threats, since these do not provide constant evaluation outcomes. For example, soil science considers the problem of soil erosion as a problem per se, which is the object of separate enquiries, published studies and protective measures. By contrast, we have already seen that local perceptions may well integrate soil erosion in a wider context, also encompassing soil fertility, soil diversity and genesis, and rainfall (Warren et al. 2003).

Thus, a genuine coupling of traditional knowledge and its scientific equivalent might involve a reorientation of the ontological evaluation mechanism alluded to in Fig. 14.2a, away from threat and negative values and towards humanity itself and its place in the world (Lockwood 1999; Fig. 14.2b). This feat invites us to an unprecedented revolution in our mode of relating to the world. If indeed traditional agriculture represented no more than a "performance," at once dissociated from culture and history and strictly based on improvisational skills upon confrontation with a particular situation, then the reorientation of ontological processes proposed here would be rendered even more difficult in view of the ensuing blurring of the perception that humans have of their destiny. On the contrary, taking into account the particular significance of agricultural activity and its functional link with the core of human identity leads us to a different conclusion: it is through agriculture that ontological processes may be safely reoriented, while maintaining the human ability to evaluate the world and react to it.

While modern science attempts to adhere to an ideal of despiritualization of agriculture, the time has come to bridge the widening gap between mentalities and agricultural practices. In spite of valuable initiatives such as those mentioned above and many others, there is a dearth of solid models and procedures on how to integrate scientific and local knowledge in land management practice (Winklerprins 1999). The resistance to integration may not be only passive, but also active, as some attitudes of agricultural scientists and extentionists may be viewed as a barrier to effective knowledge integration (Gupta 1989). To succeed in this integration will require the consideration of knowledge not as a product of the human experience (that is, as the result of a threat evaluation process), but as a component of it. As that of humans themselves, the organic nature of knowledge should be recognized. Just like living things, knowledge can be conceived as a dissipative structure, self-organizing while acting upon the outside world. Also, knowledge is evolving, as does life itself, in both a contingent and a chaotic manner.

A recognition of the organic nature of knowledge leads to an appreciation of the symbiosis it has established with humans. Symbiosis occurs when two or more dissimilar entities live in or on one another in an intimate relationship (Roossinck 2008). Symbionts play a critical role in generating the host phenotype. The phenotypic significance of the symbiont is modular, in the sense that it alters or determines coherent and autonomous fractions of the host phenotype, and conservative, since major evolutionary changes in the host lineage do not subvert the phenotypic effect of the symbiont on the host. The acquisition of the symbiont is a breakthrough that allows the invasion of a new adaptive zone (Sterelny 2004). Whereas it is apparent that knowledge possesses all of the above-noted properties, there is a difficulty in qualifying knowledge as a symbiont of humans since, as such, it would be endogenous to its host. It is apparent, however, that knowledge is endogenous to the human species, but not to the individual who must acquire it as an outside body shaped by oral or written transmission vehicles. Hence, the recognition of the symbiotic nature of knowledge amounts to the creation of a new, interpolated knowledge: it will be interpolated because it will not be constructed within a person, a culture or a system, but "between." Such new knowledge will be created between people (Östberg 1995), and also between mankind and other beings. A utilitarian view of knowledge, which invites to define traditional and scientific knowledge and to study the means for their reconciliation, should yield to a conception of knowledge as a symbiont of humans. The transmissible and ever-expanding component of our conscience, symbiotic knowledge is also our guide within the world.

Acknowledgments I thank Len Tsuji (University of Waterloo, Canada) for helpful comments.

References

Agrawal A (1995) Dismantling the divide between indigenous and scientific knowledge. Dev Change 26:413–439

Altieri MA (2002) Agroecology: the science of natural resource management for poor farmers in marginal environments. Agr Ecosyst Environ 93:1–24

Antle JM, Stoorvogel JJ, Valdivia RJ (2006) Multiple equilibria, soil conservation investments, and the resilience of agricultural systems. Environ Dev Econ 11:477–492

Asrat K, Idris K, Semegn M (1996) The "flexibility" of indigenous soil and water conservation techniques. In: Reij C (ed) Sustaining the soil: Indigenous soil and water conservation in Africa. Earthscan Publications, London, pp 156–162

Bennett J (2001) The enchantment of modern life: attachments, crossings, and ethics. Princeton University Press, Princeton, NJ

Bilsborrow RE (1992) Population growth, internal migration, and environmental degradation in rural areas of developing countries. Eur J Popul 8:125–148

Boven K, Morohashi J (2002) Best practices using indigenous knowledge. Nuffic, The Hague

Braimoh AK (2002) Integrating indigenous knowledge and soil science to develop a national soil classification system for Nigeria. Agric Hum Values 19:75–80

Brodnig G, Mayer-Schönberger V (2000) Bridging the gap: the role of spatial information technologies in the integration of traditional environmental knowledge and Western science. Electron J Inf Syst Dev Countries 1:1–15

Brodt S (2001) A systems perspective on the conservation and erosion of indigenous agricultural knowledge in central India. Hum Ecol 29:99–120

Brookfield H (2001) Exploring agrobiodiversity. Columbia University Press, New York

Cauvin J (1998) Naissance des divinités, naissance de l'agriculture: la révolution des symboles au Néolithique. Flammarion, Paris

Conklin HC (1957) Hanunóo agriculture: a report on an integral system of shifting cultivation in the Philippines. Food and Agriculture Organization of the United Nations, Rome

Cory GA Jr (2000) From Maclean's triune brain concept to the conflict systems neurobehavioral model: the subjective basis of moral and spiritual consciousness. Zygon 35:385–414

Crawford JW, Verrall S, Young IM (1997) The origin and loss of fractal scaling in simulated soil aggregates. Eur J Soil Sci 48:643–650

Descola P (2005) Par-delà nature et culture. Gallimard, Paris

Dion P (2008) Reconstructing soil biology. In: Nautiyal CS, Dion P (eds) Molecular mechanisms of plant and microbe coexistence. Springer, Berlin, pp 75–97

Ellen R, Harris H (2000) Introduction. In: Ellen R, Parkes P, Bicker A (eds) Indigenous environmental knowledge and its transformations: Critical anthropological perspectives. Harwood Academic Publishers, Amsterdam

Fischer MD (2005) Culture and indigenous knowledge systems: emergent order and the internal regulation of shared symbolic systems. Cybernet Syst 36:735–752
García-Barrios R, García-Barrios L (1990) Environmental and technological degradation in peasant agriculture: a consequence of development in Mexico. World Dev 18:1569–1585
Gellner E (1992) Reason and culture. Blackwell, Oxford
Glaser B, Haumaier L, Guggenberger G, Zech W (2001) The "Terra Preta" phenomenon: a model for sustainable agriculture in the humid tropics. Naturwissenschaften 88:37–41
Gowing J, Payton R, Tenywa M (2004) Integrating indigenous and scientific knowledge on soils: recent experiences in Uganda and Tanzania and their relevance to participatory land use planning. Uganda J Agric Sci 9:184–191
Gupta AK (1989) Scientists' views of farmers' practices in India: barriers to effective interaction. In: Chambers R, Pacey A, Thrupp LA (eds) Farmer first: Farmer innovation and agricultural research. Intermediate Technology Publications, London, pp 24–31
Gyasi E, Agyepong G, Ardayfio-Schandorf E, Enu-Kwesi L, Nabila J, Owusu-Bennoah E (1995) Production pressure and environmental change in the forest–savanna zone of Southern Ghana. Global Environ Chang 5:355–366
Hillel D (1991) Out of the earth: Civilization and life of the soil. University of California Press, Berkeley
Hountondji PJ (2002) Knowledge appropriation in a post-colonial context. In: Odora-Hoppers CA (ed) Indigenous knowledge and the integration of knowledge systems: Towards a philosophy of articulation. New Africa Books, Cape Town, pp 23–38
Jacob J-P (1998) Pourquoi les Winye ont-ils cessé de cultiver leurs champs permanents en 1986? (Burkina Faso). Autrepart 7:111–124
Kassogué A, Komota M, Sagara J, Schutgens F (1996) A measure for every site. Traditional SWC techniques on the Dogon plateau, Mali. In: Reij C (ed) Sustaining the soil: Indigenous soil and water conservation in Africa. Earthscan Publications, London, pp 69–79
Kayombo R, Ellis-Jones J, Martin HL (1999) Indigenous conservation tillage system in East Africa with an example of their evaluation from South West Tanzania. In: Kaumbutho PG, Simalenga TE (eds) Conservation tillage with animal traction: A resource book of the Animal Traction Network for Eastern and Southern Africa. ATNESA, Harare
Koning N, Smaling E (2005) Environmental crisis or "lie of the land?" The debate on soil degradation in Africa. Land Use Policy 22:3–11
Krogh L, Paarup-Laursen B (1997) Indigenous soil knowledge among the Fulani of northern Burkina Faso: linking soil science and anthropology in analysis of natural resource management. GeoJournal 43:189–197
Lalonde A (1993) African indigenous knowledge and its relevance to sustainable development. In: Inglis JT (ed) Traditional ecological knowledge concepts and cases. Canadian Museum of Nature and IDRC, Ottawa, pp 55–62
Leach M, Fairhead J (2000) Challenging Neo-Malthusian deforestation analyses in West Africa's dynamic forest landscapes. Popul Dev Rev 26:17–43
Legros J-P (2007) Les grands sols du monde. Presses Polytechniques et Universitaires Romandes, Lausanne
Lema AJ (1996) Cultivating the valleys: Vinyungu farming in Tanzania. In: Reij C (ed) Sustaining the soil: Indigenous soil and water conservation in Africa. Earthscan Publications, London, pp 139–144
Lockwood JA (1999) Agriculture and biodiversity: finding our place in this world. Agric Hum Values 16:365–379
Lopez-Gonzaga V (2002) Peasants in the hills. University of the Philippines Press, Quezon City
MacLean PD (1990) The triune brain in evolution: Role in paleocerebral functions. Plenum, New York
Mairura FS, Mugendi DN, Mwanje JI, Ramisch JJ, Mbugua PK, Chianu JN (2007) Integrating scientific and farmers' evaluation of soil quality indicators in Central Kenya. Geoderma 139:134–143

Malinowski B (2004) Magic, science and religion and other essays 1948. Kessinger Publishing, Whitefish, MT

Mando A, Fatondji D, Zougmoré R, Brussaard L, Bielders CL, Martius C (2006) Restoring soil fertility in semi-arid West Africa: assessment of an indigenous technology. In: Uphoff N et al (eds) Biological approaches to sustainable soil systems. CRC Press, Boca Raton, FL, pp 391–399

Mazoyer M, Roudart L (2002) Histoire des agricultures du monde. Du néolithique à la crise contemporaine, Éditions du Seuil, Paris

Moges A, Holden NM (2008) Estimating the rate and consequences of gully development, a case study of umbulo catchment in Southern Ethiopia. Land Degrad Dev 19:574–586

Netting RM, Stone MP (1996) Agro-diversity on a farming frontier: Kofyar smallholders on the Benue Plains of Central Nigeria. Africa 66:52–70

Okoba BO, De Graaff J (2005) Farmers' knowledge and perceptions of soil erosion and conservation measures in the Central Highlands, Kenya. Land Degrad Dev 16:475–487

Olatokun WM, Ayanbode OF (2008) Use of indigenous knowledge by rural women in the development of Ogun State. Indilinga Afr J Indigenous Knowl Syst 7:47–63

Östberg W (1995) Land is coming up: the Burunge of Central Tanzania and their environments. Stockholm Studies in Social Anthropology, Stockholm

Phillips-Howard KD, Lyon F (1994) Agricultural intensification and the threat to soil fertility in Africa: evidence from the Jos Plateau, Nigeria. Geogr J 160:252–265

Pichon FJ (1997) Colonist land-allocation decisions, land use, and deforestation in the Ecuadorian Amazon frontier. Econ Dev Cult Change 45:707–744

Quince C, Curtis TP, Sloan WT (2008) The rational exploration of microbial diversity. ISME J 2:997–1006

Retallack GJ (2008) Rocks, views, soils and plants at the temples of ancient Greece. Antiquity 82:640–657

Richards P (1993) Cultivation: Knowledge or performance? In: Hobart M (ed) An anthropological critique of development: The growth of ignorance. Routledge, New York, pp 61–78

Rocheleau D, Wachira K, Malaret L, Wanjohi BM (1989) Local knowledge for agroforestry and native plants. In: Chambers R, Pacey A, Thrupp LA (eds) Farmer first: Farmer innovation and agricultural research. Intermediate Technology Publications, London, pp 14–24

Roossinck MJ (2008) Symbiosis, mutualism and symbiogenesis. In: Roossinck MJ (ed) Plant virus evolution. Springer, Berlin, pp 157–164

Sandor JA, Eash NS (1995) Ancient agricultural soils in the Andes of Southern Peru. Soil Sci Soc Am J 59:170–179

Sanga G, Ortalli G (2003) Nature knowledge: Ethnoscience, cognition, and utility. Berghahn Books, New York, Oxford

Sillitoe P (1998) Knowing the land: soil and land resource evaluation and indigenous knowledge. Soil Use Manage 14:188–193

Sterelny K (2004) Symbiosis, evolvability, and modularity. In: Schlosser G, Wagner GP (eds) Modularity in development and evolution. The University of Chicago Press, Chicago, pp 490–516

Subbotsky E, Quinteros G (2002) Do cultural factors affect causal beliefs? Rational and magical thinking in Britain and Mexico. Brit J Psychol 93:519–543

Thrupp LA (1989) Legitimizing local knowledge: from displacement to empowerment for third world people. Agr Hum Values 6:13–24

Tsuji LJS, Ho E (2002) Traditional environmental knowledge and Western science: in search of common ground. Can J Native Stud 22:327–360

Vyse SA (1997) Believing in Magic: the psychology of superstition. Oxford University Press, Oxford

Warren A, Osbahr H, Batterbury S, Chappell A (2003) Indigenous views of soil erosion at Fandou Béri, southwestern Niger. Geoderma 111:439–456

Whitfield J (2008) Across the curious parallel of language and species evolution. PLoS Biol 6:e186. doi:10.1371/journal.pbio.0060186

Winklerprins AMGA (1999) Local soil knowledge: a tool for sustainable land management. Soc Natur Resour 12:151–161

Young IM, Crawford JW (2004) Interactions and self-organization in the soil–microbe complex. Science 304:1634–1637

Zimmerer KS (1996a) Changing fortunes: Biodiversity and peasant livelihood in the Peruvian Andes. University of California Press, Berkeley

Zimmerer KS (1996b) Environmental discourses on soil degradation in Bolivia: sustainability and the search for socio-environmental "middle ground". In: Peet R, Watts M (eds) Liberation ecologies. Routledge, Oxon, pp 107–124

Index

R

The manufacturer's authorised representative in the EU is Springer Nature Customer Service Centre GmbH, Europaplatz 3, 69115 Heidelberg, Germany. If you have any concerns regarding our products, please contact ProductSafety@springernature.com

Printed and bound by CPI Group (UK) Ltd, Croydon, CR0 4YY
01/10/2026
02225970-0012